T0190242

Studies on Binocular Vision

Archimedes

NEW STUDIES IN THE HISTORY AND PHILOSOPHY OF SCIENCE AND TECHNOLOGY

VOLUME 47

Archimedes has three fundamental goals; to further the integration of the histories of science and technology with one another: to investigate the technical, social and practical histories of specific developments in science and technology; and fi nally, where possible and desirable, to bring the histories of science and technology into closer contact with the philosophy of science. To these ends, each volume will have its own theme and title and will be planned by one or more members of the Advisory Board in consultation with the editor. Although the volumes have specifi c themes, the series itself will not be limited to one or even to a few particular areas. Its subjects include any of the sciences, ranging from biology through physics, all aspects of technology, broadly construed, as well as historically-engaged philosophy of science or technology. Taken as a whole, *Archimedes* will be of interest to historians, philosophers, and scientists, as well as to those in business and industry who seek to understand how science and industry have come to be so strongly linked.

More information about this series at http://www.springer.com/series/5644

Dominique Raynaud

Studies on Binocular Vision

Optics, Vision and Perspective
from the Thirteenth to the Seventeenth
Centuries

 Springer

Dominique Raynaud
PPL
Université Grenoble Alpes
Grenoble
France

ISSN 1385-0180   ISSN 2215-0064   (electronic)
Archimedes
ISBN 978-3-319-82645-5   ISBN 978-3-319-42721-8   (eBook)
DOI 10.1007/978-3-319-42721-8

Printed on acid-free paper

This Springer imprint is published by Springer Nature
The registered company is Springer International Publishing AG Switzerland

Preface

The aim of this book is to elucidate the question of the interrelationship between optics, vision and perspective before the Classical Age. In the Middle Ages and the Renaissance, the concept of *Perspectiva*—the Latin word for optics—encompassed many areas of enquiry that had been viewed since antiquity as interconnected, but which afterwards were separated: optics was incorporated into the field of physics (i.e., physical and geometrical optics), vision came to be regarded as the sum of various psycho-physiological mechanisms involved in the way the eye operates (i.e., physiological optics and psychology of vision) and the word 'perspective' was reserved for the mathematical representation of the external world (i.e., linear perspective).

However, this division, which emerged as a result of the spread of the sciences in classical Europe, turns out to be an anachronism if we confront certain facts from the immediately preceding periods. It is thus essential to take into account the way medieval scholars posed the problem—which included all facets of the Latin word *perspectiva*—when exploring the events of this period. What we now recognize as a 'nexus' between optics and perspective was at the time in fact seen as a single science. I submit that the earliest developments in linear perspective cannot be elucidated without reinserting them into the web of ideas that originally constituted *perspectiva*.

The central focus of this book is the theory of binocular vision, which has been virtually ignored in the field of perspective studies. This theory generated one of the most puzzling alternatives to linear perspective in the history of representation—two-point perspective which could be regarded as a 'heterodox system' inasmuch as linear perspective is taken to be the norm. However, linear perspective was not at all the standard until the late sixteenth century (Cinquecento). Before then many other systems were used, such that one would be justified in asking whether it would not be better to admit that different, parallel systems of perspective existed as late as the Renaissance. Since the norm was still to come, it was common to find painters and architects testing new methods that lay at the margins of linear perspective. As a result, there is no way to demonstrate that painters and architects as a whole were applying the rules of perspective from Brunelleschi's time onward. Up until the end of the Cinquecento the word 'perspective' referred to a series of free and uncoordinated systems, with debates being conducted in scholarly and artistic circles on the merits of each.[1]

In Chap. 1 we will seek to define more clearly the similarities and differences between perspective and *perspectiva*, i.e., medieval optics. One of the main differences was the gradual trend to decouple linear perspective from medieval optics, the course of which included an entire chapter on the formation of binocular images.

Errors—Chap. 2 investigates the emergence of perspective as a geometric science and seeks to separate what is fact from what is fiction regarding the birth of perspective in Quattrocento Italy. Events that were codified into what may be regarded as the mythology of perspective are discussed, including Brunelleschi's untraceable *tavoletta*, Alberti's *costruzione legittima*, and the perspective in Masaccio's fresco of the *Holy Trinity* in the Church of Santa Maria Novella in Florence. This chapter will show how access to knowledge could change practices; it establishes, for instance, that the solutions found by draftsmen to the problem of how to draw the perspective view of a circle varied, depending on their degree of familiarity with optics and geometry. Chapter 3 provides a classification of the types of errors that may arise in perspective constructions, deepening our understanding of the problem by presenting several examples of works that depart from the rules of perspective. Chapter 4 scrutinizes a blatant example of mistaken judgment regarding the correctness of one specific case of perspective—the interpretation by Erwin Panofsky of Masaccio's *Trinity*. Although celebrated as a milestone in the history of perspective, this fresco is not a correct example of central perspective due to the many errors—both random and systematic—that can be found in its geometric construction. These results undermine the commonly held idea that linear perspective became the unspoken rule in Brunelleschi's time, with all other alternatives being gradually abandoned. Linear perspective was neither clearly defined nor followed as a general rule in these early stages, and there was not yet a sufficient consensus to limit alternative representational systems.

Theory—Chap. 5 outlines the theory of binocular vision presented by Ibn al-Haytham in *Kitāb al-manāẓir* and discusses the innovations and limitations of this medieval Arab scholar's work in the light of modern physiological optics. Chapter 6 seeks to retrace the impact of Ibn al-Haytham's theory on Latin medieval optics. There is evidence that the study of key sections of *Kitāb al-manāẓir* and the commentaries written by European scholars ensured the wide dissemination of his theory of binocular vision. Chapter 7 focuses on certain contemporary documents

---

[1]The present book includes revised content from several papers, mostly in French, published in academic journals. Chap. 1: *Nel Segno di Masaccio*, ed. F. Camerota, Firenze, 2001, pp. 11–13. Chap. 2: *Les Espaces de l'homme*, eds. A. Berthoz and R. Recht, Paris, 2005, pp. 333–354. Chap. 3: *L'Hypothèse d'Oxford*, Paris, pp. 62–85. Chap. 4: *Nuncius* 17 (2003): 331–344. Chap. 5: *Arabic Sciences and Philosophy* 13 (2003): 79–99. Chap. 8: *Oriens/Occidens* 5 (2004): 93–131. Chap. 9: *Sciences et Techniques en Perspective* 2-1 (1998): 3–23. Chap. 10: *Zeitschrift für Kunstgeschichte* 67/4 (2004): 449–460. Chap. 11: *Physis* 45 (2008): 29–55. Appendix A: *L'Œuvre et l'artiste à l'épreuve de la perspective*, eds. M. Dalai Emiliani et al., Rome, 2006, pp. 411–430. The other parts of the book are new.

that explicitly condemned the practice of 'two-point perspective.' These texts, which were written by members of the earliest Italian academies and of the *Académie Royale de Peinture* in France, inform us that the theory and practice of monocularity continued to encounter strong resistance during the Renaissance and well into the classical period.

Sifting the Hypotheses—Applying standard techniques of error analysis, Chap. 8 and Appendix 1 address the methodological issue of how to eliminate or reduce the errors that may be introduced during the *ex post* reconstruction of a perspective view. An in-depth analysis is presented of *The Saint Enthroned*, a fresco by Giusto de' Menabuoi that illustrates the use of two-point perspective. The same methodology is then applied to 30 works produced in Italy between the Duecento and the Cinquecento in which the use of two-point perspective has been identified. The error analysis is supplemented by a reconstruction of the geometric plans and elevations in these paintings, working backward from the perspective views. This analysis based on a large number of works allows us to eliminate a series of alternative forms of representation, and the sifting of the different representational systems proves that binocular vision might have provided the foundations for the construction of these medieval and Renaissance perspectives.

However, the hypothesis that early works of perspective were constructed on the basis of binocular vision can be accepted only if all the competing assumptions are successfully rebutted. We therefore carried out an evaluation, one by one, of the various theses that currently dominate discussions of the history of perspective. In Chap. 9 we demonstrate the inconsistency on both logical and empirical grounds of the Hauck–Panofsky conjecture regarding 'curvilinear perspective.' Similarly in Chap. 10 we disprove the White–Carter conjecture regarding 'synthetic perspective' by pointing out a mathematical property that renders this system unlikely. Chapter 11 examines Andrés de Mesa Gisbert's conjecture that medieval perspective was the result of an arithmetic method of construction, a solution that, while elegant, poses some serious difficulties.

All the competing assumptions having been disproved, I conclude that binocular vision and two-point perspective constituted a genuine alternative to linear perspective from the late Duecento onward. In this way a strong interdependence between optics and perspective is established that accords with the original meaning of the word *perspectiva* and opens up the possibility for a better understanding of how perspectives were constructed in the early modern period. I submit that binocularity represents a key juncture point between the history of art and the history of science.[2]

---

[2]From this perspective, the binocular system makes a genuine difference with the foreshortening rule, which could have been derived from Euclid's *Optica*, postulate 5, as well as from practical geometry, in particular the "Turris altitudinem metiri" section included in many treatises. See for instance Stephen K. Victor, *Practical Geometry in the High Middle Ages*, Philadelphia, 1979; Hubert L.L. Busard, "The 'Practica geometriae' of Dominicus de Clavasio," *Archive for the History of Exact Sciences* 2 (1965): 520–575; and Cosimo Bartoli's *Del modo di misurare*, Venezia, 1564.

The intent of this book is to explore the various explanations and past modes of rationalizing the phenomenon of vision that can be derived from the matrix of *Perspectiva*, thus contributing to the rewriting of an important chapter in the history of optics and perspective from an angle that takes into account the criticisms that have been brought to bear on linear perspective in the past, and that is more sensitive to the precarious balance that characterizes the early stages in any process of innovation.

I express gratitude to Lisa C. Chien, who translated several chapters from the French and diligently revised the whole text.

Saint-Martin   Dominique Raynaud
June 2015

Contents

Part II   Theory

5   Ibn al-Haytham on Binocular Vision 71

6   The Legacy of Ibn al-Haytham 95

7   The Rejection of the Two-Point Perspective System 115

Part III   Sifting the Hypotheses

8   The Properties of Two-Point Perspective 133

Chapter 1
Perspectiva Naturalis/Artificialis

Abstract Perspective, as a system of visual representation, draws its name from the medieval Latin term perspectiva which means 'optics.' We owe this linguistic connection to the fact that certain principles of perspective developed from theories of vision. Between the two sets of notions one can find relationships of both continuity and discontinuity. A study of textual parallels has established this continuity. However, there are clear distinctions between perspectiva and perspective. Apart from the close relationship between science and technique that characterized them both, medieval perspectiva was a tripartite science embracing optica, catoptrica and dioptrica, whereas perspective would focus exclusively on direct vision; perspectiva postulated the binocular vision whereas linear perspective would adopt the conditions of monocular vision. These were the two main bifurcations that led to the development of perspectiva artificialis.

The system of representation that we call today "perspective" derives its name from *perspectiva*, the term used in the Middle Ages to designate the science of optics (ὀπτική in Greek and *al-manāẓir* in Arab). This connection can be explained by the fact that certain principles useful to painters and architects are based on geometrical optics, beginning with the law that objects appear to diminish in size as a function of distance:

Alhacen: Perception of size is due only to a correlation of the base of the visual cone encompassing the size to the angle of the cone at the center of sight and to the length of the cone, which represents the magnitude of the distance of the visible object.[1]

---

[1] "Comprehensio magnitudinis non est nisi ex comparatione basis piramidis radialis continentis magnitudinem ad angulum piramidis qui est apud centrum visus et longitudinem piramidis, que est remotio magnitudinis rei vise," *Opticae Thesaurus Alhazeni Arabi libri septem*, New York, 1972, p. 58; A. Mark Smith, *Alhacen's Theory of Visual Perception*, Philadelphia, 2001, vol. I, p. 185.

© Springer International Publishing Switzerland 2016
D. Raynaud, *Studies on Binocular Vision*,
Archimedes 47, DOI 10.1007/978-3-319-42721-8_1

Bacon: There can be no determination of the magnitude of an object in accordance with the size of the angle, but it is necessary that the angle be considered and the length of the pyramid.[2]

Pecham: Perception of the size [of an object] derives from perception of the radiant pyramid and comparison of the base to the length and to the size of the angle.[3]

These should not be viewed as isolated observations. A number of studies[4] conducted over the past two decades have established the debt that *perspectiva artificialis* owes to *perspectiva naturalis*. This chapter will discuss some of the connections and divergences between these two sets of ideas.

1.1   Perspective in the Classification of the Sciences

The medieval classification of the sciences[5] can help us to understand the links that existed between perspective, geometry and arithmetic. According to the classification by al-Fārābī, which was transmitted to the Latin world through the translations of Gerard of Cremona and Dominicus Gundissalinus,[6] *perspectiva* is three-fold, consisting of *optica* (direct rays), *catoptrica* (reflected rays), and *dioptrica* (refracted rays). Pictorial perspective is tied only to *optica*. The relations between the sciences were understood through the Aristotelian concept of

---

[2]"Non potest esse certificatio magnitudinis rei secundum quantitatem anguli, sed oportet quod consideretur angulus et longitudo pyramidis," *The 'Opus majus' of Roger Bacon*, ed. A.G. Little, reprint, Frankfurt am Main, 1964, pp. 115–116.

[3]"Comprehensionem quantitatis ex comprehensione procedere pyramidis radiose et basis comparatione ad quantitatem anguli et longitudinem distantie," David C. Lindberg, *John Pecham and the Science of Optics*, Madison, 1970, p. 146.

[4]For example, Emma Simi Varanelli, "Dal Maestro d'Isacco a Giotto. Contributo alla storia della 'perspectiva communis' medievale," *Arte medievale* 2. Ser. 3 (1989): 115–143; Luca Baggio, "Sperimentazioni prospettiche e ricerche scientifiche a Padova nel secondo Trecento," *Il Santo*, 34 (1994): 173–232; Francesca Cecchini, "Artisti, committenti e perspectiva in Italia alla fine del Duecento," in *La prospettiva. Fondamenti teorici ed esperienze figurative dall'Antichità al mondo moderno*, ed. R. Sinisgalli, Fiesole, 1998, pp. 56–74; *Eadem*, "Ambiti di diffusione del sapere ottico nel Duecento," in *L'Œuvre et l'artiste à l'épreuve de la perspective*, eds. M. Dalai Emiliani, M. Cojannot Le Blanc, P. Dubourg Glatigny, Rome, 2006, pp. 19–42.

[5]James A. Weisheipl, "Classification of the sciences in medieval thought," *Mediaeval Studies* 27 (1965): 54–90; *Idem*, "The nature, scope, and classification of the sciences," ed. D.C. Lindberg, *Science in the Middle Ages*, Chicago, 1978, pp. 461–482; Graziella Federici Vescovini, "L'inserimento della 'perspectiva' tra le arti del quadrivio," *Actes du IVe Congrès international de Philosophie médiévale*, Montréal/Paris, 1969, pp. 969–974.

[6]Henri Hugonnard Roche, "La classification des sciences de Gundissalinus et l'influence d'Avicenne," in *Études sur Avicenne*, eds. J. Jolivet and R. Rashed, Paris, 1984, pp. 41–63; Jean Jolivet, "Classification des sciences," in *Histoire des sciences arabes*, ed. R. Rashed, Paris, 1997, 3, pp. 255–270.

subalternation.[7] There is subalternation when a superior science (*scientia subalternans*) provides the *propter quid* of a fact presented by an inferior science (*scientia subalternata*). Ever since Aristotle's *Posterior Analytics*, optics has been subordinate to geometry, which has led either to its outright absorption into geometry, as in Boethius' *De Trinitate*, or to its insertion among the geometrical sciences, as in Dominicus de Clavasio's *Questiones super perspectiva*.[8]

Many classification systems made a clear distinction between the theoretical and the practical sciences, as in Isidorus of Seville's *Etymologiae* or the *Didascalicon* by Hugh of St Victor. In contrast, Arabic scholars saw a continuous gradation from the speculative sciences to the practical sciences.[9] Along the lines of al-Fārābī, Dominicus Gundissalinus named seven mathematical sciences as having both theoretical and practical aspects, including optics (*de aspectibus*), statics (*de ponderibus*), and engineering (*de ingeniis*). Drawing on this same tradition, Roger Bacon devoted an entire chapter of *Communia mathematica* to "Geometria speculative et practica"[10] and Fra' Luca Pacioli expounded on the "parte principale de tutta l'opera de Geometria, in tutti li modi theorica e pratica."[11] Such connections explain why perspective was so heavily dependent on the geometrical sciences and why, although a practical art, it benefitted from the contributions of speculative geometry and *perspectiva naturalis*.

Last but not least, it must be mentioned that the mathematical sciences were divided based on their subject matter—arithmetic was the science of discrete quantities (πλῆθος) while geometry was the science of continuous quantities (μέγεθος). This dichotomy remained in place from the time of Aristotle, Proclus,

---

[7]Aristotle rejects the mixing of genres during the course of a demonstration but admits the subordination of the sciences under certain conditions; Aristotle, *Posterior Analytics*, ed. H. Tredennick, Cambridge, 1966, I, IX, 66–69 and I, XIII, 88–90. He recognized, for example, that optics was subordinate to geometry, I, XIII, 88–90. Later, metaphysical considerations sometimes contributed to emancipate optics from pure mathematics. On *subalternation scientiae* in the Middle Ages, see Steven J. Livesey, "Science and theology in the fourteenth century: the subalternate sciences in Oxford commentaries on the sentences," *Synthese* 83 (1990): 273–292.

[8]"It is known that the mathematical sciences are five—namely arithmetic, geometry, music, astronomy and perspective—which differ, as was seen in the first conclusion/Est sciendum quod quinque sunt scientiae mathematicae, scilicet arismetrica, geometria, musica, astrologia et perspectiva quae differunt secundum quod visum in prima conclusione," Dominicus de Clavasio, *Quaestiones perspectivae*, Florence, BNCF, San Marco, Conv. Soppr. J X 19, quaest. 1, ff. 44r-v; Graziella Federici Vescovini, *Studi sulla prospettiva medievale*, Turin, 1964, p. 210.

[9]The inclusion of the practical sciences in the overall classification of the sciences seems to have begun with the ancient Greeks. Pappus reports that Heron's disciples divided mechanics into two parts: (i) the theoretical, which included geometry, arithmetic, astronomy, and physics, and (ii) the manual, which included architecture (οἰκοδομική), ironworks (χαλκευτική), carpentry (τεκτονική), and painting (ζωγραφική), Pappi Alexandrini *Collectionis quae supersunt*, ed. F. Hultsch, Berlin, 1876–8, pp. 1022.3–1028.3 (VIII, praef. 1–3).

[10]Roger Bacon, *Communia mathematica Fratris Rogeri*, ed. R. Steele, Oxford, 1940 (I, 3, 2).

[11]Luca Pacioli, *Summa de aritmetica, geometria, proportione et proportionalita*, Venice, 1494, fol. 75r.

and Geminus[12] up to the Italian Renaissance treatises that identified devices *per numero* and *per linea*.[13] In the light of these categories we can better understand why optics as a geometrical science guided the earliest experiments on perspective.

1.2   The Phases in the Development of Optics

If one examines the literature on the history of the classification of the sciences,[14] one finds that the boundaries of optics were particularly labile and the place it occupied on the tree of scientific knowledge was subject to marked fluctuations. The only scientific classification systems that even mention optics before the advent of modern science were those of Aristotle in the *Nicomachean Ethics* (ca. 340 BC), al-Fārābī in his work *Kitāb iḥṣā' al-'ulūm* (*Opusculum de scientiis*, ca. 950), and the English friar Robert Kilwardby in *De ortu scientiarum* (ca. 1250).

If one compares the chronology of optical treatises to these milestones, one immediately notes that the introduction of optics into the classification of the sciences coincided with those periods in which research in this area was most prolific. This correlation should not surprise us for it is when new knowledge emerges that the need arises to assign it a place reflecting its importance. The first period of intense activity was seen in antiquity, with the work of Euclid (ca. 300 BC), Hero of Alexandria (ca. 70 AD), Damianus (ca. 100), Ptolemy (ca. 127), and Theon of Alexandria (before 405).[15] The second period corresponded to the study of optics in the Arab world, which it would perhaps be more accurate to refer to as the science of optics in the Arabic language, given the significant contributions of Greek, Nestorian and Persian savants who expressed themselves in this language. The best known texts are those of al-Kindī (ca. 846), Ḥunayn ibn Isḥāq (ca. 857), Qusṭā ibn

---

[12]Bernard Vitrac in Euclide, *Éléments*, vol. 2, pp. 19, 22.

[13]Luca Pacioli, *Divina proportione*, Venice, 1509; Pietro Cataneo, *L'Architettura*, Venice, 1567; Andrea Palladio, *I Quattro Libri de architettura*, Venice, Domenico dei Franceschi, 1570. On the devices *per numero* and *per linea*, see Samuel Gessner, *Les Mathématiques dans les écrits d'architecture italiens, 1545–1570*, Paris, 2006, pp. 109–144.

[14]James A. Weisheipl, "Classification of the sciences in medieval thought," *Mediaeval Studies* 27 (1965): 54–90; Graziella Federici Vescovini, "L'inserimento della 'perspectiva' tra le arti del quadrivio," *Arts libéraux et philosophie au Moyen Âge*, Paris/Montréal, 1969, pp. 969–974; Jean Jolivet, "Classification des sciences" in *Histoire des sciences arabes*, eds. Roshdi Rashed and Régis Morelon, Paris, 1997, vol. 3, pp. 255–270.

[15]*Euclidis opera omnia*, vol. VII: *Optica... Catoptrica cum scholiis antiquis*, ed. J.L. Heiberg, Leipzig, 1895; Wilfred R. Theisen, "Liber de visu: The Greco-Latin Tradition of Euclid's Optics," *Mediaeval Studies* 41 (1979): 44–105; *Heronis Alexandrini opera quae supersunt omnia*, vol. II. *Mechanica et Catoptrica*, eds. L. Nix and W. Schmidt, Stuttgart, 1900; *Damianos Schrift über Optik*, ed. R. Schöne, Berlin, 1897; Albert Lejeune, *L'Optique de Claude Ptolémée dans la version latine d'après l'arabe de l'émir Eugène de Sicile*, Leiden, 1989; *Euclidis opera omnia*, vol. VII: *Opticorum recensio Theonis*, ed. J.L. Heiberg, Leipzig, 1985.

Lūqā (ca. 860), Aḥmad Ibn 'Īsā (after 860), Ibn Sahl (ca. 985), and above all Ibn al-Haytham, known in the Latin world as Alhacen (d. after 1040).[16]

The third great period in the history of optics was that of the thirteenth century in Europe and the most significant contributions are associated with the names of Robert Grosseteste (ca. 1235), Roger Bacon (ca. 1266), Witelo (ca. 1277), John Pecham (ca. 1279), and their fourteenth-century epigones, including Egidius de Baisiu (ca. 1300), Dietrich de Freiberg (ca. 1304), Dominicus de Clavasio (before 1362) and Biagio Pelacani da Parma (ca. 1390).[17]

Issuing from this intense activity, *perspectiva* could already lay claim to being a synthesis of physical optics (treating such problems as the multiplication of species, the instantaneous versus the temporal propagation of light), geometric optics (the images reflected in mirrors, the source of the moon's light, the theory of the rainbow), physiological optics (the anatomy of the eye, the phenomenon of the persistence of vision, the conflicting theories of intromission and extramission of visual rays), and psychological optics (used, for example, to explain optical illusions).[18]

1.3   The Similarities Between *Perspectiva* and Perspective

Before embarking on a discussion of the relationship between *perspectiva* and perspective, it should be pointed out that this relationship falls into the category of "a necessary but not sufficient condition." Not sufficient because there were many determining factors in the emergence of perspective—not only the development of theories of vision, but also the support of medieval theologians for iconography and

---

[16]Elaheh Kheirandish, *The Arabic Version of Euclid's Optics: Kitāb Uqlīdis fī ikhtilāf al-manāẓir*, New York, 1999; Roshdi Rashed, *Optique et mathématiques*, Aldershot, 1992; *idem, Géométrie et Dioptrique au Xe siècle: Ibn Sahl, al-Qūhī et Ibn al-Haytham*, Paris, 1993; *idem, Œuvres philosophiques et scientifiques d'al-Kindī: L'optique et la catoptrique*, Leiden, 1996; *idem, Geometry and Dioptrics in Classical Islam*, London, 2005; Abdelhamid I. Sabra, *The Optics of Ibn al-Haytham, Books I-III: On Direct Vision*, London, 1989; *idem, The Optics of Ibn al-Haytham, Books IV-V: On Reflection and Images Seen by Reflection*, Kuwait, 2002; A. Mark Smith, *Alhacen's Theory of Visual Perception*, Philadelphia, 2001; *idem, Alhacen on the Principles of Reflection*, Philadelphia, 2006; *idem, Alhacen on Image-Formation and Distortion in Mirrors*, Philadelphia, 2008; *idem, Alhacen on Refraction*, Philadelphia, 2010.

[17]Ludwig Baur, "Die philosophischen Werke des Robert Grosseteste," *Beiträge zur Geschichte der Philosophie des Mittelalters* 9 (1912): 1–778; David C. Lindberg, *Roger Bacon and the Origins of Perspectiva in the Middle Ages*, Oxford, 1996; Witelo, *Opticae Thesaurus*; David C. Lindberg, *John Pecham and the Science of Optics*, Madison, 1970; José-Luís Mancha, "Egidius of Baisiu's theory of pinhole images," *Archive for History of Exact Sciences*, 40 (1989): 1–35; Maria Rita Pagnoni-Sturlese, Rudolf Rehn and Loris Sturlese, *Dietrich von Freiberg. Opera Omnia, IV. Schriften zur Naturwissenschaft*, Hamburg, 1985; Graziella Federici Vescovini, "Les questions de 'perspective' de Dominicus de Clivaxo," *Centaurus* 10 (1964): 236–246; Blaise de Parme, *Questiones super perspectiva communi*, eds. G. Federici Vescovini et al., Paris, 2009.

[18]Gérard Simon, *Le Regard, l'être et l'apparence dans l'optique de l'Antiquité*, Paris, 1988.

the use of imagery as a mnemonic technique,[19] the desire for social mobility on the part of artisans (which motivated them to reduce the gulf between the mechanical arts and the liberal arts by demonstrating that their work was based on knowledge of the *quadrivium*),[20] etc.

The number of underlying factors can be greatly reduced if one focuses on the study of the textual parallels linking Renaissance treatises on perspective with the treatises on optics written in previous epochs. These parallels, which sometimes bordered on outright copying, can be found in the writings of Lorenzo Ghiberti,[21] Leon Battista Alberti, Piero della Francesca, and Leonardo da Vinci. A passage typical of such undeclared borrowings can be found in the *Codex Atlanticus*, fol. 543r:

> English translation: Light produces an impression in the eye that is directed toward it. This result is proved by an effect, for when the eye sees brilliant lights, it suffers and endures pain. Also, after a glance [at bright lights], images of intense brightness remain in the eye, and they cause a less illuminated place to appear dark until the traces of the brighter light have disappeared from the eye.

> Leonardo: La luce operando nel uedere le chose contra se conuerse alquanto le spezie di quelli ritiene. Questa conclusione si pruoua per li effetti perche la uista in uedere luce alquanto teme. Ancora dopo lo sguardo rimangano nel locchio similitudine della chosa intensa e fanno parere tenebroso il luogo di minor luce per insino che dallocchio sia spartito il uestigio de la impression de la magiore luce.[22]

> Pecham: Lucem operari in uisum supra se conuersum aliquid impressiue. Hec conclusio probatur per effectum, quoniam uisus in uidendo luces fortes dolet et patitur. Lucis etiam intense simulacra in oculo remanent post aspectum, et locum minoris luminis faciunt apparere tenebrosum donec ab oculo euanuerit uestigium luminis maioris.[23]

This is the literal translation of a paragraph from John Pecham's *Perspectiva communis* on which, however, Leonardo does not elaborate. Research that I have devoted to these parallel texts allows me to formulate certain conclusions regarding the history of perspective.[24]

---

[19]Alain Besançon, *L'Image interdite. Une histoire intellectuelle de l'iconoclasme*, Paris, 1994; Emma Simi Varanelli, "Arte della memotecnica e primato dell'imagine negli ordines studentes," *Bisancio e l'Occidente: arte, archeologia, storia*, Rome, 1996, pp. 505–525.

[20]Robert E. Wolf, "La querelle des sept arts libéraux dans la Renaissance, la Contre-Renaissance et le Baroque," *Renaissance, Maniérisme, Baroque*, Paris, 1972, pp. 259–288.

[21]Klaus Bergdolt, *Der dritte Kommentar Lorenzo Ghibertis. Naturwissenschaften un Medizin in der Kunsttheorie der Frührenaissance*, Weinheim, 1998.

[22]Leonardo da Vinci, *The Notebooks of Leonardo da Vinci*, ed. Jean Paul Richter, New York, Dover, 1970, vol. I, p. 24.

[23]David C. Lindberg, *John Pecham and the Science of Optics*, p. 62.

[24]For a detailed study of the textual parallels, see Raynaud, *L'Hypothèse d'Oxford*, Paris, 1998, pp. 163–209; *idem*, "L'ottica di al-Kindī e la sua eredità latina. Una valutatione critica," in *Lumen, Imago, Pictura*, Atti del convegno internazionale di studi (Rome, Bibliotheca Herziana, 12–13 April 2010), eds. S. Ebert-Schifferer, P. Roccasecca and A. Thielemann, Rome (in press); *Idem*, "An unknown treatise on shadows referred to by Leonardo da Vinci," in *Perspective as Practice. An International Conference on the Circulation of Optical Knowledge in and Outside the Workshop*, eds. S. Dupré and J. Peiffer, Max Planck Institut für Wissenschaftsgeschichte (Berlin,

1. This research shows first of all that there was a marked continuity between the study of optics and the study of perspective, thus greatly reducing the credibility of the classic thesis that a major rupture took place during the Renaissance in Italy. This was doubtless true on certain levels, but curiously enough the treatises on perspective seemed to form an exception to the rule.
2. Another finding is that the treatises on optics most often cited during the Renaissance were not those of antiquity but texts from the Arab and Latin Middle Ages. This appears to be quite strange given the fact that the Renaissance has been characterized by scholars as the period of the "rediscovery of the antique."
3. Among the medieval authors, the ones most frequently cited belong to the tradition of the 'Perspectivists,' principally Alhacen and his Western successors. But here again is another source of surprise: Witelo, who had close connections with the papal court in Viterbo, is rarely quoted, and Biagio Pelacani da Parma hardly more often.
4. In *L'Hypothèse d'Oxford. Essai sur les origines de la perspective*, I proposed that these anomalies could be understood by introducing a socio-historic factor. *The texts on optics that the perspectivists of the Renaissance consulted were likely to have been the ones that were most accessible in terms of the number of manuscript copies in circulation.* The hierarchy between these texts can be reconstructed from their distribution: in libraries across Europe a total of 65 manuscripts by Bacon and 64 by Pecham can be counted, compared to 25 by Witelo and 16 by Biagio Pelacani da Parma.[25] Thus, the frequency with which authors borrowed from Bacon and Pecham could be due to the exceptional diffusion of their texts during the course of the thirteenth and fourteenth centuries.

(Footnote 24 continued)

12–13 October 2012), Berlin (in press); *Idem*, "Application de la méthode des traceurs à l'étude des sources textuelles de la perspective. Auteurs, traités, manuscrits," in *Vision and Image-Making: Constructing the Visible and Seeing as Understanding*, Actes du colloque international, Centre d'Études Superieures de la Renaissance et Le Studium CNRS, Orléans (Tours, 13–15 September 2013).

[25]David C. Lindberg, *A Catalogue of Medieval and Renaissance Optical Manuscripts*, Toronto, 1975. With regard to the invention of perspective, links have also been drawn to the abacus, the cartographic projections of Ptolemy, the use of the astrolabe, or a combination of all of these sources, Birgitte Bøggild-Johanssen and Marianne Marcussen, "A critical survey of the theoretical and practical origins of the Renaissance linear perspective," *Acta ad Archaelogiam et Artium Historiam Pertinentia* 8 (1981): 191–227. This knowledge probably contributed to the development of the perspective system, but in the Quattrocento their influence remained secondary to that of optics: (1) if perspective had been based on cartography, contemporaries would probably have spoken of "the cartography of painters" rather than "the perspective of painters"; (2) the identification of certain sources appears to be conjectural because they are not supported by a study of parallel texts (Raynaud, *L'Hypothèse d'Oxford*, pp. 165–167); (3) the notion of a "source" depends on one's point of view. Simply because knowledge appears to us on logical grounds to be 'pertinent' to a subject does not necessarily mean that it would have been utilized.

5. A study of the holdings in Italian libraries sheds light on the context in which these borrowings unfolded.[26] For example, a comparative analysis of Florentine inventories before the middle of the Quattrocento shows that there were no treatises on optics in the Badia Fiorentina or the Medici library, but that they could be found in convent libraries. While the Dominicans of Santa Maria Novella had no manuscripts on *perspectiva*, the Augustinian order of the Basilica of Santo Spirito possessed one (*Perspectiva magistri Vitellonis*) and the Franciscans of Santa Croce no less than six (Robert Grosseteste, *De luce seu inchoatione formarum*; Bartholomew of England, *De proprietatibus rerum*; John Pecham, *Tractatus de perspectiva* and *Perspectiva communis*; Bartholomeus de Bononia, *De luce*; and Petrus Aureolus, *Scriptum in II Sententiarum*).

6. The large number of treatises on optics to be found in the libraries of the Franciscan convents during the Middle Ages can be explained by the conjunction of two factors: (1) a homophilic bias, that is, the preference of a religious community for authors belonging to the same order (thus, Dominican authors were over-represented in the libraries of Dominican convents, Franciscan authors in the collections of Franciscan libraries, and so on),[27] and (2) the strong commitment of Franciscans to the writing and copying of manuscript treatises on *perspectiva*. A tally beginning with the *Catalogue of Optical Manuscripts* shows that among 310 manuscripts from the thirteenth and fourteenth centuries preserved in European libraries, 92 % were redacted by clerics and of these 80 % were written by friars belonging to the mendicant orders. A total of 71 % (220 MSS) were the work of Franciscan friars, of which 66 % (205 MSS) were by just three authors—Grosseteste, Bacon and Pecham.[28]

The interest of the Franciscans in the subject of optics, joined to the principle of homophily, properly explains the presence of Franciscan 'best-sellers' in Italian libraries. In *Optics and the Rise of Perspective* I used this data to show that the diffusion of optics was one of the pre-conditions for the development of linear perspective during the Renaissance.

The purpose of this book is different. It will test the hypothesis that there were close ties between optics and perspective, but from a different angle; that is, by asking whether long-abandoned medieval notions of optics may have left traces in the way perspective was envisaged in later epochs. From such traces—if they do exist—it should be possible to furnish proof of how close the relationship was between *perspectiva* and perspective. Since my aim here is more to lay out and conduct a scientific test than a discourse on culture, I will begin by reviewing the most salient differences between *perspectiva* and perspective.

---

[26]Raynaud, *L'Hypothèse d'Oxford*, pp. 301–349.

[27]See the statistical tables in Raynaud, *L'Hypothèse d'Oxford*, p. 329.

[28]Raynaud, *Optics and the Rise of Perspective*, Oxford, 2014, chapter 3, especially pp. 64–65.

1.4   The Differences Between *Perspectiva* and Perspective

The existence of correspondences between *perspectiva* and perspective does not negate the possibility that differences exist between medieval optics and Renaissance perspective. In addition to the fact that their relationship was one of theory to practice, or of science to technology, two other bifurcations marked the passage from one to the other.

First of all, *perspectiva* as it was understood and taught during the Middle Ages was a tripartite science that comprised the study of direct rays (*optica*), reflected rays (*catoptrica*), and refracted rays (*dioptrica*).[29] By comparison, Renaissance treatises on perspective covered a much narrower field of investigation, ignoring for example the study of burning mirrors and such natural phenomena as the rainbow, the halo of the moon, and the apparent twinkling of the stars. An entire facet of *perspectiva* thus disappeared as scholars concentrated on direct vision.

Secondly, all medieval treatises on *perspectiva* speculated at length on the central conundrum of binocular vision—how do the separate images received by the two eyes come to be fused?[30] And yet modern summaries, as well as the sources of the period, continually underline the close ties that link the invention of perspective and the postulate of monocular vision. These presuppositions have been laid out by most historians of perspective. At the beginning of the twentieth century, Erwin Panofsky observed that in order to construct a perspective it is necessary to grant, "First, that we see with a single and immobile eye."[31] Thirty years later Gioseffi declared in his turn that *monocular vision* was the condition that guaranteed the integrity of the system of perspective.[32] In his account of the history of perspective, Laurent expounded on this point: "The two eyes of binocular vision are reduced to a single one (monocular vision) called the eye and placed at the summit of the visual cone."[33] The historical sources are no less prolix. The postulate of monocular vision figures prominently in Manetti's *Vita* of Brunelleschi, in which the *tavoletta* of the baptistery of San Giovanni in Florence is described: "It is necessary that the painter postulate beforehand a single point from which his painting should be viewed/Il dipintore bisognia che presuponga un luogo solo

---

[29]"Otherwise, vision is fundamentally triple, depending upon whether it is made of straight, refracted or reflected rays/Aliter vero triplicatur uisio secundum quod fit recte, fracte et reflexe," *The 'Opus majus' of Roger Bacon*, ed. Little, p. 162.

[30]Alhacen, *Opticae Thesaurus*, pp. 76–87; Smith, *Alhacen's Theory of Visual Perception*, vol. II, p. 562–582; *The 'Opus majus' of Roger Bacon*, pp. 92–99; Lindberg, *John Pecham and the Science of Optics*, pp. 116–118; Witelo, *Opticae Thesaurus... Item Vitellonis Thuringopoloni libri decem*, pp. 98–108.

[31]Erwin Panofsky, "Die Perspektive als symbolische Form," *Vorträge der Bibliothek Warburg* 4 (1924/5): 258–331, *Perspective as Symbolic Form*, New York, 1991, p. 29.

[32]Decio Gioseffi, *Perspectiva artificialis*, Trieste, 1957, p. 8.

[33]Roger Laurent, *La Place de J.-H. Lambert (1728–1777) dans l'histoire de la perspective*, Paris, 1987, p. 37.

d'onde s'a a uedere la sua dipintura."[34] The same condition is formulated in the commentary to De visu by Grazia de' Castellani: "And you put a single eye at point C where there is a small hole/E tu ponj un solo occhio al punto.c. doue è uno picholo bucho."[35] As the vanishing point is the orthogonal projection of the eye onto the picture plane, the monocular postulate imposes the uniqueness of the vanishing point in a central linear perspective.

In contrast, the theory of monocular vision was much less developed in medieval optics and, it seems, was always seen in relation to the size of an object. This was illustrated by the classic experiment of the hand and the wall, which is cited in turn by Alhacen, Pecham and Alberti:

> Alhacen: For instance, if an observer looks at a wall that lies at a moderate distance from the eye, and if he accurately determines the distance and size of that wall, and if he accurately determines the magnitude of its breadth, then, if the observer places his hand in front of one of his eyes between the center of sight and the wall *and closes the other eye*, he will find that his hand will cover a considerable portion of that wall.[36]

> Pecham: If a *one eyed man* looks at a large wall and, after certifying its size, places his hand before his eye, the hand will appear under an angle equal to or larger than that under which the wall is seen; nevertheless, the hand will appear to him smaller than the wall because it is less distant.[37]

> Alberti: I say that the part of the rod that lies between C and B goes as many times into the distance that is between B and D, i.e., between *your eye* and the foot of the rod, as many times as the height of the tower goes into the distance that is between your eye and the foot of the tower.[38]

[34]Antonio di Tuccio Manetti, *The Life of Brunelleschi, by Antonio di Tuccio Manetti/Vita di Filippo di Ser Brunelleschi*, eds. H. Saalman and C. Engass, University Park, 1970, p. 43.

[35]Gino Arrighi, "Un estratto dal 'De visu' di M° Grazia de' Castellani," *Atti della Fondazione Giorgio Ronchi* 22 (1967): 44–58, p. 47; Filippo Camerota, "Misurare 'per perspectiva'," *La prospettiva. Fondamenti teorici ed esperienze figurative dall'Antichità al mondo moderno*, Fiesole, 1998, pp. 293–308.

[36]"Verbi gratia, quod quando visus aspexerit parietem remotum a visu remotione mediocri, et certificaverit visus remotionem illius parietis et quantitatem eius, et certificaverit quantitatem latitudinus eius, deinde apposuerit aspiciens manum uni visui inter visum et parietem *et clauserit alterum oculum*, inveniet tunc quod manus eius cooperiet portionem magnam illius parietis," Alhacen, *Opticae Thesaurus*, p. 52; Smith, *Alhacen's Theory of Visual Perception*, vol. I, p. 171.

[37]"Si *monoculus* aspiciat aliquem parietem magnum et quantitatem eius certificet deinde oculo suo manum anteponat, ipsa manus uidebitur sub eodem angulo uel sub maiori quam paries uisus est, nec tamen tanta ei apparebit quantus paries apparet quia minus distat," Lindberg, *John Pecham and the Science of Optics*, p. 146.

[38]"Dico che la parte del dardo quale sta fra C et B entra tante volte nella distanza quale sta fra B e D cioè fra *l'occhio vostro* e il piè del dardo, quante volte l'altezza della torre entra nella distanza quale è fra l'occhio vostro et il piè della torre," Alberti, *Ex ludis rerum mathematicarum*, Cambridge, Mass., MS. Houghton Typ 422.2, fol. 1v.

Reducing the field of investigation to the study of direct vision (*optica*) and substituting the postulate of binocular vision for that of monocular vision would appear to be the two principal bifurcations—a consequence of the compartmentalization of the sciences—that set the seal on the continued evolution of *perspectiva artificialis*.

But there are just as many questions to be posed regarding the origins of the new theory of perspective, because the differences between *perspectiva* and perspective could have resulted from a lack of knowledge of the texts on optics, or a rejection of theoretical optics in favor of other sources such as the use of the astrolabe or practical geometry, or even the draconian selection from the textual sources of only those elements that were compatible with the development of linear perspective. The method used by Brunelleschi to depict the *tavoletta* of the baptistery could be viewed as part of a historical continuum, a logical consequence of the dearth of sources on monocular vision available during the Middle Ages. It could equally well be seen as an application of practical geometry to the measurement of inaccessible sizes,[39] thus favoring the discontinuity thesis. How might this issue be resolved?

The path that I will follow in this book differs from the one adopted earlier in *L'Hypothèse d'Oxford* and in *Optics and the Rise of Perspective*. If the origins of perspective are to be found *preponderantly* in medieval optics, then one should be able to identify some of its vestiges in the earliest experiments on perspective, which were conducted between the end of the Duecento and the second half of the Cinquecento. The uniformity of the procedure for creating perspective views was a consequence of its being taught as a regular part of the curriculum in the academies, beginning with the Accademia del Disegno (established in Florence in 1563) and the Accademia di San Luca (founded in Rome in 1577).[40] Before this time neither the concepts of perspective nor its methods were fixed and perspectivists, not being constrained to follow a definite set of rules, came up with a number of approaches that would all be regarded as "heterodox systems" once the rules for the representation of perspective were fixed and adopted. The period from the Duecento to

---

[39]This exercise was included by many authors in their treatises on geometry, from Euclid to Johannes of Muris, and from Dominicus of Clavasio to Cosimo Bartoli; Euclid, *Liber de visu*, ed. W. Theisen, p. 72; Stephen K. Victor, *Practical Geometry in the High Middle Ages*, Philadelphia, 1979, p. 295; Hubert L.L. Busard, *Johannes de Muris. De Arte mensurandi*, Stuttgart, 1998, p. 145; idem, "The Practica Geometriae of Dominicus de Clavasio," *Archive for the History of Exact Sciences* 2 (1965): 520–575, p. 539; Cosimo Bartoli, *Del modo di misurare le distantie, le superfitie, i corpi*, Venezia, 1564, fol. 19v, 24r.

[40]Marica Marzinotto, "Filippo Gagliardi e la didattica della prospettiva nell'accademia di San Luca a Roma, tra XVII e XVIII secolo," *L'Œuvre et l'artiste à l'épreuve de la perspective*, Rome, 2006, pp. 153–177.

the Cinquecento therefore offers an ideal field of investigation to explore whether the medieval principles of optics inspired systems of representation other than linear perspective.

We have characterized the passage from *perspectiva* to perspective in terms of two bifurcations: (i) the reduction of tripartite *perspectiva* (*optica, catoptrica, dioptrica*) to direct vision alone, and (ii) the adoption of the postulate of monocular vision. As *catoptrica* and *dioptrica* do not seem to have left any mark on the new system of perspective, the central axis of this book will consist in exploring whether the postulate of *binocular vision* could have inspired the many and varied systems of representation that were conceived beginning in the Duecento.

Part I
Errors

Chapter 2
Knowledge and Beliefs Regarding Linear Perspective

Abstract The aim of this chapter is to deconstruct the notion that linear perspective formed a stable system of representation beginning in the Quattrocento. Doubts must be raised because the history of perspective is in fact quite conjectural due to the many lacunae scattered along its path; one crucial example is the exact nature of the contributions of Brunelleschi, Alberti, and Masaccio. A second obstacle is the fact that a multiplicity of approaches were in use from the end of the Duecento to the Cinquecento, when the academies formally introduced the teaching of perspective techniques. Between these two time points perspectivists explored numerous systems of perspective, introducing errors and variations that can be explained by the uneven distribution of knowledge regarding the laws of optics and geometry.

The challenge facing the practitioner in representing space may be summed up as follows: how can one apprehend and capture the three-dimensionality of a solid in the two dimensions of a plane? Among the strategies commonly used, some consist in decomposing the object into a series of partial views—the horizontal plane, elevation, and profile—from which one can, with a little practice, mentally reconstruct the spatiality of the object. Other strategies instead attempt to provide a visual synthesis that is capable of immediately evoking the three-dimensionality of the solid. Parallel axonometric projections (isometric, dimetric, trimetric) and oblique projections (cavalier and military), both of which conserve the parallelism of an object's straight lines, fall into this category. Perspective itself rejects the property of parallelism for the principle of a gradual reduction in size, reproducing as closely as possible the conditions of natural vision: i.e., two straight lines that are not confined to the frontal plane converge toward a vanishing point. Linear perspective is just one of the systems that respects this principle (since it holds true for the curvilinear and synthetic perspectives as well), but it is the version that is generally considered in discussions of perspective *tout court*. I will conform to this usage by discussing only the case of linear perspective here.

The argument that can be advanced is that the characterization of perspective space as a unitary, coherent and stable representation is not sufficient because it fails to take into account the wide range of practices that are known to have existed.

© Springer International Publishing Switzerland 2016   15
D. Raynaud, *Studies on Binocular Vision*,
Archimedes 47, DOI 10.1007/978-3-319-42721-8_2

Linear perspective constitutes an *open* rather than a closed *system*, one that reflected the mobilization over time of specific intellectual resources.

In the first part of this chapter, it will be shown that the work of Italian craftsmen at the beginning of the Quattrocento did not lead to a codified and homogeneous set of perspective practices (illustrating, in sociological terms, the effects of belief). In the second part it will be shown that the diversity of perspective conceptions in circulation can be explained differences in the optical-geometric resources available to the perspectivists (the effects of knowledge).

2.1 The Myth of Perspective

To begin, it will be useful to examine the supposedly stable nature of the perspective system. It is true that one finds, from Euclid[1] to Gibson,[2] unvarying expressions of the law of diminution in size as a function of distance. But the solidity of this principle has sometimes served as a pretext to impose the uniqueness of the perspective system and to reify it, particularly as far as the Renaissance is concerned, when in fact research on perspective often took the form of disparate and uncoordinated initiatives. Let us examine the contributions of Brunelleschi, Alberti and Masaccio, to whom have been attributed the invention, codification, and first major realization of the concept of perspective, respectively.

2.1.1 Filippo Brunelleschi

Filippo Brunelleschi (1377–1446) is usually credited with having realized the first rigorous work of perspective, in Florence around the year 1413. The documentation is scarce, but the artist apparently conducted an ingenious demonstration of the accuracy of his construction. He stood at a distance of three *braccia* (arm's lengths) from the main portal of the cathedral of Santa Maria del Fiore facing the Baptistery of San Giovanni, holding a mirror in one hand and a panel painting of the octagonal-shaped building in the other in such a way that he could observe, through a small hole pierced in the panel, the image of the painting reflected in the mirror. From his position he could see at the same time the image and the actual building, and thus judge the accuracy of his perspective drawing. The first difficulty regarding this experiment is that no material trace of it has survived. In particular, the

---

[1] "Objects of equal size unequally distant appear unequal and the one lying nearer to the eye always appears larger/Aequales magnitudines inaequaliter expositae inaequales apparent et maior semper ea quae propius oculum adjacet," *Optica*, ed. J.L. Heiberg, Leipzig, 1895.

[2] James J. Gibson, *The Ecological Approach to Visual Perception*, Boston, 1979.

tavoletta (panel painting) has been lost and there is no way of knowing what perspective method was used by the architect.

The only description that has come down to us is a second-hand account attributed to Antonio di Tuccio Manetti, who was born a decade after the experiment took place. What is more, his account does not grant Brunelleschi's display the status that is generally ascribed to it of an experiment in optics. In fact, Manetti never employs the word "experiment" although the term is amply attested to both in medieval Latin and in the Italian vernacular.[3] He does couch his description in very concrete terms: "[Brunelleschi] put into practice" (*misse inatto*), "he displayed a panel" (*mostro una tauoletta*), "he made a painting" (*fecie una pittura*) ... but from this one cannot strictly speaking infer either an experiment of a public nature conducted before eyewitnesses, nor the existence of an experimental set-up of any kind. Hence, there is no concrete proof that Brunelleschi carried out a demonstration in the doorway of the cathedral of Florence. What Manetti's biographical account does offer is a fairly detailed description of his painting of the baptistery.

Let us identify the crucial points relating to perspective in this account, which are conditions A (the vantage point of the viewer), B (the scene depicted) and C (the size of the eyehole). These three conditions as described by Manetti in fact contradict one another. It is a simple matter to calculate the theoretical field of vision based on conditions A and B: the point of view chosen for the viewer ("some three *braccia* inside the central portal of Santa Maria del Fiore/*dentro alla porta del mezzo di Santa Maria del Fiore qualche braccia tre*") and the painted scene ("up to the arch and the corner of the sheep [market] ... up to the corner of the straw [market]/*insino all uolta e canto de Pecori ... insino al canto alla Paglia*") dictate a theoretical field of vision of 54°. The actual field of vision can be calculated from condition C: Manetti stated that the diameter of the eyehole at the end facing the observer was 5 mm ("a lentil bean/*una lenta*"), widening to 30 mm at its posterior end ("a ducat, or a bit more/*uno ducato o poco piu*"). For the eyehole to form a truncated cone ("it widened conically like a straw hat/*si rallargaua piramidalmente come fa uno capello di paglia*"), the minimal thickness of the panel must have been about 15 mm. In this case the actual field of vision based on the distance of the crystalline lens from the anterior end of the eyehole[4] would have been between 13°

---

[3]One finds numerous references in the Latin and Italian translations of Ibn al-Haytham's *Kitāb al-manāẓir* (Alhacen's *De aspectibus/De li aspecti*). The terms that are attested to in Arabic, Latin and Italian are: *i'tibār* > *experientia-experimentatio* > *sperimento-sperimentatione; i'tabara* > *experimentare* > *sperimentare; mu'tabir* > *experimentator* > *sperimentatore*; cf. Abdelhamid I. Sabra, "The Astronomical Origin of Ibn al-Haytham's Concept of Experiment," *Actes du XIIe Congrès International d'Histoire des Sciences*, Paris, 1971, tome III.A, pp. 133–136.

[4](A) "In order to paint it, it seems that he stationed himself some three braccia inside the central portal of Santa Maria del Fiore/E pare che sia stato a ritrarlo dentro alla porta del mezo di Santa Marie del Fiore qualche braccia tre...," (B) "In the foreground he painted that part of the piazza encompassed by the eye, that is to say, from the side facing the Misericordia up to the arch and corner of the sheep [market], and from the side with the column of the miracle of St. Zenobius up to the corner of the straw [market]/Figurandoui dinanzi quella parte della piaza che ricieue l'occhio cosi uerso lo lato dirinpetto alla Misericordia insino alla uolta e canto de Pecorj cosi da lo lato della

and 19°, i.e. only one-fourth to one-third of the expected theoretical value. When Brunelleschi's "experiment" was reproduced in situ in April 1995, it was found that conditions A, B, and C were in fact mutually exclusive. The field of vision carves out a square measuring 7–8 m on each side corresponding precisely to the door of the Baptistery. Since all the lines lie in the frontal plane containing the façade, this is not a perspective image.[5] The results of a second experiment conducted in May 2001 as part of the 4th *ILabHS* were no more convincing as a demonstration of perspective.[6] Despite the many positive analyses of this episode that continue to appear, all serious attempts to reconstruct Brunelleschi's experiment have failed and for one simple reason: it is *physically impossible* to reproduce the tableau based on the conditions described by Manetti.

If one adds to this the fact that the only work of perspective extant that can be attributed with any probability to Filippo Brunelleschi—an engraving on a silver plaque of *Christ Casting Out a Demon* (Louvre)—does not follow the rules of linear perspective,[7] one is forced to conclude that Brunelleschi's contribution has been considerably overestimated. The doubts raised here do not concern his involvement in the development of perspective, which is incontestable, but the exact nature of this contribution, about which we know nothing. In truth only three pieces of evidence exist on the role played by the artist.

The first is a letter written by Domenico da Prato to Alessandro Rondinelli on 10 August 1413, in which Filippo Brunelleschi is described as "an ingenious man on perspective/*prespettiuo ingegnoso uomo*," but this reference could simply attest to the fact that the architect took a general interest in the subject of optics (*perspectiva* in Latin); rigorously speaking it certainly does not allow a *terminus ante quem* to be fixed for the invention of perspective.

(Footnote 4 continued)

colonna del miracolo di Santo Zanobi insino al canto alla Paglia…," (C) "The hole was as tiny as a lentil bean on the painted side and it widened conically like a woman's straw hat to about the circumference of a ducat, or a bit more, on the reverse side/El quale buco era piccolo quanto una lenta da lo lato della dipintura et da rouescio si rallargaua piramidalmente come fa uno cappello di paglia da donna quanto sarebbe el tondo d'uno ducato o poco piu…," Antonio di Tuccio Manetti, *Vita di Filippo di Ser Brunelleschi*, eds. H. Saalman and C. Engass, University Park, 1970, p. 43ff. The first reassessment of this account was made by Martin Kemp, "Science, non-science and nonsense: The interpretation of Brunelleschi's perspective," *Art History* 1 (1978): 134–161.

[5]The field of vision is fixed by the distance between the centre of the crystalline lens and the anterior opening of the eyehole, that is, $a_0 \approx 15\,\text{mm}$ in the case of an exophthalmic eye and $a_1 \approx 22.2\,\text{mm}$ in the case of a normal eye. This information allows us to calculate $\alpha = \arctan(d/a)$: $13°\,05' < \alpha < 18°\,54'$; see Raynaud, *L'Hypothèse d'Oxford*, pp. 132–150.

[6]Filippo Camerota, "Brunelleschi's panels," *The 4th International Laboratory for the History of Science*, Florence, 25 May 2001 and personal communication; *Idem*, "L'esperienza di Brunelleschi," *Nel segno di Masaccio*, Florence, 2001, pp. 32–33: "Ma date le dimensioni, non consentiva di vedere tutto il dipinto, bensì solo una porzione piuttosto limitata della facciata del Battistero" [that is, nothing else than the door].

[7]Raynaud, *L'Hypothèse d'Oxford*, pp. 73–75.

Secondly, around 1461 Filarete wrote in his treatise on architecture, "I *believe* this is the way that Pippo di Ser Brunellesco found this perspective, which had not been used before,"[8] a declaration that must be taken for what it is worth, as a statement of belief rather than an assertion of fact.

Finally, around 1480 Manetti asserted that: "[Brunelleschi] himself put into practice what painters today call perspective, because it is part of that science [i.e. optics],"[9] but this claim was based on an inappropriate interpolation of the text, and he makes no mention of an "inaugural experiment" nor does he provide a method that would permit the reconstruction of his perspective.

None of these references can be regarded as unambiguous and beyond them, the rest remains conjecture. It is necessary therefore to retain a more nuanced picture of the contribution of Brunelleschi; his role in the development of perspective is in fact quite obscure.

2.1.1.1 Leon Battista Alberti

In *De pictura*, Leon Battista Alberti (1404–1472) sets out what is generally recognized to be the first codified procedure for the representation of perspective. Even today his method is often qualified as *costruzione legittima*, a term that gained wide currency thanks to Erwin Panofsky, who wrote: "Trecento pictures after the Lorenzetti became, so to speak, progressively more false, until around 1420, when *costruzione legittima* was (as we may well say) invented."[10] The expression is replete with meaning, because it implies the existence of a law for the representation of space that is universally true. As a consequence, it imposes the notion of a unified vision of perspective that formed at the beginning of the Quattrocento and still holds today. And yet any law, to be legitimate, must meet two conditions: it has to be based on a rational order, and it must be applied. Let us examine these two points.

With regard to the foundations of the rule of perspective, on re-reading *De pictura* it becomes clear that Alberti's only intention in this text is to describe a series of *empirical* operations. He makes no attempt to justify these operations, either in terms of their correspondence to reality (perspective as the tracing of a visual experience) or their logical consistency (perspective as a system whose validity could be demonstrated).[11] The approach adopted by Alberti was strictly

---

[8]"Credo che Pippo di Ser Brunellesco trovasse questa prospettiva, la quale per altri tempi non s'era usata," Antonio Averlino detto Il Filarete, *Trattato di architettura*, eds. A.M. Grassi and L. Finoli, Milano, 1972, p. 653.

[9]"Misse innatto luj propio quello che dipintorj oggi dicono prospettiua perche ella e una parte di quella scienza...," Manetti, *Vita*, p. 43.

[10]Erwin Panofsky, "Die Perspektive als symbolische Form," *Vorträge der Bibliothek Warburg* 4 (1924/5): 258–331, *Perspective as Symbolic*, New York, 1991, p. 62.

[11]This question would not be raised in studies on perspective until much later. In 1585 Giovanni Battista Benedetti demonstrated that Alberti's construction was correct; Judith V. Field, "Giovanni

procedural, and the rational foundations for a *costruzione legittima* cannot be deduced from his text.

As for the eventual application of this rule, two facts must be pointed out. First, the study of a large body of perspective paintings from the Quattrocento shows that artists utilized various approaches in drawing their perspectives that were frequently erroneous, and did not follow the principles laid out by Alberti.[12] Secondly, recent research has shown that the use of the expression *costruzione legittima* to describe the pictorial representations of the Renaissance is in reality an anachronism. Not finding any trace of this term in texts from the Quattrocento, scholars initially believed that its first appearance could be identified in a treatise on perspective published by Pietro Accolti in 1625.[13] But in terms of occurrences this constitutes an approximation, because Accolti qualified as legitimate only the "planes" or "operations" that contribute to the construction of a perspective.[14] Only recently has it been discovered that *costruzione legittima* was in fact translated from the German term *legitime Verfahren*, an interpolation by Heinrich Ludwig, who employed it for the first time in 1882 in his edition of Leonardo da Vinci's *Trattato della pittura*. The expression was then taken up by Winterberg (1899), Kern (1915), and Panofsky (1924), with the consequences that we now know.[15] One cannot therefore liken *costruzione legittima* to a rule that was "applied" by Alberti's contemporaries.

Hence, as with Brunelleschi it is necessary to draw a more nuanced picture of the contribution of Alberti to the development of linear perspective. Three points emerge regarding the actual meaning that should be ascribed to the expression that

---

(Footnote 11 continued)

Battista Benedetti on the mathematics of linear perspective," *Journal of the Warburg and Courtauld Institutes* 48 (1985): 71–99.

[12]"There is not a single verified example of a painting done with the 'construzione legittima'," James Elkins, *The Poetics of Perspective*, Ithaca, 1994, p. 86: "We should note that in a large number of Renaissance paintings the perspective turns out to be incorrect in mathematical terms," Judith V. Field, "Alberti, the Abacus and Piero della Francesca's proof of perspective," p. 72; Raynaud, *L'Hypothèse d'Oxford*, pp. 49–120.

[13]Luigi Vagnetti, "La posizione di Filippo Brunelleschi nell'invenzione della prospettiva lineare," *Filippo Brunelleschi. La sua opera e il suo tempo*, Florence, 1980, p. 305; Field, "Alberti, the Abacus and Piero della Francesca's proof," p. 69.

[14]Pietro Accolti, *Lo inganno degl' occhi*, Firenze, 1625, pp. 19, 57–58: "... to find, by means of their *legitimate* respective plans, the true perspective design... among the two previous operations, only this can be *legitimate*, in which we estimate ourselves to be deceived/trouare, mediante le loro *legittime* respettiue Piante, il proprio, e vero prospettiuo disegno... delle due sudette operazioni, questa sola poter essere *legittima*, nel che noi stimiamo ingannarsi.".

[15]Leonardo da Vinci, *Das Buch von Malerei*, ed. H. Ludwig, Wien, 1882, vol. 3, p. 177. In particular, consider Pietro Roccasecca, "Punti di vista non punto di fuga," *Invarianti* 33/99 (1999): 41–49, who writes: "Finally, we would like to point out that the existence of a discussion of different procedures... dispels the myth that perspective in the first half of the Quattrocento was the work of a lonely hero/Vorremmo infine segnalare che l'esistenza di una discussione tra diverse procedure... sfata il mito che la prospettiva della prima metà del Quattrocento sia l'opera di un eroe solitario," p. 48. See also Carlo Pedretti, "Leonardo 'discepolo della sperientia'," *Nel segno di Masaccio*, Florence, 2003, p. 170.

has been used, ever since Panofsky, to describe Alberti's work: (1) *costruzione legittima* did not exist in the Quattrocento; (2) the first attempts to codify perspective emerged in the sixteenth century, as a matter of course to meet the requirements of academic teaching; and (3) the unitary conception of perspective space probably dates to no earlier than the end of the nineteenth century.

2.1.1.2 Masaccio

The third milestone in this process—the first application in a large-scale work of the laws of perspective invented by Brunelleschi and codified by Alberti—has been attributed to Tommaso di Ser Giovanni, detto Masaccio (1401–1428). The *Trinity* fresco, which was painted around 1425–1427 in the church of Santa Maria Novella in Florence, has traditionally been viewed as an exemplary application of the laws of perspective. It has formed the object of universal praise ever since the declaration by Panofsky: "... at any rate, Masaccio's *Trinity* fresco is already exactly and uniformly constructed."[16] Modern studies have led to a reassessment of his conclusion, although a handful of scholars can still be found who assert that the fresco conforms to the canons of true perspective.

Doubtful of judgments that were in reality impressions based on a simple visual examination of the work, Field, Lunardi and Settle[17] undertook the first rigorous study in which the lines of perspective in the fresco were measured in situ. They presented proof that in constructing his perspective Masaccio introduced some serious accidental errors, beginning with the coffered barrel vault whose receding lines converge only approximately toward a central vanishing point. In addition, the method used to convey the perspective view of the abacuses above the corner columns was found to be faulty. Finally, the positioning of the longitudinal ribs of the vault in relation to the construction lines presents numerous irregularities. The authors concluded that Masaccio probably drew on the conceptions of Brunelleschi, Alberti and Donatello, but his work does not possess the mathematical rigor that some have ascribed to it.

Other studies of the fresco conducted using photogrammetry and computer reconstructions[18] have yielded varying results. Based on the analysis of a photogrammetric outline, we recently showed that this fresco contains not only accidental errors, as pointed out by Field, Lunardi and Settle, but also fundamental *errors of principle* in the method of reduction applied. Many studies have attempted to determine where the viewer was supposed to stand and look at the picture, based on the assumption that the perspective line was correctly drawn. Such an exercise can be contemplated because, once the depth scale has been calculated, there is a

---

[16]Panofsky, *Perspective as Symbolic Form*, p. 62.

[17]Judith V. Field, Roberto Lunardi, and Thomas B. Settle, "The perspective scheme of Masaccio's Trinity fresco," *Nuncius* 4 (1989): 31–118.

[18]Volker Hoffmann, "Masaccios Trinitätsfresko: Die perspektivkonstruktion und ihr Entwurfsverfahren," *Mitteilungen des Kunsthistorischen Institutes in Florenz* 40 (1996): 42–77.

single vantage point from which the painting should ideally be seen. An anomaly that ought to have raised doubts much earlier is that among fifteen studies conducted between 1913 and 1997, the calculated distance between the viewer and the painting varied from 210.5 to 894.2 cm.[19] How can such a dispersion of values be explained? If the fresco by Masaccio adhered to the rules of linear perspective, these values should have diverged only slightly, following a normal distribution and being solely attributable to errors in the graphic reconstruction. Such is not the case, which leads to the conclusion that the rules of linear perspective were not strictly applied in the *Trinity* fresco. The error has since been identified in the receding lines of the coffered vault, the only part of the fresco that allows an evaluation of the reduction method used. Between the perspective drawn by Masaccio and the theoretically correct one there is a disparity of more than 12 cm in the positioning of the orthogonals corresponding to the ribs of the vault. None of this can be ascribed to an accidental error or to an "adjustment" in the line for purely aesthetic reasons; it reflects instead an indifference to the principles of perspective reduction.

The conclusions that can be drawn at this point are: (1) the *Trinity* fresco does not follow the rules of linear perspective; (2) it does not represent an application of *costruzione legittima*; and (3) the search for the ideal vantage point, which only makes sense in the case of a linear perspective, is therefore destined to remain an exercise without a solution.[20]

The data as reviewed above regarding the work of Brunelleschi, Alberti and Masaccio (and many other contributions of the same nature) expose the fact that the modern interpretation of the development of perspective is founded, in the final analysis, on *ideological premises*.[21] The received version according to which linear perspective was invented by Brunelleschi based on his initial experiment and codified by means of *costruzione legittima* into a set of laws by Alberti, which were then applied by Masaccio, all share a point in common: they accentuate the thesis of a revolution (with its attendant components) that led to a complete change in the method of representing three-dimensional space.

The divergence between the historical evidence and the modern reconstruction of this process offers a case study in the sociology of knowledge, raising in a fresh context the question of the nature of beliefs and why they are adhered to. Many factors appear to be responsible for the firm attachment to the notion that perspective was codified and applied *ab origine* in a literal fashion:

– The "effect of authority," which has long conditioned the reading of Vasari's *Lives of the Most Excellent Painters, Sculptors, and Architects*. But how much

---

[19]Hoffmann, "Masaccios Trinitätsfresko," p. 75.

[20]See Chap. 4.

[21]If we set aside all adventitious hypotheses, there remain two possible inaugural dates: from a practical point of view, the first correct perspective was the fresco in Assisi *Christ among the Doctors*, attributed to the atelier of Giotto (ca. 1315); the first theoretical treatment was the demonstration of perspective in Piero della Francesca's *De prospectiva pingendi* (ca. 1475).

evidential weight can be given to a work that was written to glorify the reign of the Medici?

- The quest to define the pillars of Western civilization, which has led us to identify (and, subconsciously, to venerate) the key milestones in its evolution;
- Economic interests in the art world, which may inadvertently be maintaining mythologized or hagiographic versions of the history of art.

To these may be added factors specific to the Italian context:

- The historic rivalry between Florence and Rome for the title of cultural capital of Italy, leading to centuries of contention between the two cities;
- A separate ministry for the conservation of Italy's cultural heritage was not established until 1998, before which time the *Soprintendenze per i beni culturali ed ambientali* were subsumed under the Ministry of Tourism.

These are all elements that could help to explain the persistence of "the myth of perspective" in the face of rational arguments and obvious lacunae in the chain of evidence. The abiding belief in the centrality of the Renaissance has been ensured up to now by a complex set of disposition and communication effects, in particular what sociologists refer to as "relay effects."

2.2 Perspective and Knowledge

Once it is allowed that the unitary vision of perspective is in fact a myth, it remains to explain the many variations in the conception and techniques of perspective that emerged during the course of two centuries. I will argue that these variations can be linked to the specific notions of optics and geometry available to the individuals concerned.

2.2.1 Geometry and the Perspective of the Circle

To show how a knowledge of geometry could influence conceptions of perspective, it suffices to consider the case of the perspective of the circle. It is a simple matter to extract from Apollonius of Perga's *Conics* the property that the perspective view of the circle is an ellipse (proposition I, 13).[22] Nevertheless, the theory of conic

---

[22]*Apollonii Pergaei quae graece extant cum commentariis antiquis*, t. I–II, ed. J.L. Heiberg, Leipzig, 1891–3. After showing in Book I, prop. 9, that no oblique section of a cone is a circle, Apollonius sets out the three cases of the parabola (prop. 11), the hyperbola (prop. 12) and the ellipse (prop. 13). He then demonstrates that the projection of a circle onto a plane that is not parallel to the circle will be an ellipse or a conic section. Exact copies of the figure on which he based his argument were used by Commandino, Benedetti and others in the sixteenth and seventeenth centuries.

sections would not be applied systematically to the problem of perspective until the late sixteenth and early seventeenth centuries with the work of Commandino, Benedetti, Guidobaldo del Monte and Aguilonius (Fig. 2.1).[23]

This explains the proliferation of empirical methods for depicting the circle in perspective before this date—from gibbous to ovoid figures, from an oval with four centers to a rectangle flanked by two semicircles, etc. Let us study the differences between those who adopted such approximations and those who sought to draw a true ellipse.

The application of the ellipse as the perspective view of a circle seems to have been unknown to Masaccio; one need only examine the astragals of the capitals in the *Trinity*. Lorenzo Ghiberti was equally unaware of this geometrical notion and in the bas-relief *Joseph* on the Florence Baptistery's Gates of Paradise he used a semicircle (*EG*) and two gibbous forms (*AB* and *CD*) to create an ellipse (Fig. 2.2).

The case of Albrecht Dürer is more atypical, because he correctly identified the ellipse (*die linie ellipsis*) but gave it an ovoid shape.[24] In contrast, one finds the true ellipse in certain works by Piero della Francesca[25] (*Chalice 1758A*), in the notes of Leonardo da Vinci (*Ring, Codex Atlanticus*, 263ra) and in the circle of Antonio da Sangallo (*Mazzocchi 830A, 831A, 832A*) (Fig. 2.3).

What do these differences stem from? The response would not be the same for all artists. Let us attempt to differentiate between them.

Piero della Francesca (1420–1492) can be distinguished from his predecessors because of his mathematical approach to perspective; it was he who first demonstrated that a perspective could be constructed based on the distance point using

---

[23]In his edition of *Claudii Ptolomaei liber de analemnate*, Rome, 1562, Francesco Commandino studied the projection of the circle and showed that the perspective view often appeared in the form of an ellipse. Giovanni Battista Benedetti, *Diversarum speculationum mathematicarum et physicarum liber*, Turin, 1585, proved the theorem that the intersection of a cone by two parallel planes will produce similar conic sections. Guidobaldo del Monte, *Perspectivae libri sex*, Pesaro, 1600, studied the projection of a circle on an inclined plane and the similar problem of the figure that casts the shadow of a sphere on a plane. Franciscus Aguilonius, *Opticorum libri sex*, Anvers, 1613, studied the shadow of the sphere and determined geometrically the position of the axes of the elliptical projection of a circle.

[24]"The Ancients showed that one could cut a cone in three ways and obtain thusly three different sections... The erudite called the first section an ellipse: it cut the cone obliquely and drew nothing from the base of the cone," Albrecht Dürer, *Géométrie*, ed. J. Peiffer, Paris, 1995, p. 174. Dürer then described point by point the construction of the "egg line or ellipse" (sic) by means of a double projection, that is, by transferring onto its face the points of intersection recorded on the plane and profile.

[25]*Mazzocchi* 1756A, 1757A, 1758A has traditionally been attributed to Paolo Uccello, but there is no evidence to support this apart from an inconclusive reference in Giorgio Vasari's *Vite de' più eccellenti pittori*, stating that he possessed "... a mazzocchio drawn only with lines, so beautiful that nothing save the patience of Paolo could have executed it/un mazzocchio tirato con linee sole, tanto bello, che altro che la pacienza di Paulo non lo avrebbe condotto" (sic).

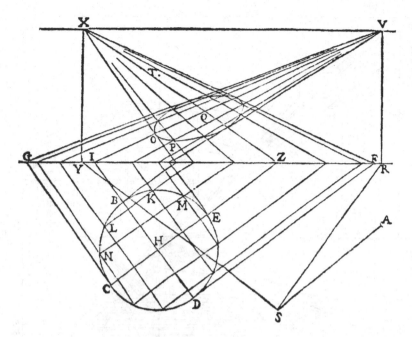

Fig. 2.1 Perspective of the circle, Guidobaldo del Monte, *Perspectivae libri sex*, Pesaro, 1600, IV, 24, p. 217

similar triangles. Only three treatises by Piero della Francesca have come down to us: *Trattato d'abaco*, *Libellus de quinque corporibus regularibus* and *De prospectiva pingendi*.[26] *Libellus* of course introduces the five regular polyhedra—the tetrahedron, the cube, the octahedron, the icosahedron, and the dodecahedron—but this material only takes up Book I. Book III also treats problems in stereometry (how to measure the volumes of solid figures), including sixteen exercises involving the sphere and the cone. These assume a knowledge of conics, which

[26]Piero della Francesca, *Trattato d'abaco*, ed. G. Arrighi, Pisa, 1970; *idem*, *De prospectiva pingendi*, ed. Nicco-Fasola, Florence, 1984; *idem*, *Libellus de quinque corporibus regularibus*, ed. F. P. di Teodoro, Florence, 1995. On Piero della Francesca's mathematics, see Marshall Clagett, *Archimedes in the Middle Ages*, vol. 3, Philadelphia, 1978; Menso Folkerts, "Piero della Francesca and Euclid," *Piero della Francesca tra arte e scienza*, eds. M. Dalai Emiliani and P. Curzi, Venice, 1996, pp. 293–312; Judith V. Field, *Piero della Francesca, A Mathematician's Art*, New Haven, Yale University Press, 2005; Pietro Roccasecca, "Dalla prospettiva dei pittori alla prospettiva dei matematici," *Enciclopedia italiana di scienze, lettere ed arti. Il Contributo italiano alla storia del pensiero*, Roma, Istituto della Enciclopedia italiana, 2013, pp. 137–144.

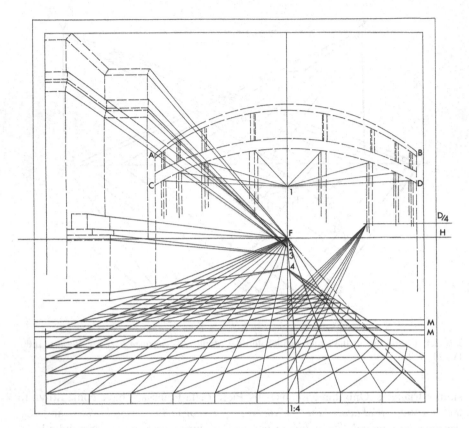

Fig. 2.2 The circle depicted as a gibbous figure, Lorenzo Ghiberti, *Storia di Giuseppe,* 1425–52 (Firenze, Museo dell'Opera di Santa Maria del Fiore), author's reconstruction

Piero della Francesca must have possessed at least through Archimedes' work *On Conoids and Spheroids,* for it is known that he owned a copy of the treatise (which subsequently passed into the possession of the dukes of Urbino).[27]

---

[27]*Archimedis de konoidalibus et speroidibus figuris* (Urbinato latino 261, fol. 44v–45r). Archimedes provides definitions at the beginning of his treatise that are comparable to those of Apollonius: "If a cone be cut by a plane meeting all the generators of the cone, the section will be either a circle or an ellipse... And if a cylinder be cut by two parallel planes meeting all the generators of the cylinder, the sections will be either circles or ellipses," *De la sphère et du cylindre, La mesure du cercle, Sur les conoïdes et les sphéroïdes,* ed. Ch. Mugler, Paris, 1970, p. 158. He then determined the area of the ellipse by comparing it to the area of a circle with the same diameter as the long axis of the eclipse. The ratio between the two areas would be equal to the ratio between the rectangle circumscribing the ellipse and the square circumscribing the circle or, in what comes to the same thing, as the ratio of the short axis of the ellipse to the diameter of the circle, pp. 166–170. The *Divina proportione* (1509) by Fra' Luca Pacioli, a part of which was copied from *Libellus,* also addresses the problem of the ellipse: if one takes a square and trans- forms it into a rectangle of the same length, whose height is equal to the diagonal of the square,

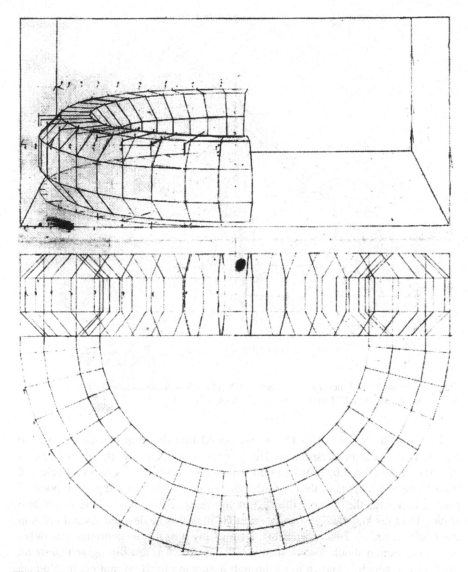

Fig. 2.3 The circle represented as an ellipse, Cerchia di Sangallo, *drawing of a mazzocchio* (Florence, Gabinetto Disegni e Stampe degli Uffizi), inv. 832A, from Christoph L. Frommel and Nicholas Adams, *the architectural drawings of Antonio Da Sangallo the younger and his circle*, Cambridge, MIT Press, 1994, vol. I, p. 150

(Footnote 27 continued)

every point of the circle inscribed in this square will correspond to a point on the ellipse that is inscribed in the rectangle. In this way Fra' Luca Pacioli obtained the *"proportioned circle/circulo proportionato"*. The same method was adopted by Dürer, who did not, however, link it to the drawing of an ellipse, Dürer, *Géométrie*, p. 70.

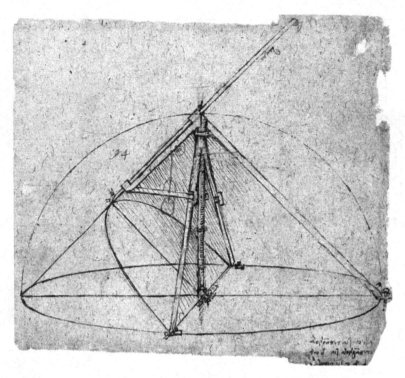

Fig. 2.4 Leonardo da Vinci, Parabolograph, 1478–1518 (Milan, Biblioteca Ambrosiana, *Codex Atlanticus*, fol. 1093r, ed. Hoepli, Florence, 1894, olim 394ra)

Leonardo da Vinci (1452–1519) understood that the ellipse must be used to represent a circle in perspective. He sometimes resorted to the discontinuous drawing of the ellipse by projecting horizontal chords obliquely across a circle and transposing the lengths of these chords onto perpendiculars raised at each point of intersection with the oblique line.[28] But his perspective view of the circle also sprang from his knowledge of conic sections, to which he devoted several notes in the *Codex Arundel*. Two instruments designed by him for the continuous drawing of a conic section should also be mentioned. The first was an ellipsograph (*seste da far l'ovato*), which is known to us through a drawing by Benvenuto della Volpaia preserved in the Biblioteca Marciana in Venice.[29] The other was a parabolograph, a sketch of which appears in the *Codex Atlanticus*[30] (fol. 349ra, Fig. 2.4).

---

[28]*Codex atlanticus*, fol. 115rb (ca. 1510), Carlo Pedretti, *Léonard de Vinci architecte*, Paris, 1983, p. 302.

[29]Venice, Biblioteca Marciana, MS. It. 5363, fol. 18r.

[30]Pierre Sergescu, "Léonard de Vinci et les mathématiques," *Léonard de Vinci et l'expérience scientifique*, Paris, 1952, pp. 73–88; Otto Kurz, "Dürer, Leonardo and the invention of the ellipsograph," *Raccolta vinciana* 18 (1960): 15–25; Gino Arrighi, "Il 'compasso ovale invention di

On comparison one notes that these are in fact exact equivalents to mathematical instruments described by al-Qūhī and al-Sijzī and by their twelfth-century successors al-Baghdādī and al-Ḥusayn. In particular, Leonardo's ellipsograph exactly corresponds to al-Sijzī's description of "the perfect compass."[31]

The *mazzocchi* associated with Antonio da Sangallo and his circle provide another interesting example of the application of geometric concepts to problems of perspective. After detailed study of the material traces of these drawings, certain historians have formulated the hypothesis that he used a method based on conic sections. Their thesis is supported by a drawing (830A) from the *Gabinetto Disegni e Stampe* of the Uffizi in which, to trace the outline of his *mazzocchio*, Sangallo derives an ellipse from the section of a conical pyramid composed of thirty-two faces (an admissible approximation of the cone). But the architect's knowledge of conics did not stop here, because drawing 1102A presents an ellipsograph (*sesto per fare avovati*) that is equivalent in every way to the one utilized by Leonardo.[32]

The most striking trait in the work on the perspective representation of the circle conducted by Piero della Francesca, Leonardo Da Vinci, and Antonio Sangallo is that they all derived correct projections from the theory of conic sections, geometric notions which their contemporaries were either ignorant of (Masaccio and Ghiberti) or had misunderstood (Dürer). This demonstrates clearly how geometric concepts could be applied to resolve problems of perspective.

(Footnote 30 continued)

Michiel Agnelo' dal Cod. L.IV.10 della Biblioteca degl' Intronati di Sienna," *Le Machine* 1 (1968): 103–106; Paul L. Rose, "Renaissance Italian methods of drawing the ellipse and related curves," *Physis* 12 (1970): 371–404; Carlo Pedretti, *Studi vinciani*, Genève, 1957; *Idem, Léonard de Vinci architecte*, p. 336; *idem*, "Leonardo discepolo della sperientia," pp. 184–185; Pedretti believed that Leonardo based his knowledge of conic sections on his reading of *De rebus expetentis* by Giorgio Valla (1501). This was a possible source for the notes in *Codex Atlanticus*, fol. 394va, dated 1513–1514. All the same, an identical diagram dated ca. 1480 appears on fol. 32ra, with the note "a youthful study". Leonardo therefore drew on an earlier source that still remains to be identified.

[31]Roshdi Rashed, "Al-Qūhī et al.-Sijzī: sur le compas parfait et le tracé continu des sections coniques," *Arabic Sciences and Philosophy* 13 (2003): 9–43. The close parallels that exist between the *seste da fare l'ouato* of Leonardo and al-Sijzī's compass cast doubt on the thesis that the ellipsograph was an Italian invention. A borrowing by way of a Latin translation constitutes a credible hypothesis, Raynaud, "Le tracé continu des sections coniques à la Renaissance. Applications optico-perspectives, héritage de la tradition mathématique arabe," *Arabic Sciences and Philosophy* 17 (2007): 299–345.

[32]Pietro Roccasecca, "Tra Paolo Uccello e la cerchia sangallesca," in *La prospettiva. Fondamenti teorici ed esperienze figurative dall' antichità al mondo moderno*, ed. Rocco Sinisgalli, Fiesole, 1998, pp. 133–144. A direct transcription from the Arabic cannot be excluded: Antonio da Sangallo is known to have made a drawing of an astrolabe, "Strolabio egyptizio daritto e da riverso" (Cabinet of Drawings and Prints of the Uffizi 1454A) in which the divisions on the limb, mater and rete are marked in Arabic characters.

2.2.2 Optics and Binocular Vision

The preceding discussion can be reproduced point for point with regard to the influence that a knowledge of optics had on the practice of perspective.

Let us consider the case described in some texts as "two-point perspective". Such representations are similar to those with a central linear perspective, but rather than receding towards a single vanishing point, the lines converge on two vanishing points located on the same horizon. About thirty works whose perspective is based on this principle are known,[33] including *The Trial of Pietro d'Abano*, a fresco by Altichiero in the Palazzo della Ragione in Padua (ca. 1270–1280). The groins of the vault in this painting form two sets of converging lines, with the pencil on the left extending toward point F' on the right, and the pencil on the right extending toward point F on the left. The vanishing points F and F' have been correctly placed on the same horizon line (Fig. 2.5).

These perspective traces admit a priori of more than one interpretation. The lateral walls could be slightly convergent rather than parallel to each other, occasioning two sets of lines that do not meet in a single, central vanishing point.[34] If the two pencil of lines, rather than continuing towards points F and F', ended along the central axis AF, the view would be similar to the "axial perspective" interpreted by Panofsky and White in terms of a curvilinear or synthetic perspective.[35] It is also possible that Altichiero unintentionally introduced a deviation in the lines while executing the drawing, creating two artificial vanishing points that in fact can be resolved into one, which would mean that this fresco is an example of central linear perspective.[36]

Various hypotheses may be proposed to interpret this drawing in perspective and the correct one must be chosen based on credible criteria. Further on it will be shown that such two-point perspectives do not correspond to any of the most widely known forms of perspective and, even if they bear certain points of similarity to linear perspective, they could in actuality be the result of an application of *binocular vision*.

Two-point perspective was used from the end of the Duecento up to Classical times. One can even find examples from the mid-seventeenth century. The approach has attracted little attention up to now due to the prevailing belief that the rules of

---

[33]See for instance: Giusto de' Menabuoi, *Christ among the Doctors* (1376–1378), Stefano di Sant'Agnese, *Madonna with Child* (ca. 1390), Taddeo di Bartolo, *The Last Supper* (1394–1401), Lorenzo Monaco, *Adoration of the Magi* (ca. 1421), Lorenzo Ghiberti, *Christ among the Doctors* (ca. 1415), Niccolò di Pietro, *St. Benedict Exorcising a Monk* (ca. 1420), Gentile da Fabriano, *The Crippled and the Sick Cured at the Tomb of St. Nicholas* (1425), Giovanni di Ugolino, *Madonna with Child* (1436). A corpus of about thirty works displaying two-point perspective was identified for study; the process by which they were selected is described in Chap. 8 and the results of the analysis are discussed in subsequent chapters.

[34]This hypothesis is discussed in detail in Chap. 8.

[35]See Chaps. 9 and 10.

[36]See Chap. 8.

Fig. 2.5 Perspective drawing of Altichiero's *Trial of Pietro d'Abano*, ca. 1370–80 (Padua, Palazzo della Ragione), author's reconstruction after Giampiero Bozzolato, *Il Palazzo della Ragione*, Roma, 1992, plate XLIII

linear perspective were codified and rigorously followed from the Quattrocento onward. And yet, contrary to all expectations, references to two-point perspective—which has been retroactively judged to be quite heterodox—can be found even at a relatively late date. Vignola and Danti, for example, in 1583 refuted the notion that a perspective could be constructed based on two vanishing points, arguing that because visual sensations are fused to the optic chiasma they produce a single image, which implies a single vanishing point.[37] The very fact that these authors felt it necessary to devote a long critique to the use of two-point perspective shows that the construction was still in widespread use around 1583. However heterodox they may have become, the foundations of two-point perspective were no less rational than those of single-point perspective and deserve to be examined more closely.

The text by the architect Vignola and the commentary provided by the mathematician Danti furnish in this regard a promising path of investigation which this book intends to pursue: the two-point construction of perspective could bear points

---

[37]*Le Due Regole della prospettiva pratica di M. Iacomo Barozzi da Vignola*, Roma, 1583, pp. 53–54.

of affinity with the principles of binocular vision (*hauendo l'huomo due occhi …*). In fact the postulate of monocular vision imposed by linear perspective was regularly violated, both *before and after* its presumed invention by Brunelleschi.[38] How can the fundamental principles of two-point perspective be retraced? One strategy that will be adopted here is to systematically compare knowledge of the science of optics with the perspective practices of the period. Ghiberti, for example, who utilized the perspective construction based on two principal vanishing points (Appendix B, No. 25), provides an invaluable first-hand source of information on the knowledge of optics among perspectivists in fifteenth-century Italy. His *Commentario terzo* is a compilation of the most widely read treatises of the period, and includes the works of Ibn al-Haytham (known in the West as Alhacen), Bacon, Pecham and (to a lesser degree) Witelo.[39] In this compilation Ghiberti reproduces Ibn al-Haytham's chapter on binocular vision *almost in its entirety*. Following the Arab scientist's exposition[40] of the conditions for the fusion of images, Ghiberti utilized a similar experimental set-up to examine the case of diplopia, and came to similar conclusions.[41] As we will discover, the textual sources of the period provide all of the necessary elements to construct a perspective with two vanishing points. What follows is a question that will constitute a central theme in this book: Why did Ibn al-Haytham and Egnatio Danti draw such different conclusions regarding how binocular vision operates?

This difference appears to be the result of the assimilation of different sets of knowledge. The response to the question of binocular vision and how two images come to be fused into one is not the same if the problem is approached by way of optics (as in the case of Ibn al-Haytham and his Latin successors) or anatomy (as with Egnatio Danti, a mathematician who was in this case reasoning as an anatomist). The arguments in support of the *constant* unification of visual sensations stem directly from anatomical studies of the optic tract—the *chiasma* or *decussatio*

---

[38]Although the painting by Brunelleschi has been lost, some information can be gleaned from Manetti, who wrote in his biography: "It is necessary that the painter postulate beforehand a single point from which his painting must be viewed/Il dipintore bisognia che presuponga *un luogo solo* d'onde s'a a uedere la sua dipintura," Manetti, *Vita di Filippo di Ser Brunelleschi*, fol. 207v. The passages on the monocular postulate are quoted in Chap. 1.

[39]The most comprehensive study to date of the sources is provided by Klaus Bergdolt, *Der dritte Kommentar Lorenzo Ghibertis*, Weinheim, 1988. See also Raynaud, "Le fonti ottiche di Lorenzo Ghiberti," pp. 79–81.

[40]Abdelhamid I. Sabra, *The Optics of Ibn al-Haytham*, Books 1–3: *On Direct Vision*, London, 1989, vol. 1, pp. 237–240.

[41]"All those things we have described can be demonstrated by experiment, and once the proof has been seen, take a board of light wood… When therefore the experimenter has understood these lines and individuals [the wax sticks], in fact if there is not one line down the middle [of the board], but two appear… then the one and the other will appear to be doubled/Tutte quelle cose noi abbiamo dette si possono sperimentare, e veduta la certificazione tragassi una tavola del legno leggiero… Quando adunque lo sperimentatore arà compreso queste linee e gli individui, veramente non è se non è una linea nel mezzo, ma paiono due… allora l'uno e l'altro appariranno due," Lorenzo Ghiberti, *I Commentari*, ed. O. Morisani, Naples, 1947, pp. 144–147; Bergdolt, *Der dritte Kommentar*, p. 330.

—whereas Ibn al-Haytham and his Latin commentators held the disparate images to be normal and subscribed to the thesis of a *conditional* fusion of images depending on their degree of disparity. As heirs to the geometric school of optics, the early perspectivists clearly chose to ignore the teachings of the anatomists.[42] This shows that different pieces of optical knowledge could lead to different conceptions of perspective space.

2.3 Conclusion

If one excludes a handful of general principles such as the law of the reduction in size, a unitary conception of perspective does not exist. Perspective is an open system reflecting the optical and geometric knowledge available at a given time. This observation is not without importance to the socio-historical study of systems of representation. An explanation for the birth of perspective has long been sought in social factors such as the role played by the Florentine bourgeoisie, the rivalry between Italian city-states, the humanist movement, and so on. However, *stricto sensu* none of these factors is capable of providing a satisfactory explanation for the development of perspective during the Renaissance. The rise of the bourgeoisie or the competition between city-states could have contributed to the development of the arts (through the bias of the art patron or of the type of work commissioned), but it does not explain why artists focused their attention on perspective rather than some other mode of representation. Likewise, the chronology of humanism and the rediscovery of the texts of antiquity is not consonant with the facts regarding the development of perspective. The study of the mathematical texts of the Greeks did not begin until relatively late, in the sixteenth century.

An analysis of the textual parallels in the treatises of the Quattrocento demonstrates that the optical-geometric sources most often cited were not the mathematicians of ancient Greece but those of the Middle Ages; in general Euclid was known only through medieval commentaries. In comparison to the traditional theses regarding the invention of perspective, *explaining such a disruptive innovation in terms of the mobilization of the resources most pertinent to the development of this innovation* might appear to be somewhat bland. In point of fact, however, it opens up fresh paths of investigation and raises fewer difficulties because it focuses directly on the objective phenomenon to be explained, that is, on perspective rather than on pictorial representation or the arts generally. It therefore provides a simple (and verifiable) explanation for the variety of forms observed (for example, the perspective view of a circle might be rendered as an ellipse or a gibbous figure depending on the optical-geometric knowledge available in a certain milieu).

---

[42]See Chap. 5.

The above reflections have led to a research study that lies at the intersection between sociology and the history of the sciences, and whose objectives may be summarized as follows:[43]

1. To understand the processes involved in the *appropriation of knowledge*. This requires in particular that one pay greater attention to what the *errors* committed by perspectivists can reveal about the development of that science. Such errors often denoted a misunderstanding of the problem, or of the time that might be required to master it. This is why a considerable number of false perspectives in Renaissance art have been identified.
2. To assess the *availability of knowledge* at a given time, beginning with a study of the networks through which knowledge spread. It is a fact that not all knowledge is equally accessible and studying the inhomogeneous distribution of resources as a function of distance and social milieu may help to clarify the cognitive bases on which a disruptive innovation may have been conceived and thus the relationship between the representation of perspective and its optical-geometrical foundations.

All of this leads to the conclusion that *no specific and unitary conception* of perspective space exists, precisely because so many variations can be found.

If one attempts to retrace the shared origins of the many and diverse solutions to the problem of perspective that were being explored from the late Duecento onward, two main blocks of knowledge can be identified that made possible the gradual emergence of the perspective system:

– Euclidean geometry, which was rediscovered in Europe in the twelfth century through translations of the Greek mathematician's work from the Arabic (the works of Adelard of Bath, Robert of Chester, Gerard of Cremona, etc.)[44] and the keen interest in practical geometry (for example, Abraham bar Ḥiyya Savasorda's *Liber embadorum*, Leonardo Fibonacci's *Practica geometrie*, and John of Muris's *De arte mensurandi*);[45]
– the optics of Alhacen, which was introduced into Latin Europe through a translation by the school of Gerard of Cremona, and in parallel the optics of the Latin scholars Bacon, Pecham and Witelo.[46]

---

[43]These issues are addressed in *L'oeuvre et l'artiste à l'épreuve de la perspective*, eds. M. Dalai Emiliani, M. Cojannot Le Blanc, and P. Dubourg Glatigny, Rome, 2006.

[44]See the works of Hubert L.L. Busard.

[45]Maximilian Curze, *Der Liber Embadorum des Abraham bar Chijja Savasorda in der Übersetzung des Plato von Tivoli*, Leipzig, 1902; Baldassare Boncompagni, *Leonardi Pisani Practica geometriae*, Roma, 1862; Hubert L.L. Busard, *Johannes de Muris. De Arte mensurandi*, Stuttgart, 1998.

[46]Sabra, *The Optics of Ibn al-Haytham*; David C. Lindberg, *Roger Bacon and the Origins of Perspectiva in the Middle Ages*, Oxford, 1996; *Idem, John Pecham and the Science of Optics*, Madison, 1970; *Opticae thesaurus... Item Vitellonis Thuringopoloni libri X*, New York, 1972.

This body of optical-geometric knowledge did not *immediately* coalesce into a definitive conception, which seems to have been the result on the one hand of the gradual acquisition of these texts and on the other to the codification brought about by the academic teaching of linear perspective.[47] As a result of these factors, it is improbable that perspective contributed in any way to the modern conception of space as infinite, homogeneous, and isotropic.[48]

If therefore, despite what has been shown by recent and more rigorous analyses, the present conception of perspective—which is based on a mistaken interpretation of the contributions of Brunelleschi, Alberti and Masaccio and completely ignores the alternatives to linear perspective that were in widespread use until the Cinquecento—remains so profoundly reified, it will be necessary first of all to examine the errors in perspective that can be observed in the works of the period, in order to be able to identify, characterize and classify these deviations in relation to the canons of linear perspective.

---

[47]The *Accademia*—the seat *par excellence* for the codification of perspective—reduced the plethora of uncoordinated initiatives to a single method. Therefore, the problem is not to justify the existence of heterodox practices lying on the margins of the "pure type" of linear perspective, but rather to explain the procedures by which distinct conceptions were gradually separated out and evaluated in order to allow the rise of a single orthodox conception.

[48]In *Perspective as Symbolic Form*, p. 70, Panofsky proposed the thesis that a conception of space as "infinite, homogenous and isotropic" emerged during the Renaissance. This thesis was supported by successive scholars, such as Manfredo Tafuri and Rudolf Wittkower, who transposed the concept to architecture. The thesis runs up against various logical and empirical difficulties, however. (1) The type of reasoning that characterizes perspective does not consist in working within a space considered to be of unlimited extent, but in the manipulation of figures, i.e. finite bodies. With regard to the projective geometry of Desargues, Michel Chasles wrote that it consisted in an exercise in "reasoning on the properties of figures," *Aperçu historique sur l'origine et le développement des méthodes géométriques*, Bruxelles, 1837, p. 74. (2) The concept of "perspective space" itself was anachronistic in the fifteenth century because it assumed a codification that had not yet been formulated. The notion was applied to architecture by August Schmarsow at the end of the nineteenth century, and taken up again by Panofsky, Jantzen, Frey and Badt. For more on this filiation, see Roland Recht, *Le croire et le voir*, Paris, 1999, pp. 44–45.

Chapter 3
Understanding Errors in Perspective

Abstract This chapter examines the question of errors in perspective from the viewpoint of the painter rather than the spectator, a distinction that significantly modifies the way in which the problem is approached. Perspective is therefore judged in terms of the methods used by the painter or architect who constructed it, that is to say, in terms of the goals that he set for himself and the means he had at his disposal to achieve them. We then explain the reasons why Renaissance painters accepted the three main types of error in perspective: "accidental errors" (type I), "ad hoc errors" (type II), and "systematic errors" (type III).

The evaluation of the correctness of a perspective is often based on the feeling of "uneasiness" that is evoked when looking at certain paintings. Let us examine *The Wedding at Cana* painted by Duccio di Buoninsegna around 1311. The disquiet here arises from the fact that the table top is tilted at such an angle that we half expect the plates and cutlery to slide off the tablecloth at any moment... Whatever the significance of this typification—which may appear to be both immediate and effective—the judgment that 'something is wrong' with this scene originates with the *spectator*, for it is he who feels uncomfortable and it is again he who expects to see the objects slide off... However, alongside the viewer's sensorial characterization of a painting another is possible, which invites us to approach the problem of perspective through the eyes of the painter or architect who constructed it. A given perspective is then judged in terms of the method used by its creator, that is to say, in terms of the goals which he set himself and—crucially—the means that he had at his disposal to achieve them.

These approaches reiterate the antique distinction between aesthetics[1] (αἰσθάνομαι meaning "to feel" in Greek) and poietics (ποιέω meaning "to make" in

---

[1]Panofsky often engaged in this type of aesthetic evaluation: "The represented objects *appear* to stand, for the most part, more above than on the floor"; "It is also an inconsistent space, in that objects—for example, in our panel the table of the Last Supper—*appear* to stand in front of the 'space box' rather than in it"; "Space thus *seems* to extend forward across the picture plane; indeed, because of the short perpendicular distance it *appears* to include the beholder standing before the panel," Erwin Panofsky, "Die Perspektive als symbolische Form" (1924/5), *Perspective as Symbolic Form*, pp. 55, 56, 60 (italics mine). Panofsky notes that the feeling of correctness is socially constructed, which suggests that one can only begin to evaluate a perspective by questioning the presuppositions that form the background to an axiological evaluation.

© Springer International Publishing Switzerland 2016
D. Raynaud, *Studies on Binocular Vision*,
Archimedes 47, DOI 10.1007/978-3-319-42721-8_3

Greek). The poietic evaluation raises a justifiable point regarding the aesthetic evaluation of a work of art—if Western painting has undergone a transformation, it is not because the eye of the spectator has changed, but because it has trained itself to appreciate works of art that the painter has constructed differently. To declare that the perspectives of the Renaissance are "correct" is no more satisfying than to assert the "falseness" of medieval paintings. From a technical point of view, the various errors in perspective are constructions—perhaps mistaken, but nonetheless deliberate operations.

3.1   The Classification of Errors

Whatever the apparent correctness of the final result, the only question that deserves to be asked in all such cases is: How was this perspective constructed? Let us first examine three types of error that were frequently made by painters: "accidental errors" (type I), "ad hoc errors" (type II) and "systematic errors" (type III).

Type I errors. From an operating point of view, the accidental error is characterized by the fact that it does not arise out of the artist's adherence to a logical-semantic network. For example, an isolated vanishing line that unintentionally slips off at an angle should simply be interpreted as an accidental mistake. On the other hand, if it extends to the point of concurrence with other vanishing lines, it could represent an example of a type II or type III error.

Type II errors. The ad hoc error is a conscious error that can be understood from the draftsman's point of view as the solution to a practical problem. Let us take an example: when the paving tiles in the floor of a painting are interrupted by one or two steps, it is worthwhile for the artist to draw a single perspective network and to alter the vanishing lines at the point of the steps—such a deviation from correct perspective will go unnoticed. In this case, the error is comprehensible because it obeys a pattern of logic that is *instrumental*, i.e., one designed to achieve a specific end through means which are consistent with the draftsman's technique.

Type III errors. The systematic error can be distinguished from accidental and ad hoc errors because it forms part of a coherent logical-semantic network based on the acceptance of certain rules of construction. Let us consider the *Madonna and Child* painted by Fra' Filippo Lippi around 1452, whose vanishing lines converge on a centric point coinciding with the Virgin's eye. Here we are faced with a systematic error, for we can imagine that the painter saw a relationship between the eye and the vanishing point. Each time we categorize a mistake as a type III error, we may view it in terms of the analysis of false beliefs proposed by Boudon, who reiterates Pareto's observation: "Logic tries to discover why a thought process is false whereas sociology tries to discover why it is so frequently adhered to."[2] The question is: What might have been the reasons for a painter's adherence to a given system of rules?

---

[2]Raymond Boudon, *L'Art de se persuader*, Paris, 1990, p. 7.

Furthermore, was there a *single* and unanimous consensus regarding the rules of linear perspective or were there *several* simultaneous consensuses that reflected the divisions between various artistic schools and workshops?

The question "How are perspectives constructed?" should lead to a reconsideration of our approach to the analysis of paintings. To construct a correct linear perspective the artist must carry out two series of operations. The first consists in setting the *orthogonal lines*, i.e. those whose direction is parallel to that of the viewer's gaze and will become *vanishing lines* in a perspective view. The second involves appraising the *transversal lines* that are perpendicular to the viewer's gaze. The orthogonals force the painter to fix a vanishing point, of which there can be only one in central perspective. The transversals force the painter to choose a method of *foreshortening* that will determine the intervals between the transversals (e.g., the receding horizontals of the floor plane).

Ever since Panofsky, historians of perspective have tended to view the positioning of the centric point as being more important than the method of foreshortening used, such that their evaluation of the correctness of a painting's perspective has been almost exclusively based on an examination of its system of vanishing lines. An analysis of the foreshortening technique utilized is nevertheless necessary in order to evaluate to what extent painters actually subscribed to the rules of perspective. Let us embark on such an analysis now, based on the twofold sequence of operations outlined above.

3.2 Methods of Foreshortening

While the tracing of the orthogonal lines leaves little room for imagination, the same cannot be said for transversal lines. With regard to the treatment of diminishing intervals, Renaissance painters proposed a range of empirical solutions.

How is one to represent horizontal lines that are perpendicular to the viewer's gaze? In the floor plane, the spacing of the transversal lines is regular since the paving squares are all of the same dimension. This is not so when they are depicted in a perspective view and most painters subscribed to the notion that the further away the squares are, the smaller they should appear to the naked eye. But what rule of diminution were they applying? They could not resort to the theory of perspective, because this was only invented by mathematicians in the sixteenth century. Such knowledge being unavailable to them at the time,[3] artists devised various

---

[3]The situation of a heuristic search for solutions in a context of limited information is reminiscent of the early attempts by individuals unfamiliar with mathematics to solve problems of probability; see Amos Tversky and Daniel Kahneman, "Availability: a heuristic for judging frequency and probability," *Cognitive Psychology* 5 (1973): 207–232. We must not forget that most of the basic concepts of perspective, beginning with the "vanishing point," were unknown in the Renaissance (for instance, Alberti speaks of a *punctum centricus*; the concept of *punctum concursus* still lay in the future). We owe the conceptualization of this notion to seventeenth-century mathematicians

tactics in order to apply the only concept of perspective they had ever known: the *qualitative principle* of foreshortening the apparent size of objects based on distance, a principle that was presented in every medieval treatise on optics.[4]

The *quantitative translation* of this principle could follow one of several different paths. In his inventory of the methods used by Renaissance painters to construct their perspectives, Panofsky distinguishes between a series of approaches (with the exception of the reduced distance points method, which he does not mention) that yield approximately the same geometric results even though they represent the various stages in a process of progressive simplification.[5] We find: (a) Brunelleschi's (hypothetical) method (1413); (b) Alberti's section of the visual pyramid (1435), which Panofsky mistakenly refers to as the principle of *costruzione legittima*; (c) Vignola's distance point method and his tracing of oblique lines (published posthumously in 1583); (d) Viator's simplified distance point method and tracing of the diagonal line (1505); and (e) Pietro Accolti's reduced distance points method (1625). An analysis of paintings conducted today cannot be based on this inventory, however, because a fundamental problem must first be resolved. Panofsky's list, however punctilious, only includes *correct* methods. It is quite possible, therefore, that it does not contain all the solutions that were explored during the Renaissance and it must be completed by a rational examination of all the possibilities.

Imagine finding yourself in a room paved with identical square tiles. Now open a door to the point at which it crosses a tile diagonally. You will note that the door is lined up diagonally with *all* the tiles and—as trivial as it may seem—this diagonal line is a *straight* one. What you have just experienced should also hold true in a painting. For a perspective to be correct, the diagonal line crossing the tiles must be

---

(Footnote 3 continued)

such as Guidobaldo del Monte, who was one of the first to study methods of projection; see Martin Kemp, "Geometrical perspective from Brunelleschi to Desargues," *Proceedings of the British Academy*, 70 (1985): 89–132; Judith V. Field, "Alberti, the abacus and Piero della Francesca's proof of perspective," *Renaissance Studies* 11/2 (1997): 61–88.

[4]Here is a sample: "Objects of equal size unequally distant appear unequal and the one lying nearer to the eye always appears larger/Aequales magnitudines inaequaliter expositae inaequales apparent et maior semper ea quae propius oculum adjacet," Euclid, *Optica*, sup. 5; "At distance, the same object makes a small angle in the eye which it would make great when it is close/Eadem res distans facit paruum angulum in oculo quae faceret magnum quando est propinqua," Bacon, *Perspectiva*, II, III, 3; "Among equal and equidistant sizes at unequal distances from the eye, the closer will always appear greater, but they will not be seen in proportion to their distances/Aequalium et aequidistantium magnitudinum inaequaliter ab uisu distantium propinquior semper maior uidetur, non tamen proportionaliter suis distantiis uidetur," Witelo, *Optica*, IV, 25; "Among objects of equal size, that which is most remote from the eye will look the smallest/Infralle cose dequal grandeza quella chessara piu distante dallochio si dimossterra di minore figura," Da Vinci, MS. *SKM II*, fol. 63r.

[5]Panofsky, *Perspective as Symbolic Form*, French edition, pp. 229, 232.

straight;[6] otherwise every door that opens diagonally will appear to be askew, which is materially impossible. Yet this rule was often ignored, leading to two main types of errors: the "diagonal line" might assume a concave or convex curvature. Add to this the case of a true perspective where the diagonal remains straight and one has a range of three different possibilities in terms of foreshortening.

3.2.1 Correct Foreshortening

There are a number of foreshortening methods that give rise to *straight diagonals* in which the spacing between the horizontal lines diminishes correctly as the vanishing point is approached. The first perspective method to have been described in written form—that by Alberti[7]—falls into this category. Referred to as the "intersection of the visual pyramid," it begins with the formation of a pencil of visual rays linking the eye to each of the divisions in the tiled floor. This makes up a visual pyramid that is then 'cut' by the picture plane, and the intersection between the two will fix the height of each of the receding transversal lines.

The originality of Alberti's method lies in the fact that the side and perspective views are drawn on the same sheet of paper, so that the intersection points of the visual pyramid can be easily transferred.

3.2.2 Under-Foreshortening

Foreshortening can also generate *concave diagonals* when the spacing between the horizontal lines does not diminish rapidly enough on approaching the vanishing point.

Let us first examine the "zero degree" of foreshortening, which consists in drawing a series of transversal lines that are *equidistant* from one another. This method was only employed in a small number of paintings characterized by a very shallow visual field, such as *The Death of a Saint* by Simone Martini, *Confession of the New-born Child* by Donatello, *The Last Supper* by Andrea del Castagno, etc.

---

[6]Alberti verified the correctness of the perspective construction by applying this property, but was unaware that the diagonals should converge at a distance point: "If the same line is extended in the depicted floor, it will form the diagonal of the joined squares. Indeed, among mathematicians, the diagonal of a square is a kind of straight line that is drawn from one corner to the opposite corner of the square, which divides it into two parts so that to make two triangles of it/Qui quidem quam recte descripti sint inditio erit, si una eademque recta continuata linea in picto pauimento coadiunctorum quadrangulorum diameter sit. Est quidem apud mathematicos diameter quadranguli recta quaedam linea ab angulo ad sibi oppositum angulum ducta, quae in duas partes quadrangulum diuidat ita ut ex quadrangulo duos triangulos fit," De pictura, I, 20.

[7]Leon Battista Alberti, *On Painting* and *On Sculpture*, ed. C. Grayson, London, 1972; idem, *De la peinture/De pictura (1435)*, ed. J.-L. Schefer, Paris, 1992.

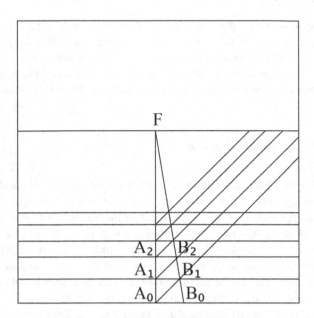

Fig. 3.1 Under-foreshortening, method 1. Author's drawing

The error in these cases is understandable, because the rule of diminishing intervals would have necessitated a drastic intervention given the lack of depth in the scene (Type II ad hoc error).

In this category we can also find the technique known as diminution by one-third. Panofsky wrote, "If we are to believe Alberti, the erroneous practice of mechanically diminishing each strip of the floor by one-third still held sway in his day."[8] Alberti in fact mentioned the *ratio vitiosa* (*De pictura*, I, 19) before applying his own method. There were a whole range of fallacies of this kind. The vertical line traced from the vanishing point (F), could divide the central square at the ½ point (Lorenzetti), the ⅓ point (van Eyck), or the ¼ point (Brunelleschi, Ghiberti). Nevertheless all of these constructions, including the *ratio vitiosa*, belong to the same category as each one is based on a division of the central square. Because this solution is justified in operating terms, it falls into the domain of type III systematic errors. Two such constructions were employed in the Renaissance, which we will analyze here.

The first (Fig. 3.1), consists in using the area of the triangle (A_0B_0F) to trace the receding transversals. From point (A_0), located on the same vertical as the vanishing point (F), draw a diagonal using the square. The diagonal will intercept the first vanishing line (B_0F) at (B_1). From that point, trace a horizontal line (A_1B_1) that will determine point (A_1) along the vertical line (A_0F). Then move the square in order to trace a parallel to (A_0B_1) from (A_1). This line will intercept point (B_2), which in turn

---

[8]Panofsky, *Perspective as Symbolic Form*, p. 62.

will fix the height of the second horizontal (A_2B_2) and so on until the horizon is reached.

This is actually a method of under-foreshortening and leads to the production of concave diagonals. Why was such an error accepted by Lorenzetti, Rogier van der Weyden, Carpaccio and other painters? Perhaps because for the Renaissance artist trained in the *abaco* school of calculation,[9] such proportional representations recalled the "Golden Rule" whereby:

$$a : b = c : d$$

Let us return to the perspective view and designate as A_iB_i and A_iA_j the apparent width and height of a square. The application of the Golden Rule gives us:

$$A_0B_0 : A_0A_1 = A_1B_1 : A_1A_2 = \cdots = A_iB_i : A_iA_j = \cdots = A_mB_m : A_mA_n$$

This relationship implies that the proportional ratio between the heights from the first interval to the last can be directly estimated by the naked eye. But in comparison to the foreshortening that a linear perspective would produce, the first intervals (A_0A_1 ...) are too short and the last (... A_mA_n) are too long. Here the error stems from a bias which consists of overweighting the sameness of the squares. Can it not be assumed that if the ratio *a:b* applies to the first intervals, then (because all the squares are identical) it should also hold true for the last ones? Despite the fact that linear perspective does not tolerate the mechanical construction of intervals using the square, painters had sound reasons to believe in the validity of the empirical rule.

Other paintings in which under-foreshortening has been identified made use of different methods from the one described above. Let us suppose that the vertical line traced from the vanishing point (F) divides the first square at the ¼ point in its width (Fig. 3.2). In order to create the spacing of the transversal lines, the height of the first square can be fixed arbitrarily by tracing the straight line (A_0P). This line will intercept the first vanishing line (B_0F) at (B_1). From this point, trace the first horizontal (A_1B_1), which will intercept the vertical line (A_0F) at (A_1). Then, from this point trace the diagonal (A_1P), which will determine point (B_2) by its intersection with the vanishing line (B_0F). Point (B_2) will fix the height of the second horizontal (A_2B_2), and so on until the horizon is reached. Compared with the previous system of construction, the diagonals are no longer parallel but converge at point (P), which is situated above the horizon.[10] This is why all the intervals

---

[9]Other connections between *abacus*—a term that encompasses arithmetic and algebra—and perspective have been proposed in order to understand the construction methods of Piero della Francesca; Judith V. Field, "Alberti, the abacus ...".

[10]This construction point (P), which foreshadows the "distance point" of linear perspective as codified in the seventeenth century, does indeed seem to have been used by painters, since we can often find it in a conspicuous place in the architectural décor; for example, at the corner of a pilaster (Brunelleschi) or on the edge of a building (Ghiberti) or on the shoulder of a figure (Donatello).

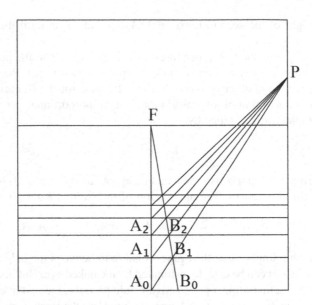

Fig. 3.2 Under-foreshortening, method 2. Author's drawing

(A_0A_1), (A_1A_2), (A_2A_3)... are longer than the ones that would normally produce exact perspective foreshortening. This method of under-foreshortening was used quite frequently, particularly by Ghiberti, Jan van Eyck and Rogier van der Weyden.

Painters who followed either one of these two (erroneous) methods did so principally because it allowed them to represent intervals that diminish the closer they are to the horizon. Within the constraints of the knowledge available to them, the goal of the *qualitative representation* of diminishing intervals was thereby achieved.

3.2.3 Over-Foreshortening

Finally, foreshortening can result in a network of *convex diagonals*. In this case the spaces between the horizontal lines diminish too rapidly as the vanishing point is approached.

We come across examples of perspective over-foreshortening (Fig. 3.3) when the point of concurrence (P) of lines (A_0B_1), (A_1B_2)..., as constructed above, is situated lower instead of higher than the horizon line. In this case the intervals of the series ... (A_0A_1), (A_1A_2), (A_2A_3) ... will tend toward a limit (A_n) at the height of (P). In a linear perspective, this point of concurrence (P) would correspond to a distance point on the horizon. However, since (P) is situated below the horizon— and assuming that the floor extends to infinity—there will always be an empty

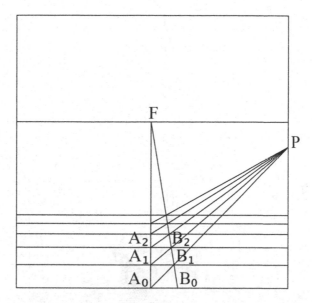

Fig. 3.3 Over-foreshortening. Author's drawing

space between this limit and the true horizon. Hence, the opposite effect to that of under-foreshortening is produced; the intervals (A_0A_1), (A_1A_2), (A_2A_3) ... are too short compared with those which an exact perspective foreshortening would produce.

This type of diminution was not very common in the corpus of Renaissance painting; it can be found on occasion in the works of Donatello and Masaccio. The preference for the under-foreshortening methods described above can be easily understood; although erroneous, they allowed the artist to fill the entire space between (A_0) and (F), which over-foreshortening does not permit.

3.3 Some Examples of Erroneous Foreshortening in Renaissance Painting

The methods of correct foreshortening, and over- and under-foreshortening described above are mutually exclusive categories. When we look for examples of these constructions in the most representative paintings of the Renaissance, it is surprising how few exhibit correct diminution apart from a handful of late works such as *Christ's Flagellation* by Piero della Francesca and *The Reflection* by the Master of the Barberini Panels (both ca. 1450). Most works from the period fail to respect the rules of linear perspective.

It is generally agreed that Filippo Brunelleschi (1377–1446) played a major role in establishing the perspective system and yet only one perspective view is actually

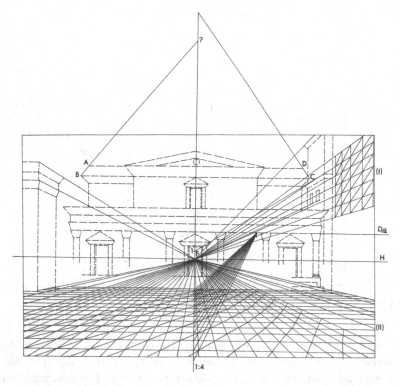

Fig. 3.4 Filippo Brunelleschi (?) *Christ Casting out a Demon*, ca. 1425, Engraved silver plate, 7 × 11 cm (Paris, Musée du Louvre, Département des Objets d'Art), author's reconstruction

attributed to him, an engraved silver plate depicting *Christ Casting Out a Demon* (ca. 1425) now conserved in the Musée du Louvre (Fig. 3.4).

Should we accept the opinion of Parronchi, according to whom, "… if one looks carefully, the perspective construction of the buildings on this plate quickly reveals the mathematical precision on which the law of proportions concerning 'diminutions' and 'augmentations' is built"?[11] It may be noted that the vertical line traced from the vanishing point (*F*) crosses the square at the ¼ point in its width. Since the network of diagonal lines on the floor is concave and point (*D*/4) is situated above the horizon, this constitutes an example of under-foreshortening. Paradoxical as it may seem, Brunelleschi—the supposed inventor of linear perspective—used a construction here that departs from the rules of correct diminution.

Let us now examine the work of Lorenzo Ghiberti (1381–1455). Art historians have often assumed that either Brunelleschi or Alberti played an advisory role in the realization of the bronze bas-reliefs for the portals of the Baptistery of San Giovanni

---

[11]Alessandro Parronchi, "Le due tavole prospettiche del Brunelleschi," *Paragone* 107 (1958), p. 16.

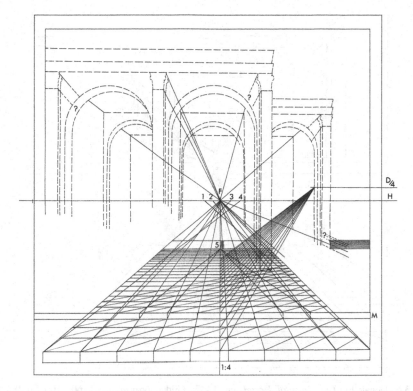

Fig. 3.5 Lorenzo Ghiberti, *The Sacrifice of Isaac*, 1437, Gilded bronze bas-relief, 79 × 79 cm (Florence, Baptistery San Giovanni), author's reconstruction

in Florence (1437). Thus Richard and Trude Krautheimer write: "In the *Isaac* and *Joseph* panels, Ghiberti applied *verbatim* the perspective construction that Alberti laid out in his text on painting."[12] John White is of the same opinion: "Here the paving in perspective, whose construction is henceforth exact, allows the figures to evolve on a platform which conforms to the new scientific principles."[13] If we study the method of diminution adopted by Ghiberti, however, it becomes clear that these conclusions do not stand up to scrutiny. The squares along the central axis are, as in Brunelleschi's engraved silver plate, shifted back by ¼ compared with the vertical line traced from the vanishing point (*F*). As the series of diagonals on the floor are concave, the system used was one of under-foreshortening. In fact, contrary to various analyses,[14] the lateral point of concurrence in both *Isaac* (Fig. 3.5) and *Joseph* is situated above the horizon. These were not accidental errors because the

---

[12]Richard and Trude Krautheimer, *Lorenzo Ghiberti*, Princeton, 1956. p. 251.

[13]John White, *The Birth and Rebirth of Pictorial Space*, (London: Faber, 1967); trans. Catherine Fraise, *Naissance et renaissance de l'espace pictural* (Paris: Adam Biro, 2003) p. 172.

[14]Martin Kemp, *The Science of Art. Optical Themes in Western Art from Brunelleschi to Seurat*, New Haven, 1990, p. 25.

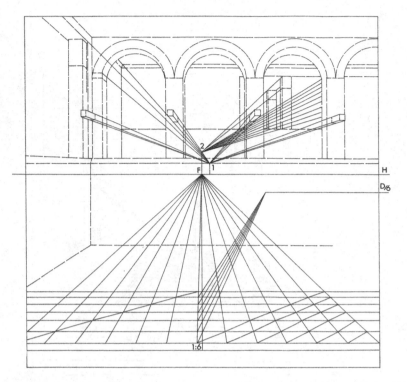

Fig. 3.6 Donatello, *The Feast of Herod*, ca. 1427, Gilded bronze panel, 60 × 60 cm (Siena, Baptistery San Giovanni), author's reconstruction

construction is identical in both panels. We must therefore refrain from concluding that Ghiberti engaged in "the strict use of artificial perspective."[15]

Let us turn to the work of Donatello (1386–1466), whose talent for perspective was praised by Cristoforo Landino: "Donato... was a great imitator of the Ancients and knew a great deal about perspective."[16] Certain historians have defended the correctness of Donatello's perspective construction. Regarding the marble relief entitled *The Feast of Herod*, now in the Musée des Beaux-Arts of Lille, Darr and Bonsanti wrote: "He constructed an architecture that was painted according to the rules of two-point perspective, which had just been codified by Leonbattista Alberti in his De pictura of 1435."[17] The fact that the network of diagonals on the floor is concave proves, however, that Donatello used a method of under-foreshortening which was inconsistent with the rules of linear perspective. Moreover, this was not

---

[15]White, *Naissance et renaissance de l'espace pictural*, p. 162.

[16]"Donato sculptore... fu grande imitatore degl'antichi, et di prospectiva intese assai," as quoted by Pietro Roccasecca, "La finestra albertiana," *Nel segno di Masaccio*, Florence, 2001, pp. 64–69.

[17]Alan P. Darr and Giorgio Bonsanti, *Donatello e i suoi*, Detroit, Firenze and Milano, 1986, p. 141.

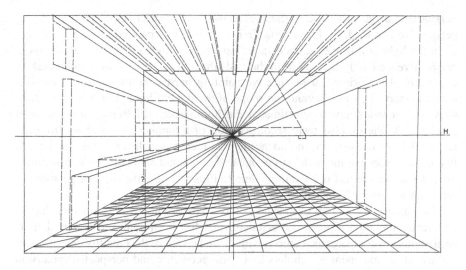

Fig. 3.7 Paolo Uccello, *Profanation of the Host: Holocaust*, 1465–9. Tempera on wood, 42 × 361 cm (Urbino, Palazzo Ducale), author's reconstruction

an episodic error on his part; in a first version of *The Feast of Herod* (Fig. 3.6), carved around 1427 for the baptistery of Siena Cathedral, Donatello had already resorted to under-foreshortening; as can be seen, point (*D*/6) is clearly situated below the horizon.

Moving on to a key figure in the history of perspective, was Giorgio Vasari justified in attributing "the perfection of this art" to Paolo Uccello (1397–1475)? As Vasari wrote:

> Paolo devoted himself, without respite, to the most difficult artistic research; he perfected the method of constructing perspective by the intersection of lines traced, using floor plans and elevations of buildings, to the very summits of cornices and rooftops. After fixing the point of view of the eye higher or lower, according to his desire, he foreshortened them and made them diminish towards the vanishing point.

If we retrace the geometric lines of *The Holocaust*, a scene from the *predella* decorated by Ucello with the story of the *Profanation of the Host* (1465–1469) and conserved in the Palazzo Ducale in Urbino (Fig. 3.7), the diagonals form a network of concave lines. One must therefore lay aside the judgment of Kemp[18] and acknowledge that Paolo Uccello used a method of under-foreshortening.

These brief observations regarding some of the greatest artists of the Italian Renaissance allow us to draw a simple—if not entirely surprising—conclusion: Not a single work produced during the first half of the Quattrocento applied the rules of linear perspective to the letter, not even the plate depicting *Christ Casting Out a Demon* attributed to Brunelleschi, who has been credited with the invention of

---

[18]Kemp, *The Science of Art*, p. 39.

perspectiva artificialis. The earliest example of the strict application of linear perspective is a work by Piero della Francesca that was painted around 1450.

In the light of this critical analysis, it must be concluded that the rules of linear perspective were only applied from the mid-fifteenth century onward and therefore any study of the origins of perspective painting must continue well into the sixteenth century. This lies beyond the scope of the present book, but it is possible to show that scholars have overestimated the homogeneity of Renaissance pictorial practices. As the ex post facto reconstruction of a number of paintings from the period demonstrates, while the art academies may have played a role in the diffusion of perspective methods, at the outset they did not standardize the operations used. I will present just one example here to illustrate how the rules of perspective were not uniformly followed by painters, even after 1450.

The paintings of Vittore Carpaccio (1460–1526) are noteworthy for their highly architectonic composition, a characteristic that has been analyzed in detail. It is enlightening to compare his methods with those applied in the Quattrocento. Should we accept the judgment of scholars that "The geometric and perspective precision

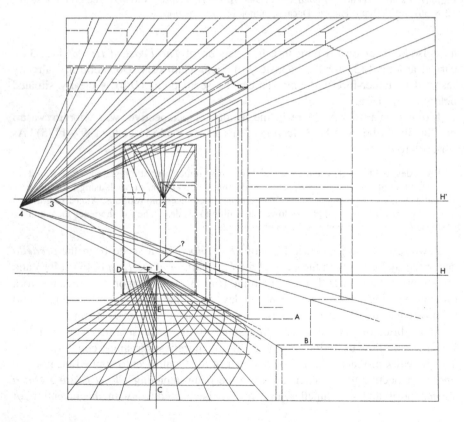

Fig. 3.8 Vittore Carpaccio, *The Birth of the Virgin*, 1504, Oil on canvas, 126 × 129 cm (Bergamo, Accademia Carrara), author's reconstruction

of the town planner and the architect is characteristic of Carpaccio's way of thinking"?[19] Consider, for example, *The Birth of the Virgin* (Fig. 3.8), which forms part of the *Albanian Cycle* (1504).

We discover not only that the system of orthogonals does not converge towards a single vanishing point, but Carpaccio's method of diminution was erroneous. If we trace the diagonals of the squares, a concave network is obtained, which shows this construction to be an example of under-foreshortening. The vertical line traced from the vanishing point (1) cuts across the axial square at the ¼ point in its width. The fact that the oblique lines (*CD … EF*) remain parallel shows Carpaccio used a method in which the diagonals were constructed by moving the square. This constitutes, I admit, an isolated instance of the transgression of the laws of perspective on his part, but Carpaccio may not have been the only sixteenth-century painter to display such independence.

3.4   Conclusion

Let us set aside our working diagrams and step back for a moment. Some general conclusions can be drawn from this examination of the methods of constructing a perspective that were developed by artists over a period of two centuries (1297–1504). The history of perspective was marked by three key moments: (1) the use for the first time—by Giotto—of a correct method of diminishing intervals (his contribution will be discussed in detail in Chap. 8); (2) the representation by Brunelleschi of perspective based on the postulate of monocular vision, although he applied an erroneous method of diminution; and finally (3) the use of both a single vanishing point and a correct method of diminution by Piero della Francesca. However, this interpretation of the history of perspective raises the question as to what is truth and what is falsehood in painting.

Isolated examples of the application of true perspective techniques at the beginning of the fourteenth century may be found in *The Funeral of a Saint* by Simone Martini and *Jesus Among the Doctors* by the atelier of Giotto, both of which were painted around 1315–1317 and in both of which a correct foreshortening of the intervals is used. Brunelleschi, Ghiberti, Uccello, Fra' Angelico, Donatello and Masaccio were still applying erroneous procedures one and a half centuries later. What is more, with regard to the supposed affinities between artists, it is apparent that the community of architects, painters and sculptors of the Quattrocento were not unanimous in their adherence to the rules of linear perspective. From a comparison of the foreshortening methods used, it emerges that no unified conception of perspective existed. Every artist adopted the method that he believed empirically to be the most correct, and knowledge was not shared. Moreover, while Ghiberti and Masaccio displayed a certain consistency in their

[19]Vittorio Sgarbi, *Carpaccio*, Bologna, 1979, p. 17.

approach, the same cannot be said of Donatello or Ucello, who experimented with several different systems. *During the Renaissance, therefore, the approach to perspective was not singular but multifarious.* This could provide the answer to the question posed by Marisa Dalai Emiliani: "Are we faced [in the Quattrocento] with the application of a single, unique discipline or rather of a discontinuity, of variations or mutations to such an extent that we could speak of a 'personal' use of perspective on the part of each individual artist?"[20]

Perspective in the Quattrocento was characterized by a series of "uncoordinated initiatives," which is precisely what sociologists observe in the unfolding of most social movements. This analysis also raises the question as to why we are so reluctant to accept the possibility that artistic movements might exist in forms other than that of social groups behaving in a uniform manner. Perhaps lingering vestiges of *Kunstwollen* can be detected here—Alois Riegl's theory of the will to self-expression through art.[21]

---

[20]Marisa Dalai Emiliani, "La question de la perspective," *Perspective et histoire au Quattrocento*, Paris, 1979, p. 17.

[21]Early on Panofsky recognized the problems raised by the use of this holistic notion. He wrote: "The term *Kunstwollen* usually refers artistic phenomena in their entirety, to the artworks of a whole era ... whereas the term 'artistic intention' is more often used to characterize an individual work of art," Panofsky, *Perspective as Symbolic Form*, French edition, p. 200.

Chapter 4
Fact and Fiction Regarding Masaccio's *Trinity* Fresco

Abstract The present chapter discusses some new findings on the question of how Masaccio constructed the perspective in his Trinity fresco. Some scholars have attempted to reduce the anomalies in the work using photogrammetry and computer analysis. On such grounds it has been argued that Masaccio used the normal technique known as costruzione legittima. However, the aberrations discovered in situ strongly suggest that Masaccio's fresco is not a model of linear perspective. The concurrence of the vanishing lines in one point is only approximate, and the foreshortening of the intervals is flawed. Masaccio apparently used the lines of the plane joining the abaci of the capitals as a guide when he drew the lines of the coffered vault. But since the horizontal plane and the barrel vault are not coincident, he adopted an erroneous method of foreshortening—a fact that has been disregarded up to now. The thesis that Masaccio designed the fresco with the aid of a ground plan and elevations is dubious, and the search for its ideal viewing point is destined to remain an unending quest.

The *Trinity* fresco in the Basilica of Santa Maria Novella in Florence, painted by Masaccio around 1425–1428, is a work of large dimensions (667 by 317 cm) that allows one to make perspective tests under very good conditions. As a consequence it has formed the subject of a wide range of publications in the areas of art history and the history of perspective techniques.[1] Many art historians, including Panofsky, have assumed the correctness of the fresco's perspective, probably following the lead of Giorgio Vasari. In his *Vite* Vasari writes: "At Santa Maria Novella... there is something even more beautiful than the figures: it is a barrel vault, drawn in perspective and divided into coffers filled with rosettes whose proportions decrease

---

[1]Before 1989, the main works on the construction of perspective were: G.J. Kern, Das Dreifaltigkeitsfresko von S. Maria Novella. Eine perspektivisch-architektur geschichtliche Studie, *Jahrbuch der königlich preussischen Kunstsammlungen* 24 (1913): 36–58; P. Sanpaolesi, *Brunelleschi*, Milan, 1962; H.W. Janson, "Ground plan and elevation of Masaccio's Trinity fresco" in F. Douglas et al., eds., *Essays in the History of Art presented to Rudolf Wittkower*, London, 1967, pp. 83–88; J. Polzer, "The anatomy of Masaccio's Holy Trinity," *Jahrbuch der Berliner Museen* 13 (1971) 18–59. Further references are given in the article by Field, Lunardi and Settle cited in note 5 below.

© Springer International Publishing Switzerland 2016
D. Raynaud, *Studies on Binocular Vision*,
Archimedes 47, DOI 10.1007/978-3-319-42721-8_4

with foreshortening so that the wall appears to be hollowed out." Panofsky believes that this work is "exactly and uniformly constructed."[2] Parronchi considers "Masaccio's *Trinity* at Santa Maria Novella, which is a work of genius and well ahead of its time" to be an example of the strict application of the rules of linear perspective.[3] John White agrees: "The foreshortening of the architecture, in accordance with the principles of artificial perspective, is accurate both in the diminution of the coffering and in the single vanishing point which lies slightly below the plane on which the donors kneel."[4]

The exacting studies on the geometric aspects of Masaccio's fresco conducted by Field, Lunardi and Settle led to somewhat unexpected conclusions. They noticed many geometrical anomalies,[5] in direct contrast to the judgment of previous art historians.

4.1   Recent Research

Following the key publication by Field et al. in 1989, several other studies appeared.[6] In a remarkable paper presented to the *4th ILabHS* in 2001, Volker Hoffmann sought to analyze and compare all the major publications concerned with the geometric analysis of the fresco. He then set out the results of his own analyses.

---

[2]E. Panofsky, "Die Perspektive als symbolische Form," *Vorträge der Bibliothek Warburg* 4 (1924/5): 258–331; *Perspective as Symbolic Form*, p. 62.

[3]Alessandro Parronchi, "Le fonti di Paolo Uccello: I perspettivi passati," *Paragone* 89 (1957), p. 7.

[4]John White, *The Birth and Rebirth of Pictorial Space*, London, 1967; transl. *Naissance et renaissance de l'espace pictural*, p. 146.

[5]Judith V. Field, Roberto Lunardi, Thomas B. Settle, "The perspective scheme of Masaccio's Trinity fresco," *Nuncius* 4/2 (1989): 31–118. They write (p. 34): "The very success of the *Trinity* fresco in presenting space that seems as real as the figures that inhabit it may explain why so many scholars have taken the perspective scheme for granted." This observation has been reiterated on many occasions, see note 7.

[6]Martin Kemp, *The Science of Art. Optical Themes in Western Art from Brunelleschi to Seurat*, London, 1990, draws up an inventory of at least six errors but supports the idea of the use by Masaccio of a ground plan and elevations. A sophisticated system has been proposed by Jane A. Aiken in "The perspective construction of Masaccio's Trinity fresco and medieval astronomical graphics," *Artibus et Historiae*, 31 (1995): 171–187. She postulates that Masaccio obtained the diminution of the vault ribs with the help of an astrolabe and stereographic projection. Nevertheless, it is highly questionable whether the orthographic and stereographic projections of the astronomers were "readily available sources to Masaccio and Brunelleschi," p. 173. The length and complexity of the procedure shows an evident lack of proportion between the ends and the means, so that one would be justified in wondering whether so sophisticated a technique had ever actually been used. See Volker Hoffmann, "Masaccios Trinitätsfresko: Die perspektivkonstruktion und ihr Entwurfsverfahren," *Mitteilungen des Kunsthistorischen Institutes in Florenz* 40 (1996): 42–77. Rona Goffen, ed., *Masaccio's Trinity*, Cambridge, 1998. "The Trinity of Masaccio: perspective construction—isometric transformation—coordinate system," *Art, Science and Techniques of Drafting in the Renaissance*, 4th ILabHS, working paper, Florence, 24 May-1 June 2001. Cristina Danti, ed., *La Trinità di Masaccio. Il restauro dell'anno duemila*, Firenze, Edifir, 2002.

In line with Field et al., Hoffmann was primarily concerned with the issue of the foreshortening of the intervals, which has not received the same attention as the convergence of the vanishing lines towards the vanishing point. The restoration of the *Trinity* fresco undertaken in the year 2000 also yielded a wealth of information on the perspective drawing, most notably about the incisions left in the *intonaco*.

Three issues are worthy of discussion here: (1) the convergence of the vanishing lines, (2) the foreshortening of the intervals, and (3) the determination of the ideal viewing point for the fresco.

The vanishing lines incised in the *intonaco* do not converge precisely in one point, as confirmed by the 2000 restoration of Masaccio's fresco. One-third of the lines deviate appreciably from the vanishing point F. This can be checked by superimposing the incised lines (broad lines) and their extensions (narrow lines) over a photogrammetric image of the fresco. The lines that deviate most from the correct perspective are lines 1, 2, 7, 8, 15, 19, 24 and 25 (Fig. 4.1). This confirms the earlier conclusion by Judith V. Field that "There are serious departures from mathematical correctness in Masaccio's *Trinity* fresco," a statement on which she further expanded at the *4th ILabHS*: "The lines of the edges of the ribs do not meet so neatly... So the ribs do not really provide strong evidence for Masaccio having understood the properties of what Alberti, writing about ten years later, was to call the centric point".[7]

But the major problem with Masaccio's fresco is much more serious: none of the authors who have attempted a perspective reconstruction agree on the distance from which the work should be viewed, proposing values that range from 210.5 to 894.2 cm and which strictly speaking are baseless according to the rules of linear perspective.

Hoffmann lays out the problem in these terms: "There are only two explanations for such a chaotic scientific situation: [either] the fresco has *not* been constructed with the help of linear perspective or the methods of analyzing the perspective are not worth anything."[8] He then opts for the hypothesis that the reconstruction methods were at fault, and carries out a scrupulous analysis of the *Trinity* fresco based on photogrammetry and computer analysis. Using a four-point reconstruction, based on the assumption that Masaccio took the lines of the horizontal plane joining the *abaci* of the capitals as his guide for the foreshortening of the coffered vault, he claims to have obtained a good fit between the lines incised in the *intonaco*

---

[7]Danti, *La Trinità di Masaccio*, pp. 89–94 and, most importantly, plate VII. Judith V. Field, in *The Invention of Infinity: Mathematics and Art in the Renaissance*, 1997, p. 72; "What mathematical analysis can tell us about a fifteenth-century picture," in *Art, Science and Techniques of Drafting in the Renaissance, 4th ILabHS*, working paper, Florence, 24 May-1 June 2001.

[8]Volker Hoffmann, "Brunelleschi's invention of linear perspective: The fixation and simulation of the optical view," *Art, Science and Techniques of Drafting in the Renaissance, 4th ILabHS*, working paper, Florence, 24 May-1 June 2001, p. 2.

Fig. 4.1 Photogrammetric image of Masaccio's *Trinity* fresco showing some of the vanishing lines that deviate most from the correct perspective. Author's drawing adapted from Field et al. and Danti

(broad lines) and his perspective scheme (narrow lines). This result as well was confirmed by the information gathered during the 2000 restoration (Fig. 4.2).[9] For that reason, Hoffmann reverts to the traditional judgment of the correctness of Masaccio's perspective: "All in all, he writes, there is enough reason to believe that

---

[9]After having unsuccessfully tested a three-point reconstruction (*Entwurf/Dreipunkte-Rekonstruktion*), the author experiments with a four-point reconstruction (*Ausführung/Vierpunkte-Rekonstruktion*) that corresponds to a superimposition of the photogrammetric and perspective drawings based on a fixed congruence of the following points: *F* (vanishing point), *A* and *B* (the upper corners of the abaci of the ionic capitals in the foreground) and *C'* (the top corner of the abacus of the ionic capital in the background on the left), Hoffmann, "Masaccios Trinitätsfresko," p. 45. Danti, *La Trinità di Masaccio*, plate VII.

Fig. 4.2 Photogrammetric image of Masaccio's *Trinity* fresco showing the partial correspondence between the incised lines and the foreshortening of the barrel vault. Author's drawing adapted from Field et al. and Danti

Masaccio constructed the perspective of the Trinity according to the normal case of costruzione legittima."[10]

What could have formed the basis for such a discrepancy of views between Hoffmann and Field et al.? The present chapter will attempt to answer this question, focusing on the issue of the method of foreshortening used by Masaccio, and the ensuing question regarding the point from which the fresco should be viewed. Approaching Masaccio's fresco geometrically,[11] Hoffmann maintains a threefold thesis:

---

[10]Hoffmann, "The Trinity of Masaccio," p. 8 (italics mine).

[11]"As the construction of linear perspective first and foremost centers around questions of projective geometry… the demonstration will have to be of a geometrical nature," Hoffmann in "The Trinity of Masaccio," p. 2.

(1) We must rethink the problem and propose a new scheme to explain the construction of the vault ribs in the fresco.
(2) This leads to the conclusion that Masaccio applied *costruzione legittima*.[12]
(3) Using plans and elevations, we can then calculate the viewing distance (452 cm).

Let us examine each of these points in more detail.

4.2   The Construction of the Vault Ribs

Within this circumscribed context, the main contradiction between Hoffmann and Field et al. concerns the geometric construction as such. In Fig. 4.3 we have retraced Hoffmann's diagram, using a different lettering system for the convenience of discussion. Lines AF, JF, KF... BF (images of the orthogonals) converge at the vanishing point F. Lines AB, A_1B_1, A_1B_1... $C'D'$ (images of the transversals) exhibit perfect foreshortening, because the diagonal BC' covers all of the intersections of the orthogonals with the transversals.

Hoffmann's results depart from those of Field et al., which were based on the gathering of in situ measurements of the arch divisions and led to the conclusion: "[Masaccio] seems to have changed his mind about the importance of this simplification [i.e. equal arcs], because the lowest division... gives an arc that subtends about 30° at the center of the circle. The three remaining arcs, which are equal to one another, each subtend about 20°."[13] In Hoffmann's reconstruction, the length of the arcs AJ', $J'K'$, $K'L'$, $L'M'$ depends on the perspective reduction of the inner divisions of the square $ABC'D'$. It is then possible to calculate the values of the angles subtending these arcs as a function of the perspective parameters. Consider the orthogonal coordinate system (O, x, y). If we assume that F is the point with coordinates $(0, 0)$, M (i.e., the center of the circle of the front arch) is the point with coordinates $(311.2, 0)$, and R is the radius of the circle $(R = 105.8)$, then the equation of the circle is: $(x - 311.2)^2 + y^2 = 105.8^2$.

---

[12]Hoffmann understands by *costruzione legittima* either the "viewing beam method" of Alberti or the "distance point method" of Piero della Francesca. It is common knowledge that both of these approaches lead to the same results, but Hoffmann makes some attempt to show that the second method fits in better with Masaccio's perspective scheme. We will leave aside the semantic difficulty that emerges when one applies the word *costruzione legittima* to an early Quattrocento painting. Scholars have written at great length on this anachronism; for instance Field, *The Invention of Infinity*, and Pietro Roccasecca, "Il 'modo optimo' di Leon Battista Alberti," *Studi di Storia dell'Arte* 4 (1993): 245–262.

[13]Field et al., "The perspective scheme," pp. 49–50.

Fig. 4.3 Hoffmann's scheme
for Masaccio's *Trinity* fresco.
Author's drawing

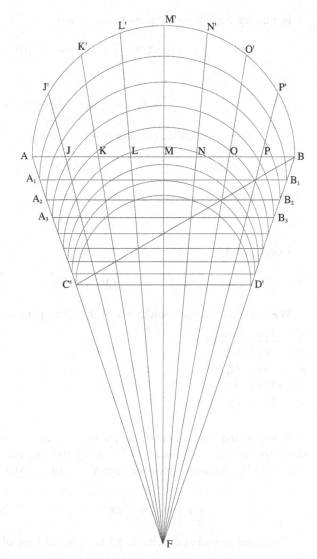

On the other hand, given that *A, J, K, L, M* are equidistant points, the equations for the straight lines *FA, FJ, FK*... are easy to determine. For instance, *FJ* is $y = 0.225x$.

Since the points being searched for—*J', K', L'*...—lie both on the circle and on straight lines, their coordinates represent the solutions to a system of two equations. The point $J'(x_{J'}, y_{J'})$ has to fit:

$$\begin{cases} (x - 311.2)^2 + y^2 = 105.8^2 \\ y = 0.255x \end{cases}$$

Replacing y in the first equation, we obtain:

$$(x-311.2)^2 + (0.255x)^2 = 105.8^2$$

Then:

$$1.065x^2 - 622.4x + 85651.8 = 0$$

The positive solution to this second-degree equation is:

$$x_{J'} = 362.6$$

The second equation yields:

$$y_{J'} = 92.5$$

Consequently:

$$J'(362.6, 92.5)$$

We can arrive at the coordinates of the other points by the same method:

A   (311.2, 105.8)
J'   (362.9, 92.5)
K'   (393.2, 66.8)
L'   (411.5, 33.7)
M'   (417.0, 0)

Knowing the coordinates of the points A, J', K'... of the vault, we can calculate the length of the chords AJ', J'K'... using the formula for calculating Euclidean distances. For instance, from the coordinates of A and J' we obtain the formula:

$$d_{AJ'} = \sqrt{(x_A - x_{J'})^2 + (y_A - y_{J'})^2} = 53.1 \, \text{cm}$$

Given the formula of a chord, we have in the case of $\alpha = \angle AMJ'$:

$$\alpha = 2 \arcsin\left(\frac{d_{AJ'}}{2R}\right) = 29°04'$$

We thus can fix with precision the angles subtended by all the chords according to Hoffmann's scheme. Let us now present the results in tabular form:

		Field et al.	Hoffmann
α	∠AMJ′	∼30°	29° 04′
β	∠J′MK′	∼20°	21° 46′
γ	∠K′ML′	∼20°	20° 36′
δ	∠L′MM′	∼20°	18° 34′

We observe that the angular values calculated according to Hoffmann's scheme match quite closely the angular values that were independently measured by Field et al. on the fresco itself. Given that the maximum difference between the theoretical and observed values is one and a half degrees,[14] and that Field et al. only cite their measurements in approximate terms, e.g. "about 30°… about 20°,"[15] we have every reason to believe Hoffmann is correct when he maintains that Masaccio used the lines FA, FJ, FK… FB to set up the vault ribs.[16] This is the most convincing part of his analysis.

4.3   Masaccio's Use of the So-Called *Costruzione Legittima*

Hoffmann then moves towards the conclusion that Massacio used the *costruzione legittima* in the composition of his fresco, a thesis that is highly questionable.

In order for the diagonal BC′ in Fig. 4.3 to make sense, the two sets of lines AB, A_1B_1, A_2B_2… and J′F, K′F, L′F… should be located on the same plane, which is in fact not the case. On this point Hoffmann's thesis lacks consistency, because he confuses the operations carried out on the horizontal plane and those carried out on the cylinder. He writes: "The points created by the intersection of the lines running parallel to AB with the vertical [MF] become the centers of semicircles, the endpoints of which rest on AC′ and BD′. Semicircles and orthogonals create curvilinear trapezoids: these are the coffers of the barrel vault."[17]

Let us suppose first that the transversals AB, A_1B_1, A_2B_2… C′D′ are placed on the horizontal plane ABC′D′, while the orthogonals J′F, K′F, L′F… P′F belong to the cylinder of the barrel vault. In that context the diagonal BC′ has neither a precise geometrical meaning nor a specific spatial location.

---

[14]The mean difference between the four angular values, reported in situ, represents 22 mm on the arch line.

[15]Field et al., "The perspective scheme," pp. 49–50.

[16]So much so that we no longer need an "artistic" explanation for the difference between angle AMJ′ and the three other angles: "The best explanation for the 'incorrect' positioning of the outermost ribs would seem to lie in a consideration of the surface geometry of Masaccio's picture. As painted, the ribs link Christ's hands with the volutes of the columns. Moreover, their closeness to the receding edges of the front abaci allows the eye to run easily along these lines, whereas the short receding edge might otherwise have been rather lost against the pattern of the vault," Field et al., "The perspective scheme," pp. 50–51. Masaccio could have followed the simple pattern enhanced by Hoffmann.

[17]Hoffmann, "The Trinity of Masaccio," p. 6.

Suppose now the diagonal *BC'* to have a geometrical signification. Lines *J'F,*
K'F... *P'F* are then the same as lines *JF, KF*... *PF* on the horizontal plane *ABC'D'*.
In this context the lines *JF, KF*... *PF* are baseless, because they are not visible in
the fresco. We should note here that the transversal and orthogonal lines of the
coffered vault are the *only visible lines* of the fresco that can provide us with
information regarding Masaccio's method of foreshortening.[18] Other vanishing
lines being absent from both the fresco and its *intonaco*, one could draw the lines
more or less anywhere one chooses in order to demonstrate the correctness of the
perspective. Thus the reconstruction is more in line with an assumption than a
demonstration, and each of the possible alternatives leads to incongruities.

The curvilinear trapezoids that Hoffmann claims represent the coffers of the
barrel vault are arranged in a fashion that defies both common architectural patterns
and the rules of linear perspective in various ways.

On the one hand, if we assume that the perspective in the *Trinity* is correct, then
it must be concluded that the coffers of the vault are of variable width. Let the circle
be of perimeter *C* and radius *R* = 105.8 cm; then, given the angular values: $\alpha = 29°$
04', $\beta = 21° 46'$, $\gamma = 20° 36'$, $\delta = 18° 34'$, we may deduce the length of all the
corresponding arcs by means of the relationship:

$$\overset{\frown}{XY} = \frac{\alpha}{2\pi} C, \quad \alpha = \angle XMY$$

Beginning with the first arch, the arcs would measure:

AJ'   53.6 cm
J'K'   40.2 cm
K'L'   38.0 cm
L'M'   34.3 cm

But such a pattern is unheard of with respect to the models of Renaissance
architecture and others extending back in time to ancient Rome and forward to the
neo-classical period. The coffers that we come across throughout the history of
architecture are always of regular shape and indeed are almost always square.[19]

---

[18]With the exception of the lines drawn from the *abaci* of the capitals, but these run in a somewhat
erratic manner, so that the right and left *abaci* provide us with a distance point varying from simple
to double (see Sect. 4.5).

[19]Listed here are some examples of such coffered vaults. Type I (square-coffered barrel vault):
Thermae in Rome, *Nympheaum of Cicero's Villa* in Formia, *S. Andrea* in Mantova by Alberti, *S.*
Pietro in the Vatican by Bramante and Maderno, *Gesù* in Rome by Della Porta. Type II (flat
ceiling with square coffering): *Basilica of the Palace* in Treves, *S. Maria Maggiore* in Rome.
Type III (dome with square coffering): *Pantheon* in Rome, *S. Maria in Campitelli* in Rome by De
Rossi, *Library project* by Durand. Type IV (other Baroque geometrical patterns): *S. Carlo ai*
Catinari in Rome (circles), *S. Andrea al Quirinale* in Rome by Bernini (hexagons), *S. Carlo alle*
Quattro Fontane in Rome by Borromini (hexagons, circles, ovals and crosses).

Let us suppose, on the other hand, that the coffered vault in Masaccio's *Trinity* is regular (and, we would repeat, there is not a single example in architecture of variable rectangular coffers). Then we must impute to Masaccio a serious error of construction or—at the very least—a high degree of "independence" in the light of how perspectives were usually composed. This discrepancy is probably the point that most convincingly invalidates the idea that Masaccio designed his perspective with the help of plans and elevations.[20] If he had done so, he would never have drawn the coffers of the barrel vault with a variable width. There is, in fact, only one method (known from Euclid's *Elements*, Book III, 30) of drawing the longitudinal division of a vault in *correct perspective*: "Dividing a semicircular arc into eight equal parts, or a quadrant into four, is a very simple mathematical task, merely involving repeated bisections. Having divided his semicircle, or quadrant, Masaccio could then have joined the points marking the divisions to the centric point, thus obtaining the center lines of the longitudinal ribs."[21]

As we can see in the comparison presented in Fig. 4.4, the above construction is quite different from the one assumed by Hoffmann.

It is apparent that Masaccio did not use this basic—but nonetheless correct—method. Indeed, in correct perspective the angles AMJ'', $J''MK''$, $K''ML''$ ... $P''MB$ are all equal to $\frac{\pi}{8} = 22°30'$:

		Hoffmann	Exact values
α	∠AMJ'	29° 04'	22° 30'
β	∠J'MK'	21° 46'	22° 30'
γ	∠K'ML'	20° 36'	22° 30'
δ	∠L'MM'	18° 34'	22° 30'

If C is the perimeter of the circle corresponding to the front arch, we may calculate the length of an arc XY in correct perspective:

$$\stackrel{\frown}{XY} = \frac{1}{16}C = 41.5\,\text{cm}$$

The maximum difference between the measured values and the values calculated using Hoffmann's model is reached for the outermost points $J'J''$ and $P'P''$, for which we have: $J''J' = P'P'' = 12.1$ cm. This metric difference is far too great to be

---

[20]Hoffmann, "The Trinity of Masaccio," pp. 11–12, admits that Masaccio used ground plans when creating this perspective: "It is fairly obvious that one of those plans had to be the ground plan... To this [Masaccio] added either the frontal elevation or the side elevation. But there is reason to believe that Masaccio actually used both the frontal elevation and the side elevation." Nothing is obvious here, except the need to establish such an assumption.

[21]Field et al., "The perspective scheme," p. 49.

Fig. 4.4 Correct scheme for
Masaccio's *Trinity* fresco.
Author's drawing

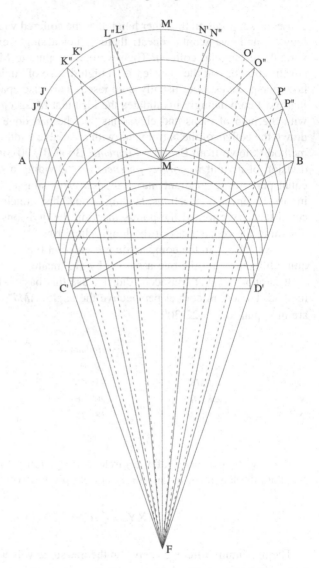

considered an accidental error; the method as such must be wrong. Therefore, by ruling the coffers of the vault based on the inner divisions of the square *ABC'D'*, Masaccio took an erroneous shortcut which was in a way much more complicated than the method he actually should have applied to resolve the problem of perspective presented. And this is, in fact, the conclusion we are in a position to deduce from Hoffmann's most interesting analysis.

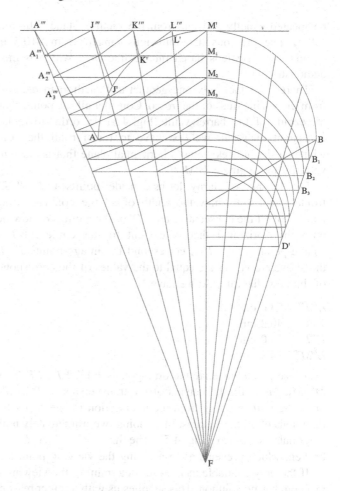

Fig. 4.5 Discrepancy between Masaccio's scheme and the true perspective of the barrel vault. Author's drawing

4.4   The Determination of the Viewing Distance

After he became convinced that Masaccio was applying the so-called *costruzione legittima*, Hoffmann adopted a fresh approach to the difficult problem of how to determine the viewing distance. To fix the viewing point for the fresco it is necessary to assume that Masaccio used plans and elevations to construct his perspective, but this has not been established. Setting aside this fact, the pattern of the coffered vault is the only way to calculate the distance from which one the fresco should be viewed.[22] Based on Hoffmann's model, the viewing point would have to

[22]Field et al., "The perspective scheme," pp. 42–43, nevertheless tried in Appendix 6 to determine the viewing distance by another method, using the vanishing lines of the abaci of the capitals. But they did not manage to achieve a viable result, because there is too great a discrepancy between the distance obtained from the upper front abacus (594.4 cm) and from the lower back abacus (345.7 cm).

be located exactly 452 cm from the picture plane. This intermediate value—re-calling that previous determinations ranged from 210.5 to 894.2 cm—remains dubious because it is based on the line BC', which is the weakest point in his demonstration.[23]

There is in fact only one geometrical method to determine the viewing point from the visible lines of the fresco. Consider all the points that define the curvilinear trapezoids of the barrel vault (Fig. 4.5). In order to replace the line BC' by a consistent diagonal, we first have to transfer all the points $A_n, J'_n, K'_n, L'_n \ldots$— whatever the indices—onto a horizontal plane that is tangent to the cylinder of the vault by the segment $M'M_1$.[24]

For the sake of clarity, let us consider points A, J', K', L', M' belonging to the frontal arch. We know the width of all the coffers—which corresponds to the lengths of all the consecutive arcs XY of the vault. We now need to unroll the circle onto the horizontal that is tangent to the circle $AM'B$ at point M'. To the source-points J', K', $L' \ldots$ correspond the image-points J''', K''', $L''' \ldots$ Nonetheless, the above arc values are equal to the values of the corresponding straight segments of the new diagram. So we have:

$A'''J'''$ 53.6 cm
$J'''K'''$ 40.2 cm
$K'''L'''$ 38.0 cm
$L'''M'''$ 34.3 cm

These points generate the orthogonals FA''', FJ''', $FK''' \ldots$ while the arc summits M', M_1, $M_2 \ldots$ fix the heights of the transversals $A'''B'''$, $A'''_1 B'''_1$, $A'''_2 B'''_2 \ldots$ We can draw the grid resulting from the intersection of the two sets of lines and trace the diagonals of all the squares. In so doing, we immediately notice that the network of diagonals is *convex* (Fig. 4.5). The lines $A'''_2 K'''$, $A'''_3 L''' \ldots$ are not straight but broken, which prevents us from fixing the viewing point.

If the only available method for determining the viewing distance fails, then the problem has no solution. This supplies us with another reason to think that, once we have solved the perspective problem, Masaccio's fresco cannot be considered a model of linear perspective. And this is why, when Hoffmann writes "There are only two explanations for such a chaotic scientific situation: [either] the fresco has *not* been constructed with the help of linear perspective or the methods of analyzing the perspective are not worth anything,"[25] we are forced to choose not the second, but the first alternative.

---

[23]Hoffmann, "Masaccios Trinitätsfresko," plates 1–6.

[24]Figure 4.4 has been traced from the previous photogrammetric reconstruction (Fig. 4.1) to avoid any distortion of the image of the fresco.

[25]Hoffmann, "Brunelleschi's invention of linear perspective," p. 2.

4.5   Conclusion

Taken together, the arguments laid out in this chapter strongly support the view that Masaccio designed the vaulted space of the *Trinity* in a somewhat empirical manner. As Hoffmann has noted, Masaccio apparently took the lines of the horizontal plane joining the *abaci* of the capitals as a guide for drawing the vanishing lines of the coffered vault. But in doing so, he followed an erroneous method of perspective construction. The painting of course conveys a very convincing sense of depth, but it is not a linear perspective whatsoever because the fresco was created with very little mathematical rigor. Consequently, it is not necessary to assume that the painter worked with plans and elevations. Moreover, the search for the viewing point is destined to remain an unending problem, because it is definable only if the rules of linear perspective are applied *à la lettre*.

The present analysis contributes further to—and at times go beyond—the understanding of the *Trinity* fresco previously proposed by Field, Danti, and others. In contrast, I think that some of the views defended by Hoffmann overestimate the correctness of the *Trinity* fresco's perspective. Attention should be drawn to this fact before we take our next step forward in the rush to apply the latest technology.

Part II
Theory

Chapter 5
Ibn al-Haytham on Binocular Vision

Abstract Early modern physiological optics introduced the concept of correspondence to the study of the conditions for the fusion of binocular images. The formulation of this concept has traditionally been ascribed to Christiaan Huygens (in a work published posthumously in 1704) and to an experiment often attributed to Christoph Scheiner (1619). Here it will be shown that Scheiner's experiment in fact had already been conceptualized, first in antiquity by Ptolemy (90–168 AD), then in the Middle Ages by Ibn al-Haytham (Latinized as Alhacen) (d. after 1040), and the extent of the latter's knowledge of the mechanisms of binocular vision will be analyzed. It will then be explained why Ibn al-Haytham, who was a mathematician but addressed this problem as an experimentalist, succeeded in discovering the theoretical horopter (the locus of points in space that yields single vision) and yet failed to recognize that the horizontal line of the horopter could be described as a circular plane around the viewer's head, credit for which must instead go to Vieth (1818) and Müller (1826). Nevertheless, through his experimental studies Ibn al-Haytham established the notion of corresponding points, explored what the cases of homonymous (direct) and heteronymous (crossed) diplopias could reveal about the mechanisms of vision, and prepared the ground for the discovery by Panum of the fusional area. The influence on Western science of al-Haytham's pioneering treatise *Kitāb al-manāẓir (Book of Optics)* is examined, beginning with his successors in the Latin-speaking world, and in particular Italy.

A key chapter in the physiology of optics considers the conditions for the fusion of the quasi- or displaced images generated by the two eyes. Interestingly, the ancient Greeks did not explore the questions raised by binocular vision in any depth. Euclid only devotes three propositions to this problem (*Optica*, prop. 26–28)[1] and limits his analysis to what is seen of a sphere in binocular vision. If the diameter of the sphere is less than, equal to, or greater than the inter-pupillary distance, then the two eyes will perceive a spherical cap that is greater than, equal to, or less than the

---

[1] Elaheh Kheirandish, *The Arabic Version of Euclid's Optics*, New York, 1999, pp. 80–90; Wilfred R. Theisen, "Liber de visu," *Mediaeval Studies* 41 (1979), pp. 78–80.

© Springer International Publishing Switzerland 2016
D. Raynaud, *Studies on Binocular Vision*,
Archimedes 47, DOI 10.1007/978-3-319-42721-8_5

hemisphere, respectively. Nowhere does Euclid address the question as to how disparate visual stimuli are integrated.

Galen attempted to rectify this lacuna by deriving a definition of the disparity between the quasi-images produced by the two eyes from one of Euclid's propositions (*Optica*, prop. 30). Unfortunately, Galen's exposition on binocular vision is extremely brief[2] and its primary aim is to provide an anatomical description of the ocular paths.

Ptolemy's treatment of binocular vision is far more comprehensive (*Optica*, II, 27–46, III, 25–62),[3] providing the framework for Ibn al-Haytham's research on the same subject. A comparison of the texts by the two savants confirms that many of the results obtained by Ibn al-Haytham were based on the thorough study and critical analysis of his predecessor's work. Nevertheless, it would appear to be more useful to take the work of Ibn al-Haytham rather than Ptolemy as our departure point to study the development of the binocular theory of vision. This choice can be justified on two grounds.

Firstly, while Ptolemy may have furnished the original matrix for Ibn al-Haytham's theory, his ideas are presented in *Optica* in anything but a systematic manner, with the propositions regarding binocular vision straddling Books II and III. What is more, certain results appear to be quite tentative, a fact that earned him the criticism of Ibn al-Haytham: "[Ptolemy] said that when the eye gazes at the middle object assumed in the middle of the ruler at the point intersection of the two diameters, then the two lines or diameters representing the visual axes will be seen as a single line that coincides with the common axis... But it is an error attested both by reasoning and experience."[4] In the final analysis therefore, Ptolemy can contribute little to our understanding of the history of binocular vision, because at each turn we find ourselves having to explain elements that require modification or amendment.

Secondly, unlike his work *Almagest*, which circulated widely and was universally read between the Duecento and the Cinquecento, Ptolemy's treatise on optics never achieved the stature of an authoritative text in Europe. In contrast, it appears to have been better known in the Arab world; we find it cited two times in Greek and five times in Arabic compared to just twice in Latin treatises from the Middle Ages. Historians have suggested that the eclipse of Ptolemy's *Optics* can be explained by the appearance of Ibn al-Haytham's *De aspectibus*, which was read by

---

[2]*De usu partium*, X, 12–13; *Oeuvres anatomiques, physiologiques et médicales de Galien*, ed. Charles Daremberg, Paris, 1854, pp. 638–645.

[3]Albert Lejeune, *L'Optique de Claude Ptolémée dans la version latine d'après l'arabe de l'émir Eugène de Sicile*, Leiden, 1989, pp. 26–34 and 102–118. On the perspective derived from Ptolemy's optics, see the excellent article by Laura Carlevaris, "La prospettiva nell'ottica antica: il contributo di Tolomeo," *Disegnare* 27 (1989), pp. 16–29.

[4]Ibn al-Haytham, *Al-Shukūk 'alā Batlamyūs*, p. 65. Ibn al-Haytham's critique of Ptolemy has not always been correctly assessed in the literature; see Craig R. Aaen-Stockdale, "Ibn al-Haytham and psychophysics," *Perception* 37 (2008), pp. 636–638, which criticizes Ian P. Howard's appraisal of the Arab scholar, "Alhazen's neglected discoveries of visual phenomena," *Perception* 25 (1996), pp. 1203–1217.

the Latin perspectivists and would orient their work. As Smith wrote: "With the continued dissemination of Ibn al-Haytham's treatise and the proliferation of Perspectivist works during the late thirteenth and early fourteenth centuries, Ptolemy's *Optics* was bound to lose its status as a legitimate source in optics."[5] In fact, another reason not to take Ptolemy as our reference point is because Latin authors retained that the explanation of binocular vision was not to be found in Ptolemy's *Optics* but in Ibn al-Haytham's *De aspectibus*.

It is therefore to the Arab world that we must turn in our search for the key to a correct reading of the further development of the theory of binocular vision. As we begin to re-trace this history, one of the first texts that deserves mention is the *Book of the Causes of the Diversity of Perspectives in Mirrors* by the Melkite physician and scientist Qusṭā Ibn Lūqā (820–912), who settled in Baghdad and translated many ancient Greek texts into Arabic. Qusṭā Ibn Lūqā draws the distinction between pathological diplopia and physiological diplopia and poses the question:

> By what cause is one single thing seen twice or more? And in how many ways is this possible? We have said earlier that the sense of sight perceives visible [things] if the visual ray falls upon them. The visible will be seen as unique if only one ray falls upon it; if two visual rays fall on it, it will be seen as double ... If it happens that the two cones emerging from the two eyes separate from each other, such that a radiant cone from each of the two eyes falls on the same visible [thing], then the same thing will be seen as double.[6]

He then constructs a classification of the causes—both natural and artificial—which could result in the separation of the visual cones and cause diplopia; among these are strabismus and the image of two objects located at different distances:

> It may happen that one single thing will be seen as double ... when a man fixes his pupil on a close thing and on another thing in the direction of that [thing] which he has fixed on but which is further than it from the eye; it is then that he will see one single thing as two. In effect, when he looks at the closer [thing] and fixes on it, one of the two rays [from the two eyes] will bend in relation to the other, and the two rays will fall on the more distant visible [object] in this manner; as a consequence he will see two things.[7]

The term "artificial cause" used by Qusṭā Ibn Lūqā, and the fact that he classifies among these causes both pathological conditions (such as strabismus) and accidental circumstances (e.g., pressure exerted on the eyeball) could give rise to confusion. However, he clearly states that diplopia may occur under normal as well as abnormal conditions. The disparity arising from the observation of objects situated at different distances does not necessarily imply that there is a problem with the visual apparatus; it is *normal physiological diplopia* and one cannot apply the theory of neutralization here to explain the integration of two sets of visual sensations.

Ibn al-Haytham (d. after 1040) was the second Arab savant to turn his attention to binocular vision. He is unanimously recognized for his work in three fields:

---

[5]A. Mark Smith, *Ptolemy's Theory of Visual Perception*, Philadelphia, 1996, p. 60.

[6]Roshdi Rashed, *Oeuvres philosophiques et scientifiques d'al-Kindī*, vol. 1: *L'Optique et la catoptrique*, Leiden, 1996, p. 584.

[7]Rashed, *L'Optique et la catoptrique*, p. 586.

mathematics (*Commentary on Euclid's Premises; On the Completion of Conics; Exhaustive Treatise on the Figures of Lunes; On the Regular Heptagon; On the Measurement of the Paraboloid;* and *On the Measurement of the Sphere,* a short tract on what came to be known in number theory as Wilson's theorem, although al-Haytham was the first to state it), astronomy (*On the Determination of the Meridian from One Solar Altitude; On the Visibility of Stars; Doubts on Ptolemy; The Resolution of Doubts on the First Book of the Almagest; The Resolution of Doubts on the Winding Movement*), and optics (*On Optics; The Discourse on Light; On Spheric Burning Mirrors; On Parabolic Burning Mirrors; On the Burning Sphere; On the Light of the Moon; On the Halo and the Rainbow; On the Formation of Shadows; On the Shape of the Eclipse;* etc.). His studies on binocular vision are less well known.

Ibn al-Haytham drew inspiration from Ptolemy,[8] Ḥunayn Ibn Isḥāq[9] and perhaps also Qusṭā Ibn Lūqā,[10] building on their work but approaching the same questions with greater rigor.[11] For example, he set up and conducted an *experimental study* on the conditions for the fusion of the quasi-images generated by two eyes, as we will describe below, producing an impressive series of results. Not only did he identify the existence of corresponding points on the recipient surfaces of the two eyes and differentiate between the cases of homonymous (direct) and heteronymous (crossed) diplopia; he also conceived a credible model for the horizontal horopter and anticipated the discovery of Panum's fusional area,[12] despite the fact that the medieval world was not yet in possession of the Keplerian concept of the "retinal image".[13]

---

[8]Lejeune, *L'Optique de Claude Ptolémée dans la version latine d'après l'arabe de l'émir Eugène de Sicile,* p. 109. We are certain of the existence of this link based on a passage in *Doubts on Ptolemy,* in which Ibn al-Haytham explicitly mentions the experiments of his predecessor; Ibn al-Haytham, *Al-Shukūk 'alā Batlamyūs (Dubitationes in Ptolemaeum),* eds. A.I. Sabra and N. Shehaby, Le Caire, 1971, p. 65. See the observations of Abdelhamid I. Sabra, "Ibn al-Haytham's criticism of Ptolemy's Optics," *Journal of the History of Philosophy,* 4 (1966): 145–149.

[9]Ḥunayn Ibn Isḥāq, *The Book on the Ten Treatises on the Eye Ascribed to Hunain ibn Is-hâq (809–877 AD),* transl. Max Meyerhof, Cairo, 1928, p. 26. Cf. Pierre Pansier, *Collectio ophtalmologica veterum auctorum,* fasc. 7: Ḥunayn Ibn Isḥāq, *Liber de oculi;* Galen, *Littere Galieni ad corisium de morbis oculorum et eorum curis,* Paris, 1909–1933.

[10]Roshdi Rashed, *Oeuvres philosophiques et scientifiques d'al-Kindī,* vol. 1: *L'optique et la catoptrique* Leiden, 1997, p. 584.

[11]Ptolemy provides only a brief description of the instrument devised by him to study binocular vision. In contrast, Ibn al-Haytham's text allows us to reproduce the experiments and understand his results without referring to other sources; Ibn al-Haytham, *Al-Shukūk 'alā Batlamyūs,* p. 65f. See also Abdelhamid I. Sabra, *The Optics of Ibn al-Haytham, Books I–III: On Direct Vision,* London, 1989, II, p. 125. Ibn al-Haytham's corrections to his Greek predecessor's work were passed on to successive generations, confirming that the medieval Latin authors followed his lead rather than Ptolemy's.

[12]I discovered the article by Ian P. Howard after publishing "Ibn al-Haytham sur la vision binoculaire," *Arabic Sciences and Philosophy,* 13 (2003): 79–99. The objectives of our two studies were quite different, with Howard devoting three pages to Ibn al-Haytham's studies of binocular vision; Ian P. Howard, "Alhazen's neglected discoveries of visual phenomena," pp. 1210–1212.

[13]Johannes Kepler, *Ad Vitellionem paralipomena, quibus astronomiae pars optica traditur,* Frankfurt, 1604. It should be recalled that the retinal image is not sharp on all parts. As with any

The experiments conducted by Ibn al-Haytham on binocular vision are presented in his seven-volume treatise on optics, *Kitāb al-manāẓir* (III, 2). Chap. 2 is divided into six sections. In paragraphs 2.1–2.24 the author defines various terms and concepts and the qualitative approach that he would adopt in analyzing the problem of binocular vision. Paragraphs 2.25–2.50 present the results of a series of experiments conducted using a board that he referred to as a 'binocular ruler' on which small columns of wax could be positioned. Paragraphs 2.51–2.54 compare the binocular and monocular perception of diagonal lines drawn on this tablet. Paragraphs 2.55–2.61 describe similar experiments, in which the columns of wax are replaced by words written on a piece of parchment. Paragraphs 2.62–2.85 explore the visual perception of objects whose direction is gradually displaced further and further from the anterior-posterior axis. Paragraph 2.86 presents Ibn al-Haytham's conclusions, and states that an analysis of "visual errors"[14] will be carried out in Chap. 3.

5.1 The Cases of Homonymous and Heteronymous Diplopia

There are two types of diplopia. The first is referred to as heteronymous or crossed diplopia, and the second as homonymous or uncrossed diplopia. The following experiment is often used to illustrate how the two forms of physiological diplopia manifest themselves:

(Footnote 13 continued)

centered optical system, the eye is not subject to the conditions of strict stigmatism. The deviation from the paraxial approximation, which is written as a sum of six factors, characterizes the various types of geometric aberration (spherical aberration, astigmatism, coma, field curvature and distortion). The two main factors that distort the image are spherical aberration and astigmatism. It suffices to fix one's gaze ahead to realize that an object seen through peripheral vision is indistinct; therefore the retinal image only meets the conditions of approximate stigmatism.

[14]*Deceptiones uisus* can be translated as "visual illusions." Ibn al-Haytham employed the term *aghlāṭ al-baṣar*, which has the same root as 'error' *(ghaliṭa: errare/decipi)*, but sometimes used the word *īhām (illusio)* instead. He was not the only scientist to consider (physiological) diplopia as normal. Bacon also did so; in his exposition of errors in vision, he mentioned only strabismus, the effects of cold or warm temperatures, passion, nervous derangement, problems with the vitreous humor, compression of the eyeball, obstruction of the lens, and the double pupil. See *Perspectiva*, II, I, 3, *The 'Opus majus' of Roger Bacon*, ed. by J.H. Bridges, Frankfurt am Main, 1964, p. 88–91; David C. Lindberg, *Roger Bacon and the Origins of Perspectiva in the Middle Ages*, Oxford, 1996, p. 170–176. What occurs when one fixes one's gaze on objects situated at different distances is discussed elsewhere by Bacon. His study of physiological diplopia is entitled "In quo ostenditur duobus diuersis experimentis et diuersis figurationibus, quomodo unum uideatur duo," *Perspectiva*, II, II, 2, ed. Bridges, p. 94. John Pecham no longer treated diplopia in the context of errors in vision, *Perspectiva communis*, I, 80, David C. Lindberg, *John Pecham and the Science of Optics*, Madison, 1970, p. 150.

The phenomenon of physiological diplopia can be made to happen by means of an experiment devised long ago by Scheiner. That is to say, [take] a small wooden ruler about 50 cm in length, one end of which is placed at the tip of one's nose, and which has been pierced by three pins at 30, 40 and 50 cm. If one gazes fixedly at the middle pin, the two other pins will be seen as doubled.[15]

When the ruler is pricked with a black-headed pin and a white-headed pin and one fixes one's gaze on the black pinhead: (a) if the white pinhead is located closer to the eyes than the black pinhead ... there is crossed diplopia; (b) if the white pinhead is located further from the eyes than the black pinhead ... there is direct diplopia.[16]

This demonstration has traditionally been attributed to the astronomer and mathematician Christoph Scheiner but it actually goes back much further[17] because both Ptolemy and Ibn al-Haytham devised similar experiments and came to the same conclusions. This ignorance of the historical facts brings up the issue of tacit borrowings.[18] How was it that a seventeenth-century Jesuit came to be credited with an experiment carried out long before by a scientist in ancient Alexandria and another in medieval Cairo? Ptolemy's work on binocular vision has been analyzed in depth by Lejeune,[19] so we will focus on the research of the Arab savant.

Ibn al-Haytham begins by describing the instrument that he designed for the study of binocular vision. It consists of a flat, rectangular piece of wood *ABCD* (*lawḥ, tabula*), one cubit in length (45–50 cm) and four fingers wide (6–7 cm.). At one end (*AB*) is a shallow depression (*MHN*) which serves as a nose rest. Lines are drawn across the ruler (in different colors to make them more visible); the median lines *HZ* and *KT* (*khaṭṭān, linee recte*) bisect the length and width of the ruler, and

---

[15]Henry Saraux and Bertrand Biais, *Physiologie oculaire*, Paris, 1973, 2nd ed. 1983, pp. 390–391.

[16]Annette Spielmann, *Les strabismes*, Paris, 1991, p. 116. The only modern optician to mention Ibn al Haytham's contribution to the study of physiological diplopia was Yves Legrand, *Optique physiologique*, tome 3: *L'espace visuel*, Paris, 1956, p. 210.

[17]Christoph Scheiner, *Oculus, hoc est fundamentum opticum...* Inspruck, 1619; London, 1652, pp. 32–49.

[18]The question of the borrowing of ideas is always a delicate one, and for many reasons: (1) it is difficult to conduct an exhaustive and systematic study of a specific discovery because it will depend on the interest and expertise of the researcher in a given domain of knowledge; (2) it is hampered by the fact that ideas stemming from certain key texts often spread through a multitude of canals that may be almost indistinguishable to us today; (3) an exact appreciation is required of the role that a borrowed element may have played in a scientific work. Research on the borrowing of ideas can nevertheless contribute—and this is its principal interest—to our understanding of the development of a science because it impacts on the question of scientific authorship. Overly hasty attributions could efface the *stratification* that regularly characterizes the evolution of scientific concepts. It suffices to consider the research of Ptolemy, Ibn Sahl, Ibn al-Haytham, Harriott, Snell and Descartes on the nature of refraction or that of different schools of astronomy (Marāgha, Ibn al-Shātir and Copernicus) which introduced tangent circles to explain the circular orbit of the planets.

[19]Albert Lejeune, "Les recherches de Ptolémée sur la vision binoculaire," *Janus* 47 (1958): 79–86. This article was based on a communication delivered in 1957 to the 2nd Benelux Congress on the History of Science.

the diagonals *AD* and *BC* cross it (*quṭrān, diametra*).[20] Ibn al-Haytham then fashions three small columns of wax painted in different colors and positions them at various points along these lines. With this instrument he conducts a series of experiments on the conditions for the fusion of binocular images that would serve as a model for the study of optics in the West for centuries.

5.2   The Notion of Corresponding Points in Binocular Vision

Modern specialists in physiological optics date the introduction of the notion of "corresponding points" as an explanation for binocular vision to the seventeenth century and the research of Christiaan Huygens (1629–1695).[21] As Pigassou-Albouy observed:

> The earliest notion of corresponding retinal points can be traced back to Huygens: "[...] every point at the back of the eye has a corresponding point at the back of the other, such that when one point of an object is painted in several pairs of these corresponding points, then it will appear simply as it is."[22]

> To explain the unification of the two visual sensations, Huygens and Müller denominated 'corresponding retinal points' (CRP) the photoreceptors whose stimulus simultaneously produces the sensation of a single source.[23]

The concept of "correspondence," today defined as the association between the nasal point of one retina and a temporal point in the other retina (see below Fig. 5.2, points p_L and p_R) considerably antedates the work of Huygens. One already finds the adjective *consimilis* applied by Ptolemy to visual rays. These were the "similarly arranged rays" (*radii ordine consimiles*) translated by Lejeune as "corresponding rays,"[24] which accords with the modern usage of the term but was only codified during the nineteenth century. As Helmholtz wrote:

> We will assign to them the terms *coincident* or *corresponding* points; they have also been referred to as *identical* points, in keeping with a particular theory. Since to each point in

---

[20]Ibn al-Haytham, *Kitāb al-manāẓir*, III, II, 26. *Opticae thesaurus Alhazeni Arabi libri septem*, ed. Risner 1572, reprint New York, 1972, p. 81; *The Optics of Ibn al-Haytham, Books 1–3: On Direct Vision*, ed. Abdelhamid I. Sabra, 2 vols., London, 1989, vol. I, pp. 237–238; A. Mark Smith, *Alhacen's Theory of Visual Perception*, Philadelphia, 2001, pp. 263–264.

[21]Christiaan Huygens, *Opuscula posthuma quae continent Dioptricam...* Leiden, 1704.

[22]Renée Pigassou-Albouy, *Les strabismes*, vol. 1: *Les divergences oculaires*, Paris, 1991, p. 27.

[23]Pigassou-Albouy, vol. 2: *Les convergences oculaires*, Paris, 1992, p. 7; Yves Le Grand, *Optique physiologique*, 3, p. 208.

[24]Lejeune, *L'Optique de Claude Ptolémée*, p. 104 et passim.

every visual field there corresponds a specific retinal point, one may refer to these inter-changeably as *coincident, correspondent or identical points in the two retinas.*[25]

These historical differences in terminology, which are echoed in Helmholtz's text on corresponding retinal points, are simply the result of the difficulty of finding an exact translation for the Latin word *consimilis*. Ptolemy limits his application of this adjective to visual rays, but Ibn al-Haytham uses it explicitly in his discussion of corresponding points:

> So the two forms impressed on the two points that are correspondingly situated (*fī nuqṭatayni mutashābihatay al-waḍ', in duobus punctis... consimilis positionis*) with respect to the surfaces of the two eyes reach the same point in the hollow of the common nerve, and they will be superimposed on that point so as to produce a single form.[26]

The "surface of the eyes" (*saṭḥ al-baṣar, superficies uisus*) referred to here is not the retinal surface,[27] but the anterior surface of the crystalline lens (*al-jalīdiyya,*

---

[25]Hermann von Helmholtz, *Optique physiologique*, Leipzig, 1866; Paris, 1867, p. 880. Regarding this terminology, we should specify that the expression "corresponding points," the only term still in use today, originally denoted an empirical approach to the problem (e.g., Helmholtz), in contrast to the theory based on nativism (defended by Hering) which presupposes an innate process involving the coupling of retinal points in the brain and uses instead the term "identical points." The hypothesis of corresponding points, which was first proposed by Galen in *De usu partium* IX, 12, was taken up by Johannes Müller, *Zur Vergleichenden Physiologie des Gesichtssinns*, Leipzig, 1826, who stated the "law of identity of the two retinas," and then by Ewald Hering, *Beiträge zur Physiologie*, Leipzig, 1861–1864, who outlined the "law of identical visual directions." For a discussion of these concepts, see Ch. Thomas, "La physiologie de la vision binoculaire," *Archives d'Ophtalmologie (Paris)* 31 (1971), pp. 191–192; David Stidwill and Robert Fletcher, *Normal Binocular Vision*, Oxford, 2011, p. 75.

[26]*Kitāb al-manāẓir*, III, 2, 14, *Opticae thesaurus*, p. 79, *Optics*, vol. 1, p. 234. Smith, *Alhacen's Theory of Visual Perception*, p. 256. *Ṣūra* (form) designates here the sensory image rather than the image produced by the eye, which Ibn al-Haytham denominated *al-khayāl*.

[27]Felix Platter, *De corporis humani structura et usu*, Basel, 1583, and Johannes Kepler, *Ad Vitellionem paralipomena, quibus astronomiae pars optica traditur*, Frankfurt, 1604, were the first scholars to understand the function of the retina. Kepler wrote in *De modo visionis* (Chap. 5, p. 168): "I say that vision occurs when the image of the whole hemisphere of the world that is in front of the eye, and a little more, is formed on the reddish white concave surface of the retina." The first description of the role of the crystalline lens in projecting visual images onto the retina would be provided by Scheiner, *Oculus, hoc est fundamentum opticum*. Ibn al-Haytham lists the parts of the eye (*ruṭūba al-bayḍiyya*: humor albugineus, *r. al-jalīdiyya*: crystallinus, *r. al-zujājiyya*: humor vitreus, *ṭabaqa al-'ankabūtiyya*: tela aranea, *ṭ. al-multaḥima*: consolidativa, *ṭ. al-'inabiyya*: uvea, *ṭ. al- qarniyya*: cornea), but says nothing of the retina. We may ask ourselves whether the rectification of this lacuna would have been significant; in the Middle Ages the most authoritative text on the anatomy of the eye was *The Book on the Ten Treatises on the Eye*, whose author, Ḥunayn Ibn Isḥāq, limits the role of the retina (*ṭ. al-shabakiyya*) to that of providing nutrients to the vitreous humor and the crystalline lens. On these matters, see Muṣṭafā Naẓīf, *al-Ḥasan Ibn al-Haytham* (Cairo, 1942/3), 1, pp. 205–217; Abdelhamid I. Sabra, *Optics*, vol. 2, pp. 45–51; Gül Russell, "The anatomy of the eye in 'Alī Ibn al-'Abbās al-Majūsī: A textbook case," in *Constantine the African and 'Alī Ibn al-'Abbās al-Majūsī*, eds. Charles Burnett and Danielle Jacquart, Leiden, 1994, pp. 247–265; *Idem*, "La naissance de l'optique physiologique," in *Histoire des Sciences Arabes*, ed. Roshdi Rashed, Paris, 1997, 2, pp. 319–354 (in which they do not, however, discuss the experiments of Ibn al-Haytham).

anterior glacialis), which was thought to be the seat of sensory responsiveness and which Ibn al-Haytham termed "the surface of the sensory body" (*saṭḥ al-jism al-ḥāss, superficies uisus sentientis*).[28] Thus, the role of the retina constitutes the principal point of divergence between medieval and early modern theories of physiological optics.

Setting aside this nuance, Ibn al-Haytham deserves credit for having formulated the concept of correspondence, which would serve as the starting point for the analysis of many aspects of binocular vision. It is possible that Huygens developed his own theory of corresponding points after reading Ibn al-Haytham's treatise, which was well known in classical Europe through the edition published by Risner in 1572 (*Opticae thesaurus: Alhazeni Arabis libri septem*).

5.3   The Study of Physiological Diplopia (Experiments 1 and 2)

According to early modern physiological optics, the fusion of two images into one in binocular vision takes place when the object points are 'painted' in corresponding points on both retinas. The phenomenon of *diplopia* arises in cases where the object points fall on disparate points on the two retinal surfaces. In *homonymous diplopia* every object situated *on the further side of* the point of fixation is seen as a double image, whereas in *heteronymous diplopia* every object situated *on this side of* the point of fixation is seen as a double image. Neither Ptolemy nor Ibn al-Haytham employed a scientific term when referring to diplopia (using instead *comprehendetur duo* and *yudraku ithnayni*), but they nonetheless make a clear distinction between the two types of physiological diplopia.[29] Here we describe the experiments conducted by Ibn al-Haytham using his binocular ruler.

Experiment 1[30] When columns of wax are situated along the same diagonal at points *L Q S* with the eyes fixed on *Q*, double images of the columns at *L* and *S* are

---

[28]*Opticae thesaurus*, pp. 16–17. Since the crystalline lens is made up of denser tissue than that of the cornea, light rays are in a way 'absorbed' as they pass through into this new medium; see Abdelhamid I. Sabra, "Sensation and inference in Alhazen's theory of visual perception," in *Studies in Perception*, eds. P.K. Machamer and R.G. Turnbull, Columbus, 1978, p. 164.

[29]Ptolemy writes: "If line *HTK* is drawn parallel to line *EDZ* while the two axes remain focused on *D*, an object at point *T* [which is located below the point of fixation *D*] will appear at the two locations *H* and *K* ... But if we focus both axes upon point *T*, we will see point *D* [which is now located beyond the point of fixation *T*] at points *E* and *Z*/Cum autem producta fuerit linea *HTK* equidistans linee *EDZ* et fuerint duo axes oppositi puncto *D*, res que est super punctum *T*, videbitur in duobus locis qui sunt *H*, *K*... Et si posuerimus utrosque axes oppositos puncto *T*, videbimus tunc *D* super punctos *E*, *Z*," Lejeune, *L'Optique de Claude Ptolémée*, p. 103, see Fig. 5.6. This was also the opinion of Lejeune, "Les recherches de Ptolémée," p. 82.

[30]*Opticae thesaurus*, p. 82; *Optics*, vol. 1, p. 239; Smith, *Alhacen's Theory of Visual Perception*, p. 265.

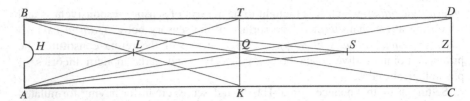

Fig. 5.1 Ibn al-Haytham, experiment 1. Author's drawing

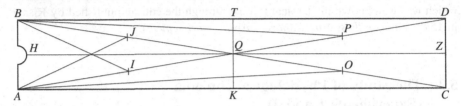

Fig. 5.2 Ibn al-Haytham, experiment 2. Author's drawing

seen (Fig. 5.1). As has already been said, the doubling of the column positioned at *S*, beyond the point of fixation, illustrates the phenomenon of homonymous diplopia, while the doubling of the column at *L* illustrates crossed diplopia.

Experiment 2[31] When columns of wax are situated along the same diagonal at points *I Q P* and the eyes are fixed on *Q*, double images of columns *I* and *P* are seen, and likewise if the columns are situated on this side of the fixation point, on *I J*, or beyond the point of fixation, on *O P* (Fig. 5.2).

5.4   The Determination of the Horopter (Experiments 3 and 4)

The concept of the *horopter* was introduced in 1613 by the Jesuit mathematician and physicist Franciscus Aguilonius.[32] He defined the horopter as the frontal plane containing the point of fixation. Following Aguilonius, and principally in the nineteenth century, the horopter formed the object of many studies, notably those of Vieth, Müller, Hering and Helmholtz.[33] Today the horopter is limited to the horizontal plane and is defined by the locus of points in space received by

---

[31]*Opticae thesaurus*, p. 82; *Optics*, vol. 1, pp. 239–240; Smith, *Alhacen's Theory of Visual Perception*, p. 266.

[32]Franciscus Aguilonius, S.J., *Opticorum libri VI*, Antwerp, 1613, Lib. II, pp. 105–150.

[33]Gerhard Ulrich Anton Vieth, "Ueber die Richtung der Augen," *Gilbert's Annalen der Physik*, 58 (1818): 233–253. Johannes Müller, *Zur Vergleichenden Physiologie*, Ewald Hering, *Beiträge zur Physiologie*, Leipzig, 1861–1864; Hermann von Helmholtz, *Optique physiologique*, Paris, 1867.

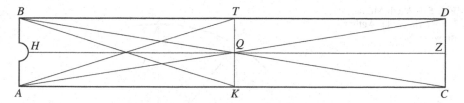

Fig. 5.3 Ibn al-Haytham, experiments 3 and 4. Author's drawing

corresponding points on the two retinas. Although both the name and the form of this geometric locus based on the notion of corresponding points were unknown to Ibn al-Haytham, his work led directly to its discovery.

Experiment 3[34] When the three columns of colored wax are situated at $T Q K$ with the eyes $A B$ fixed on Q, single images of the columns are seen (Fig. 5.3).

Experiment 4[35] The same occurs if one takes T or K as the fixation point without displacing the wax columns (Fig. 5.3).

This represents an early attempt to determine the horopter scientifically, which according to Ibn al-Haytham corresponds to the frontal line $T K$ (Aguilonius would come to the same conclusion in the seventeenth century). The historian Abdelhamid I. Sabra recognized clearly that Ptolemy and Ibn al-Haytham laid the groundwork for the discovery of the modern horopter, through which object points are re-composed as single images. He wrote:

> It would have been easy for Ptolemy and Ibn al-Haytham to generalize this conclusion further. For points in the plane of the axes, the stated conditions of single vision (taken literally) are satisfied only by points on the circumference of the circle passing through the centers of the eyes and the point of fixation (the so-called 'horopter circle' or 'horizontal horopter')... But neither Ptolemy nor Ibn al-Haytham draws this consequence. (Note, however, that Ibn al-Haytham's account is not strictly geometrical)[36]

This simple observation invites us to re-examine Ibn al-Haytham's experiments in order to determine why he did not arrive at the true form of the horopter.

5.4.1   The Theoretical Horopter

First of all, let us retrace the genesis of the Vieth-Müller horopteric circle. We will limit ourselves to a consideration of the horizontal plane of vision that allows us to

---

[34]*Opticae thesaurus*, p. 82; *Optics*, vol. 1, p. 239; Smith, *Alhacen's Theory of Visual Perception*, p. 264.

[35]*Opticae thesaurus*, p. 82; *Optics*, vol. 1, p. 239; Smith, *Alhacen's Theory of Visual Perception*, p. 265.

[36]Sabra, *Optics*, vol. 2, pp. 123–124.

Fig. 5.4 Vieth-Müller
horopteric circle. Author's
drawing

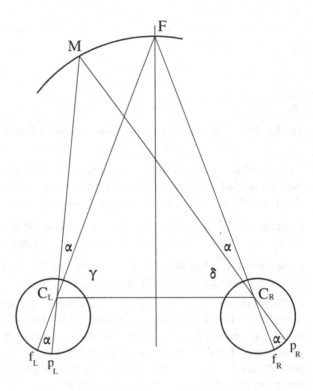

define the "longitudinal horopter". What is the geometric locus of the object points
that are seen as a single image by the two eyes?

Let C_L be the optic center of the left eye, C_R the optic center of the right eye, and
F the fixation point. Let γ designate the angle FC_LC_R and δ the angle FC_RC_L. As
they fix on point F, the rotation of the two eyes will produce quasi-images of F on
the foveae f_L and f_R (Fig. 5.4). The object point M will be received in the retinas of
the two eyes by the corresponding points p_L and p_R, that is to say, by the corresponding nasal and temporal sensory cells in the two eyes.[37]

Under these conditions the visual directions p_LM and p_RM will correspond as
well, and form equal angles (α) as they cross the arcs f_Lp_L and f_Rp_R. M then becomes
the point of intersection of the lines with an angle of $\gamma + \alpha$ and $\delta - \alpha$ in relation to
the line connecting the optic centers. One may immediately deduce from this

$$C_LFC_R = \pi - \gamma - \delta = \pi - \gamma - \alpha - \delta + \alpha = C_LMC_R$$

---

[37]The fusion of images in binocular vision comes about as a result of the architecture of the optic
fibers; information from the temporal retinas is picked up by the corresponding cerebral hemispheres, while the nasal retinas send information to the opposite hemispheres. The influx of nerve
impulses from homologous cells of the two RCPs fuse in the V area of the extrastriate visual
cortex.

Fig. 5.5 The theorem of inscribed angles. Author's drawing

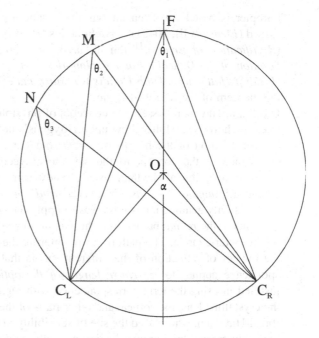

The theorem of inscribed angles teaches us that on a circle with a center O and two base points C_L and C_R, a point M describing the circle will determine an inscribed angle $C_L M C_R$ that is constant and equivalent to one half of the center angle $C_L O C_R$.

In the present case (Fig. 5.5), the object points that are seen as single images by the two eyes are points F, M, N … such that

$$\theta_1 = \theta_2 = \theta_3 = \cdots = \frac{\alpha}{2}$$

These are the grounds on which Vieth and Müller defined the theoretical horopter as the circle passing through the two optic centers C_L and C_R and the fixation point F. A given fixation point will be associated with a unique horopteric circle. Let us now suppose that F is displaced along the line OF. The horopter will then be modified and, as we continue to move F, all of the horopters will form a linear pencil of circles with C_L and C_R as their base points (these points will have a zero power in relation to all the circles of the pencil).

We now arrive at a question that is of great interest to the history of optics: having come this far, why did Ibn al-Haytham fail to construct the horopteric circle? A series of considerations may be raised in this regard.

(1) Did the Arab scholar have the mathematical knowledge necessary to realize such a construction? Yes. The property of the geometric plane on which the

horopter is based had been known since antiquity;[38] it was expounded by
Euclid (*Elements* III, 20–21),[39] Archimedes (*Book of Lemmas*),[40] and al-Kindī
(*Rectification of Errors*).[41] Ibn al-Haytham, who wrote a treatise entitled
*Solution of the Difficulties and Explanation of the Notions of Euclid's Book
(Kitāb fī Ḥall shukūk kitāb Uqlīdis wa sharḥ maʿānīh)*, was clearly aware of
the theorem of the inscribed angle.

(2) Did Ibn al-Haytham possess the concepts of physiological optics necessary to
trace the horopter? Müller drew upon Huygens's notion of correspondence in
his construction of the horopteric circle, but there is no reason to date the
emergence of the concept to the seventeenth century because, as can be
demonstrated, Ibn al-Haytham was already using it in the early eleventh
century (*fī nuqṭatayni mutashābihatay al-waḍʿ, in duobus punctis consimilis*).
Müller's only genuine innovation was to apply the concept of correspondence
to points in the *retina*, but this insight was in no way essential to the model of
the horopteric circle. The path of a ray entering the eye being determined by
the indices of refraction of the different media that it is crossing, the correspondence applies to *the entire length of the optic path*. One is therefore
justified in using the term correspondence with regard to points in the retina,
the crystalline lens, the cornea, and other parts of the eye along the optic path.
Ibn al-Haytham, who placed the site of sensibility in the crystalline lens, could
have constructed the horopter beginning with the corresponding points in the
lens.

(3) Did Ibn al-Haytham rely on an authoritative argument in constructing the
horopter? He would have found, for example, in Ptolemy: "So too, [with the
two eyes located at points *A* and *B*, and the visual axes fixed to *D*], if we draw
a line *EDZ* through point *D* perpendicular to *GD*, any object place on this line
will appear single and at its true location as long as it is aligned with point
D"[42] (Fig. 5.6).

This proposition is almost equivalent to the one advanced by Ibn al-Haytham,
but there is no reason to believe that he was merely repeating an authoritative
argument. As has been seen, in *Doubts on Ptolemy* Ibn al-Haytham rejected

---

[38] The architects of ancient Greece were equally aware of this property of the circle and realized
that, using a set square, they could apply it to produce the fluting of their Ionic columns. With a
slight rotation of the set square, the position of the two raised borders of the groove could be
marked; by repeating this step one was (in geometric terms) simply displacing the tangent represented by the right angle of the set square along the circumference of a circle.

[39] Euclid, *Elements*, vol. 1: *Livres I–IV: Géométrie plane*, ed. Bernard Vitrac, Paris, 1990, pp. 431–
433. Euclid's *Elements* was known to medieval Arabic scholars through at least three translations
produced between the eighth and tenth centuries, all circulating under the title *Kitāb al-uṣūl*.

[40] Archimedes, *Des corps flottants, Stomachion, La méthode, Le livre des lemmes, Le problème des
boeufs*, trans. and ed. by Charles Mugler (Paris, 1971), p. 513. Archimedes' works were translated
into Arabic by Thābit Ibn Qurra.

[41] Roshdi Rashed, *L'Optique et la catoptrique*, pp. 278, 312.

[42] Albert Lejeune, *L'Optique de Claude Ptolémée*, p. 103; Smith, *Ptolemy's Theory of Visual
Perception*, p. 141.

Fig. 5.6 Ptolemy, *Optics* III, 26. Author's drawing after Lejeune, *L'Optique de Claude Ptolémée*, p. 103

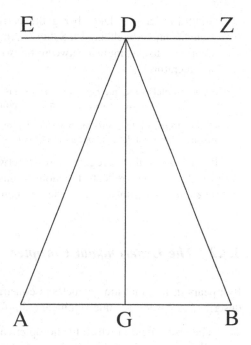

certain conclusions arrived at—supposedly on the basis of experimental evidence—by his predecessor. Since Ibn al-Haytham made such a critique of Ptolemy, it is difficult to admit that he would have borrowed another of the latter's propositions without having tested it.

(4) There is, all the same, another reason that could explain why Ibn al-Haytham was not able to arrive at the demonstration of the horopteric circle: he approached the problem as an experimentalist[43] rather than a geometer. If there was a marked gap between the theoretical and the experimental horopter, there would no longer be any reason to judge the conclusions of Ptolemy and Ibn al-Haytham as unsatisfactory in comparison to the horopteric circle of

---

[43]On the notion of experimentation and the experimental approach of the author, see Naẓīf, *Ibn al-Haytham*, 1, pp. 43–48, with comments by Roshdi Rashed, in *Optique et mathématiques*, Aldershot, 1992, pp. 235–239; Matthias Schramm, *Ibn al-Haythams Weg zur Physik*, Wiesbaden, 1963; Abdelhamid I. Sabra, "The astronomical origin of Ibn al-Haytham's concept of experiment," *Actes du XIIe Congrès international d'histoire des sciences*, Paris, 1971, vol. IIIA, pp. 133–136. Roshdi Rashed, *Entre arithmétique et algèbre* (Paris, 1984), pp. 314–315, underlines that: "It was essentially in the area of optics, with Alhazen, that one notes the emergence of this new [experimental] dimension. Everyone knows that the arrival of Alhazen marked the definitive rupture with the school that viewed optics as the geometry of vision or light." Gérard Simon arrived at similar conclusions; see "La psychologie de la vision chez Ptolémée et Ibn al-Haytham," in *Perspectives arabes et médiévales sur la tradition scientifique et philosophique grecque*, eds. Ahmad Hasnawi, Abdelali Elamrani-Jamal and Maroun Aouad, Louvain, 1997, pp. 189–190.

Vieth-Müller, the latter being the narrow interpretation adopted by Albert Lejeune and Abdelhamid I. Sabra.[44] And yet contemporary ocular physiology continues to distinguish between the two models.[45] For example, one finds in the literature:

> On a horizontal plane, the horopter corresponds *roughly* to the circle determined by the nodal points of the two eyes and the fixation point.[46]

> Different forms of the horopter have been proposed ... The empirical horopter of the horizontal correspondences appears to be more valid than the geometric horopter.[47]

The distinction that is regularly made between the 'theoretical horopter' and the 'empirical horopter' reflects the doubts that contemporary physiological optics continues to raise with regard to the geometric model of Vieth-Müller.

5.4.2 The Experimental Horopter

It appears in fact that the geometrical construction of the theoretical horopter just described is based on some significant simplifications:[48]

(1) The visual field in which the horopter can actually be tested is limited. First of all, the binocular field of vision is framed by the temporal monocular fields of the two eyes; in a normal subject the limiting nasal ray forms an angle $\alpha = 55°$ with respect to the geometric axis of straight vision. Given a fixation point

---

[44]Lejeune, "Les recherches de Ptolémée," p. 84, Sabra, *Optics*, vol. 2, p. 124.

[45]This distinction is due in part to the debate that sprang up between Hering and Helmholtz on the form of the horopter. The latter, with the collaboration of his colleagues Berthold, Bernstein and Dastich, repeated Hering's experiments but was unconvinced by the results. He wrote: "As for me, when I place myself at the distance indicated by Hering, the surface of the threads seems to me to be distinctly concave [whereas they should appear flat]," Helmholtz, *Optique physiologique*, p. 829.

[46]Spielmann, *Les Strabismes*, p. 117.

[47]Pigassou-Albouy, *Les Convergences oculaires*, p. 9.

[48]The analysis that follows will assume as given the following data regarding the anatomy of the eye: (1) thickness along the longitudinal axis: cornea 1.34 mm, aqueous humor 3.1 mm, crystalline lens 4.0 mm, and vitreous body 16.3 mm; (2) refraction index: cornea 1.38 mm, aqueous humor 1.34 mm, crystalline lens 1.42 mm, and vitreous body 1.34 mm; (3) radius of curvature: anterior surface of the cornea 7.7 mm (49 δ), posterior surface of the cornea 6.8 mm (−4 δ), anterior surface of the crystalline lens 10.0 mm (5 δ), posterior surface of the crystalline lens 6.0 mm (9 δ). The eye has a total refractive power of 59 δ, the greater part of which is furnished by the anterior corneal surface (the refraction index of air and of the cornea are very different, the anterior corneal surface is far from of the retina, and the radius of curvature of the cornea is small). The horizontal axis xx', the vertical axis yy', and the anterior-posterior axis zz' cross each other at the center of rotation of the eyeball, which is located 13.5 mm behind the anterior corneal surface, Yves Le Grand, *Optique physiologique* (Paris, 1948–1956, republished 1964–1972).

located at a distance of 25 cm (as in Ibn al-Haytham's experiment), the two eyes can only perceive objects located in a sector of $2\alpha = 110°$, which represents a center angle of 190° on the horopter.

One must also take into account Mariotte's blind spot, the physiological scotoma discovered by Edme Mariotte in 1667, which results in two zones of monocular vision within the full binocular field.

Finally, the iris does not lie in the median plane of the eye, but approximately 3.8 mm in front of the optic center. And since the iris has an average diameter of 3.5 mm, rays that penetrate the eye at an angle of less than 27° in relation to the direction of the fixation point will pass through the nodal points.[49] The more oblique incident rays will either be masked by the iris or refracted to the periphery of the crystalline lens. Their path will then be subject to spherical aberration. Under the same conditions, vision will be distinct for the points contained within a sector of 69° (cutting a center angle of 108° on the horopter).

(2)   If one attempts to correct for this situation by inducing mydriasis (increasing the diameter of the pupil to its maximum of 8 mm) the oblique incident rays will pass through the nodal points, but because the pupil is dilated most of the rays will be refracted to the periphery of the crystalline lens. In fact, the larger the diameter of the pupil, the greater the confusion created by spherical aberration.

(3)   Physiological optics utilizes a geometric model of the eye in which the retinal surface is likened to the portion of a sphere. But as Yves Le Grand acknowledged, "This geometric schema is a fiction. In reality ... the geometric axis [the anterior-posterior zz'] is not an axis of rotation; in the equatorial plane, the curvature of the ray is generally smaller on the temporal side than on the nasal side."[50] Thus, when a point issuing from the horopter reaches the retinas it will not strike two exactly corresponding points, because the corresponding rays will reach different retinal cells.

(4)   Another difficulty lies in the fact that Listing's law[51]—which states that the ocular globes are subjected to a torsion that increases with the distance of the fixation point from the horizontal or vertical axis—only holds in the case of a line of sight focusing on infinity. When the eyes move from the primary position and converge on a fixation point that is quite close, a simple movement of intorsion around the axis yy' should be produced.

---

[49]If we designate the nodal point object as N, the internal limits of the iris as I and I', and the center of the segment II' as H, then angle 2β below which a visual ray can touch the nodal point is determined by the relationship:

$$\beta = \arcsin\left(\tfrac{HI}{HN}\right)$$

Since $HI = 1.75$ mm and $HN = 3.8$ mm, then $\beta = 27° \, 25'$.

[50]Le Grand, *Optique physiologique*, p. 36.

[51]Johann Benedikt Listing, *Beitrag zur physiologischen Optik*, Göttingen, 1845.

And yet this does not take place; when the fixation point is nearer, the meridians are displaced upward, which means that the convergence leads to the outward rotation of the two eyes ... The displacements in the horizontal gaze predicted by Listing's law reach 1 degree for every 10° of convergence in the angle 2β of the lines of sight.[52]

For the left eye, the image of a horizontal segment will pivot in an anti-clockwise direction; for the right eye, the image will pivot in a clockwise direction. The distance separating the two stimulated visual cones is 4.5 μm; therefore images form on corresponding retinal points only if the torsion around the axis zz' is less than 1′ of the arc. For a fixation point situated 25 cm from the eyes (as in Ibn al-Haytham's experiment) the combined torsion of the eyeballs will be $2\alpha = 3° \, 8''$.[53] Being much greater than a torsion of 1′ of the arc, it prevents the formation of closely corresponding retinal points.

Taking into account all of these factors, the geometric locus of the object points that gives rise to corresponding retinal points *is not* the horopteric circle of Vieth-Müller. The horopter passes by the fixation point, but as the eccentricity of the retina increases, the uncertainty regarding its position increases proportionally. As Hermann von Helmholtz (1821–1894) noted in his visual analysis of frontality in three pins arranged on a wooden ruler 50 cm from the eyes, "The most favorable case is always that in which the direction of the line of pins corresponds to the direction of the tangent to the horopteric circle."[54] The conclusion drawn by the nineteenth-century physiologist was therefore identical to the one arrived at by Ibn al-Haytham based on his experiments.

On these grounds one can understand more clearly why Ibn al-Haytham, who was analyzing this problem from the perspective of an experimentalist rather than a geometer, retained the frontal line TQK as the definition of the geometric locus of object points that are seen as single images by the two eyes. The later construction of the horopteric circle is therefore based on dubious hypotheses that do not stand up on closer examination. Nevertheless, in order to grasp the rationality of Ibn al-Haytham's concept of the frontal horopter, it is necessary to study the history of the works from which the determination of the experimental horopter sprang.

---

[52]Yves Le Grand, *Optique physiologique*, vol. 1: *La dioptrique de l'oeil et sa correction*, Paris, 1965, pp. 248–249. Saraux and Biais, *Physiologie oculaire*, p. 336. Taking as his departure point a discovery by Volkmann, Helmholtz had already demonstrated that, as the fixation point is moved closer to the eye, a torsion around the anterior-posterior axis occurs, such that "the convergence leads to deviations of 2° to 2½° in the accidental image," *Optique physiologique*, p. 609.

[53]The angle of convergence 2β corresponds to the angle $C_L F C_F$ formed at the fixation point F by the visual axes of the eyes C_L et C_F. If H is the point of intersection between the median axis and the line connecting the optic centers $C_L C_R$, one has:

$2\beta = \arcsin\left(\frac{C_L C_R}{FH}\right)$

For $C_L C_R = 6.5$ cm and $FH = 25$ cm, $2\beta = 15°4'$.

Therefore, $\alpha \approx \frac{2\beta}{10} = 1°30'4''$, $\quad 2\alpha \approx 3°0'8''$.

[54]Helmholtz, *Optique physiologique*, p. 912.

5.5   Panum's Fusional Area (Experiment 5)

Taking as his departure point a critique of the Vieth-Müller circle, in 1858 Peter Ludvig Panum proposed a new conception of retinal correspondence.[55] The Danish physiologist had set out to study the degree of disparity that could be tolerated when two images are fused in binocular vision and he devised the following experiment. Given the two eyes C_L and C_R and the fixation point F, one places two small vertical bars B_1 and B_2 in proximity to F, in such a way that for one of the two eyes (say, the left eye), they are aligned along the axis $C_L B_1 B_2$. By displacing B_1 along the visual axis $C_L B_2$, one searches for the interval in which the barrettes are seen as single images by the other eye C_R (Fig. 5.7).

One then determines the external and internal limits (corresponding to homonymous and heteronymous diplopia, respectively) of the area containing the object points that are seen as simple images in binocular fusion. This is Panum's fusional area, reflecting the fact that retinal points which do not correspond exactly but are slightly disparate can nevertheless be fused into a single image.[56]

Kenneth N. Ogle[57] took up where Panum left off, devising an experiment in which the fixation point was situated 40 cm from the eyes. In this way he succeeded in determining the position of Panum's fusional area between the two arcs corresponding to the thresholds of homonymous and heteronymous diplopia (Fig. 5.8).

Ogle's abacus represents the averages of the internal and external limits—which fix the thresholds of homonymous and heteronymous diplopia—together with the Vieth-Müller circle (marked *V-M* in the figure) which diverges from Panum's fusional area with an eccentricity of more than 6°. The horopter, which from now on would be defined as the arc situated midway between the external and internal limits, satisfies the conditions for the conic section:

$$\cotan u_1 - \cotan u_2 = H$$

(given $u_1 = FC_L M$ and $u_2 = FC_R M$, using the same notation as in Fig. 5.4). As Ogle writes, the real horopter is "a curve set halfway between the Vieth-Müller circle and the tangent in F to this circle."[58] In other words, for distances on the same

---

[55]Peter Ludwig Panum, *Physiologische Untersuchungen über das Sehen mit zwei Augen*, Kiel, 1858. See the recent overview by Christian Corbé, Jean-Pierre Menu and Gilles Chaine, *Traité d'optique physiologique et clinique*, Paris, 1993, pp. 100–102.

[56]Helmholtz, *Optique physiologique*, p. 937. An explanation for the phenomenon was found only much later: "Because there is a convergence of the retinal impulses from many receptor cells toward a single peripheral ganglion cell, and because a certain degree of disparity is compatible with fusion, a point-by-point correspondence does not exist ... So that the zone of direct vision surrounding the fixation point is not just a line [i.e., the horopter], but a surface area that increases the further one moves away from the fixation point," Spielmann, *Les strabismes*, p. 117.

[57]Kenneth N. Ogle, *Researches in Binocular Vision*, Philadelphia, 1950, 2nd edition 1964.

[58]David Stidwill, Robert Fletcher, *Normal Binocular Vision*, London, 2011, p. 75.

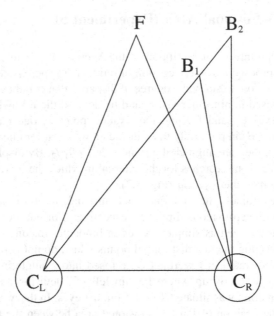

Fig. 5.7 Panum's experiment. Author's drawing

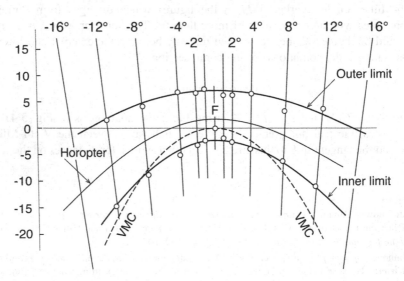

Fig. 5.8 The horopter and Panum's fusional area. Author's drawing after Ogle, *Researches in Binocular Vision*, 1950

order of magnitude as the length of Ibn al-Haytham's wooden ruler, the Vieth-Müller circle provides an upward biased model of the retinal correspondence, whereas the frontal line provides a downward biased model. The curvature of the experimental horopter changes with the distance from the fixation point. For short distances the horopter is concave. For the so-called "abathic distance,"[59] which varies from one to six meters as a function of the distance between the pupils of the two eyes, the horopter coincides with the frontal line that passes through the fixation point. Beyond the abathic distance, the experimental horopter becomes convex and has only one point of tangency with the Vieth-Müller circle.

These elements allow us to formulate a last question regarding the horopter. Could the frontal model chosen by Ibn al-Haytham simply be the result of the fact that, if one uses a sufficiently narrow ruler for the experiment (TK = 6 or 7 cm), the frontal line and the horopteric circle are virtually merged? The maximum distance between the two appears along the edge of the ruler: $TT' = KK' = 0.41$ cm.[60] Are we therefore in Panum's area, where the images of points K and K' fuse? Ogle's abacus yields, for a fixation point located 40 cm from the eyes and an eccentricity of 8° (angle QHK = 7° 28′), a tolerance of 4.6 mm (Fig. 5.8). However, the closer the fixation point, the lower the tolerance. As a consequence, in Ibn al-Haytham's experiment the tolerance should be much less than 4.6 mm and the points K K' should not be seen as fused. Repeating Ogle's experiment using Ibn al-Haytham's binocular tablet shows that a pin K' positioned at 4 mm in front of K is effectively seen as doubled.[61] One must therefore not accept the argument that the line TQK represents an approximation of the theoretical horopter, because the two would be virtually identical. For the observer, the circular horopter and the frontal horopter are *not* identical.

It may be concluded that Ibn al-Haytham adopted the hypothesis of the frontal line for a completely different reason than the one traditionally ascribed to him; from an experimental point of view it was more convincing than the model based on a circle. In fact, the object points of the frontal line TQK fall into Panum's

---

[59]Le Grand, *Optique physiologique*, vol. 3, p. 215.

[60]On the experimental ruler of Ibn al-Haytham, Q is the point of fixation. The Vieth-Müller circle passes through K' near point K, and through T' near point T. Q' is the point of intersection of the middle line HQ and the front line $T'K'$. QQ', which is the sagitta of the circle, can be calculated: $TT' = KK' = QQ' = R$ $(1 - \cos \alpha)$. Since $R = 12.75$ cm and $\alpha = 14°$ 30′, it follows that $TT' = KK' = 0.41$ cm.

[61]I reproduced these experiments with the replica of the instrument that was on display at the Galleria degli Uffizi (Florence) during an exhibition from October 16, 2001 to April 7, 2002, Dominique Raynaud, "Alhazen/Ibn al-Haytham, Tavoletta binoculare," in *Nel segno di Masaccio. L'invenzione della prospettiva*, ed. Filippo Camerota, Firenze, 2001, p. 14. In Helmholtz's experiment, which consisted of moving three sliding rulers to which pins were affixed, using the middle pin as the fixation point, the author noted that it sufficed to move the lateral pin 'a half-pin's width' (0.25 mm) to see a double image; Helmholtz, *Optique physiologique*, p. 912. Without attaining his degree of precision, we frequently observed a doubling of the image between 1 and 2 mm.

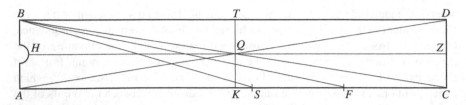

Fig. 5.9 Ibn al-Haytham, experiment 5. Author's drawing

fusional area, whereas the object points of the circle $T'QK'$ are seen as double images.

Let us return now to Panum's fusional area and the question of the approximate correspondence of images. While the experiments designed by Panum and Ogle were entirely original, the notion of measuring the degree of disparity tolerated in fusion cannot be attributed to them. In fact, one finds a first expression of this in Ibn al-Haytham's *Kitāb al-manāẓir*.

Experiment 5[62] The columns of wax being lined up once again in positions $T Q K$ and with the eyes fixed on Q, take the column in position K and move it along the side of the ruler KC. Close to K, at point S the column will still be seen as a single image. Beyond this point, for example at F, the column of wax will be seen as double (Fig. 5.9).

This observation—which remained qualitative in nature—did not allow Ibn al-Haytham to determine the extent of the space within which the object points were fused. Nonetheless it paved the way for Panum's notion of the fusional area because, by displacing the column of wax along KC, he clearly established that fusion also operates for object points that are not *strictly* contained within the horopter. It would have sufficed for Ibn al-Haytham to repeat the experiment, moving the column of wax along KC, TD, QZ and many other parallels to these lines, to arrive at a definition of the tolerances allowed by correspondence.

Ibn al-Haytham's doctrine can be summarized as follows: (1) objects will be seen as single images if they are arranged in corresponding or almost corresponding directions, that is to say, if they lie in the frontal plane that passes through the point of fixation (in modern terms, the horopter); (2) objects will be seen as single images if their position does not deviate excessively from the frontal plane (in modern terms, Panum's fusional area). If these conditions are not met, the objects will be seen as double images. Diplopia was therefore studied as a general physiological process. Ibn al-Haytham's conception of the fusion of binocular images contrasted

---

[62]*Opticae thesaurus*, p. 82; *Optics*, vol. 1, p. 240; Smith, *Alhacen's Theory of Visual Perception*, p. 266.

markedly with the optical research subsequently initiated in classical Europe, where one notes a tendency to oversimplify the problem. We will examine this in the next chapter.

Chapter 6
The Legacy of Ibn al-Haytham

Abstract Scientific ideas can have no lasting legacy unless they are accepted and transmitted by a series of successors. Our investigation will therefore focus first of all on the diffusion of texts regarding the optics of Ibn al-Haytham (known as Alhacen in the Latin West) among his peers and subsequent generations of scientists by studying the translations and commentaries that were available. Next, it is necessary that his ideas should have been applied by his successors to problems that they themselves were interested in. The conditions under which Ibn al-Haytham's theories were transferred to a new context will be examined by studying the obstacles that hindered the application of the principles of binocular vision to perspective. This analysis will provide a response to the question: Could Ibn al-Haytham's theory of binocular vision have given rise to a system of representation that was different from linear perspective?

The legacy of Ibn al-Haytham's optics in the Latin world will be studied in two parts. The first will focus on the diffusion of his work and on the availability to scholars of manuscript copies and printed editions of his texts. Since the existence of a text did not necessarily mean that it was read or its contents accepted, we will seek in the second part to clarify the conditions under which the binocular theory of vision might have provided an impetus for the work on perspective that took place between the Duecento and the Cinquecento.

6.1 The Availability of the Texts of Ibn al-Haytham in the West

As has been shown in the previous chapter, the research on optics that began in the seventeenth century (on topics such as the theory of corresponding points, direct and crossed diplopia, the construction of the horopter, and early notions of the fusional area) cannot be understood without an assessment of the work of Ibn

© Springer International Publishing Switzerland 2016   95
D. Raynaud, *Studies on Binocular Vision*,
Archimedes 47, DOI 10.1007/978-3-319-42721-8_6

al-Haytham, which laid the foundations for many of the discoveries of modern physiological optics. One can detect here a stratification of scientific concepts comparable to that seen in the studies of refraction undertaken by Ptolemy, Ibn Sahl, Ibn al-Haytham, Harriott, Snell and Descartes. Although the savants of the classical period did manage to shed considerable light on how vision functioned by explaining the role of the crystalline lens and the retina, it must be recognized that their understanding of the conditions for the fusion of binocular images was still rooted in medieval theories of optics. On this point there was no genuine break-through before the discovery of the neuro-physiological foundations of binocular vision.

To what may be ascribed the impression of a marked lack of scientific progress before the seventeenth century? An examination of how the science of optics was received allows us to formulate an explanation. In classical Europe, where translations and commentaries on most of Ibn al-Haytham's work on optics were in circulation and discussed,[1] savants might have believed the study of binocular vision to be so *saturated* with commentary that it was enough to cite the accepted authorities with no need for further analysis. Herein no doubt lies one of the main reasons for the mistaken perception that the problems of physiological optics were discovered ex novo in the seventeenth century.

Let us now examine the revival of Ibn al-Haytham's ideas on optics in more detail. His work was known through a translation from the Arab to Latin that provided the basis for the edition published by Risner in 1572. It also spread through the many commentaries written by medieval and Renaissance scholars. In particular, it is known that Ibn al-Haytham's experiments on binocular vision were repeated and abundantly discussed in the Latin world during the entire course of the Middle Ages.

If Risner's edition did not always provide a literal translation of the original text, the essential elements of the theory laid out in *Kitāb al-manāẓir* could be found in *De aspectibus*, a twelfth-century Latin translation of Alhacen's work, whose author is unknown but who probably belonged to the circle of Gerard of Cremona.[2] This text served as the source for a translation of Alhacen's work into the vernacular, *De li aspecti*, which was in all likelihood compiled in northern Italy in the middle of the

---

[1]For an overview of the influence of Ibn al-Haytham's work on optics in medieval Europe, see David C. Lindberg, "Alhazen's theory of vision and its reception in the West," *Isis* 58 (1967): 326–337; *Idem*, "Introduction to the reprint edition," *Opticae thesaurus*, New York, 1972, pp. xxi–xxx; A. Mark Smith, *Alhacen's Theory of Visual Perception*, Philadelphia, vol. 1, pp. lxxx–cxi.

[2]The Latin text can be attributed to two successive hands. The translation is quite literal up to Book III, Chapter 3, before becoming looser and sometimes resembling a paraphrase more than a translation of the original Arabic. The passages on binocular vision appear in III, 2 and therefore were the work of the first translator. Even if the presumption is that the Latin version came from Spain because of the voiceless 'c' in Alhacen appearing in the earliest manuscripts (Lindberg, *Theories of Vision*, pp. 209–210; Sabra, *The Optics of Ibn al-Haytham*, II, p. lxiv), its attribution to Gerard of Cremona remains questionable; this is discussed by A. Mark Smith, *Alhacen's Theory of Visual Perception*, pp. ix, xx–xxi.

fourteenth century.[3] The Italian edition constituted a genuine milestone that is crucial to our understanding of the spread of Ibn al-Haytham's ideas in medieval and Renaissance Italy. *De li aspecti* transmits *in extenso* the chapters dedicated to diplopia and the fusion of quasi-images.[4]

Not only could Alhacen's text be read in Latin or Italian; it also formed the subject of many scholarly commentaries.[5] Thus one finds extracts or paraphrases of passages from his work on binocular vision in the most important medieval treatises on optics: the *Perspectiva* of Roger Bacon;[6] Witelo's work of the same title;[7] *Tractatus de perspectiva*[8] and *Perspectiva communis* by John Pecham;[9] and *Questiones super perspectiva communi* by Biagio Pelacani da Parma.[10] It may be noted that the aims of each of these authors differed; Bacon elaborated a scholastic interpretation of Alhacen's theories, whereas Witelo presented a faithful paraphrasing, and Pecham condensed his ideas into a succinct aide-mémoire.

---

[3]Philological analysis has shown that MS. London, British Library, Royal 12.G.VII served as the Latin matrix for the Italian version, A. Mark Smith and Bernard R. Goldstein, "The medieval Hebrew and Italian versions of Ibn Mu'adh's 'On twilight and the rising of clouds' *Nuncius* 8 (1993): 633–639. This text, discovered by Enrico Narducci, "Nota intorno a una traduzione italiana fatta nel secolo decimoquarto del trattato d'ottica d'Alhazen," *Bollettino di bibliografia e di storia delle scienze matematiche e fisiche* 4 (1871): 1–40, was investigated in depth by Graziella Federici Vescovini, "Contributo per la storia della fortuna di Alhazen in Italia: Il volgarizzamento del ms. Vat. 4595 e il 'Commentario Terzo' del Ghiberti," *Rinascimento* 5 (1965): 17–49; *Eadem*, "Alhazen vulgarisé: Le 'De li aspecti' d'un manuscrit du Vatican (moitié du XIVe siècle) et le troisième 'Commentaire sur l'optique' de Lorenzo Ghiberti," *Arabic Sciences and Philosophy* 8 (1998): 67–96.

[4]Alhacen's exposition of the principles of binocular vision takes up folios 57r–66r of MS. Vat. lat. 4595. I wish to thank Pietro Roccasecca for allowing me to examine this manuscript. Several passages from the text are reproduced in Federici Vescovini, "Contributo per la storia della fortuna di Alhazen in Italia," pp. 31–49, and "Alhazen vulgarisé," pp. 73–79.

[5]Alhacen's *Optics* was also mentioned in treatises on other matters. Around 1230 Jordanus of Nemore refers in *De triangulis*, IV, 20 to "19 [34?] quinti perspective"; see Marshall Clagett, *Archimedes in the Middle Ages*, Philadelphia, 1964, vol. 1, pp. 668–669; and 1984, vol. 5, pp. 297–301. Around 1230–1250 Bartholomaeus Anglicus cites Alhacen in his *De proprietatibus rerum*, III, 17 as an "auctor pespective"; James Long, *Bartholomaeus Anglicus, De proprietatibus rerum, Books 3–4: On the Properties of Soul and Body*, Toronto, 1979, pp. 39–45.

[6]*The Opus maius of Roger Bacon*, ed. J.H. Bridges, Frankfurt am Main, 1964, pp. 92–99, new edition: David C. Lindberg, *Roger Bacon and the Origins of Perspectiva in the Middle Ages*, Oxford, 1996.

[7]Witelo, *Opticae libri X*, ed. Risner 1572; reprinted New York, 1972, pp. 98–108. It is worth noting that there were many more copies of the treatises on perspective by Bacon and Pecham in circulation than that of Witelo.

[8]John Pecham, *Tractatus de perspectiva*, ed. D.C. Lindberg, St Bonaventure, 1972, pp. 56–57.

[9]David C. Lindberg, *John Pecham and the Science of Optics*, Madison, 1970, pp. 116–118.

[10]Blaise de Parme, *Questiones super perspectiva communi*, eds. G. Federici Vescovini and J. Biard, Paris, 2009, pp. 178–190. Biagio's treatment of binocular vision is incomplete. He poses the question as to whether vision takes place at the chiasma, but makes no comment on the proposition by Pecham in *Perspectiva communis*, I, 80 concerning the conditions for the fusion of quasi-images.

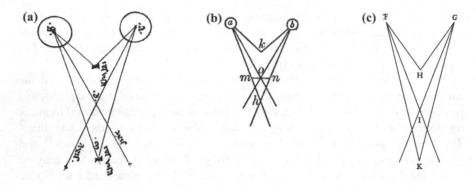

Fig. 6.1 The legacy of De aspectibus, prop. III, 2.9: **a** Alhacen, Paris, BnF lat. 7319, fol. 116v; **b** Roger Bacon, *Perspectiva* (ed. John H. Bridges, *The 'Opus majus' of Roger Bacon*, II, p. 95); **c** Author's drawing after John Pecham, *Tractatus de perspectiva*, ed. David C. Lindberg, p. 73; Permission of the Bibliothèque nationale de France; Bacon, *Opus maius*, V, II, II, 2, ed. Bridges, p. 95; Pecham, *Tractatus de perspectiva*, p. 57.)

Nevertheless, they all copied Alhacen's diagrams and arrived at the same conclusions, albeit sometimes in more simplified form. These similarities become apparent when one compares the texts and figures on some of the more important propositions regarding binocular vision.

Prop. III, 2.9. Here Alhacen distinguishes between the two cases of diplopia:

Heteronymous (crossed) diplopia: *"When that other visible object lies nearer both eyes than the visible object on which the two [visual] axes intersect/quando illud aliud visum fuerit propinquius ambobus visibus viso in quo coniunguntur duo axes"*;

Homonymous (direct) diplopia: *"When that other visible object... lies farther from both eyes than the visible object on which the two [visual] axes intersect/quando illud aliud visum... fuerit remotius ab ambobus visibus viso in quo coniunguntur duo axes"*.

Bacon and Pecham both copied Alhacen's diagram (Fig. 6.1) and came to the same conclusion: the fixation of a central point induces a doubling of the points in front of or behind it. Even so, they differed in their understanding of the problem. Roger Bacon was pursuing the same objective as Alhacen—*In which it is shown... how a single thing appears double/In quo ostenditur... quomodo unum videatur duo*—but he cut short the discussion of the figure in order to present the results of his own experiments using a binocular ruler. For his part, John Pecham interpreted binocular vision as a form of optical illusion (*errore, deceptio*) attributable to a defect in the vision or to limitations in the visual conditions,[11] which actually constitutes a misinterpretation of Alhacen's text.

Prop. III, 2.15. In this proposition Alhacen explains the fusion of two quasi-images in terms of the existence of corresponding points: *"So the two forms impressed on the two points that are correspondingly situated with respect to the*

---

[11]John Pecham, *Tractatus de Perspectiva*, ed. D.C. Lindberg, St. Bonaventure, 1972, p. 57.

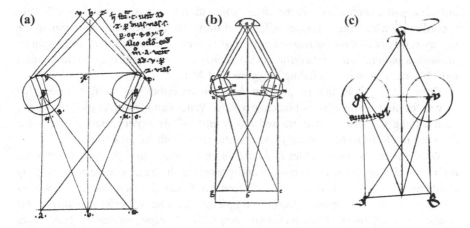

Fig. 6.2 The legacy of De aspectibus, prop. III, 2.15: **a** Alhacen, Paris, BnF lat. 7319, fol. 117r; **b** Witelo, *Perspectiva* (ed. Frederic Risner, *Opticae Thesaurus*, p. 103); **c** John Pecham, *Perspectiva communis* (Paris, BnF lat. 7368, fol. 15r). (Permission of the Bibliothèque nationale de France; Witelo, *Optica*, III, 37, ed. Risner pp. 102–103; Lindberg, *John Pecham and the Science of Optics*, p. 117.)

surfaces of the two eyes reach to that same point in the hollow of the common nerve, and they will be superimposed at that point so as to produce a single form."[12]

Latin authors such as Witelo and Pecham copied this figure (Fig. 6.2). In his text Witelo repeats Alhacen's argumentation, paraphrasing essential passages such as his notion of corresponding points (*in two points correspondingly situated/in duobus punctis consimilis positionis*). John Pecham simplifies the Arab scholar's text, arguing that vision is actually quite straightforward because, barring anomalies and accidents, the images in binocular vision are always fused at the level of the optic nerve: *"The duality of the eyes must be reduced to the unity/Oculorum dualitatem necesse est reduci ad unitatem."*[13] He schematicizes the argument by discussing the propositions on binocular vision separately (*Perspectiva communis*, I, 32 et I, 80). He does not in fact make a distinction between the two types of physiological diplopia nor does he explain the disparity in the images stemming from binocular vision and their non-corresponding positions in the two eyes until proposition I, 80 (*An object appears double when it has a sensibly different position relative to the two axes/Ex variatio sensibiliter situ uisibilis respectu duorum axium ipsum duo apparere*). Recognizable here nevertheless—apart from a few

---

[12]"Et due forme qui infiguntur in duobus punctis que sunt consimilis positionis apud superficies duorum visuum perveniunt ad illum eundum punctum concavitatis communis ipsius nervi, et superponentur sibi apud illum punctum, et efficientur una forma," Smith, *Alhacen's Theory of Visual Perception*, I, pp. 256–257.

[13]John Pecham, *Perspectiva communis*, I, 32; Lindberg, *John Pecham and the Science of Optics*, p. 116.

differences in terminology—is the same explanation that was provided by Alhacen, according to which quasi-images will fuse if they reach corresponding points in the two eyes (*impressed on the two points that are correspondingly situated/infinguntur in duobus punctis que sunt consimilis positionis*), an argument that would be used later by many authors, including Huygens and Müller.

Prop. III, 2.27. In this proposition Alhacen describes the wooden tablet that served for his experiments on binocular vision. Witelo and Bacon copied his figure, each adding the lines that illustrated the results of his experiments, whereas the drawing in *De aspectibus* simply showed the ruler with its pins (Fig. 6.3).

Witelo used the same tablet (*a board of one cubit... and four digits/tabula... unius cubiti... quattuor digitorum...*) and presented the same conclusions, slightly condensed, as those laid out by Alhacen in Chap. 5 (*then the experimenter examines..., the experimenter fixes the visual axes... if the experimenter directs his visual axes.../deinde experimentator inspiciat..., experimentator figat axes uisuales..., quod si experimentator dirixerit axes uisuales...*). Roger Bacon only retained the results of Alhacen's principal experiments (nos. 1, 3 and 5), which he presented in a series of succinct propositions. He added that one could arrive at the same conclusions without resorting to the experimental ruler (*Even without such a board, an experimenter can test many things relevant to these matters/Et experimentator potest sine tabula experiri multa in hac parte*).

While it is clear that the authors of the Latin Middle Ages broadly adopted the teachings of Alhacen (with some modifications and distortions), it may be noted that the influence of *De aspectibus* continued throughout the Renaissance.

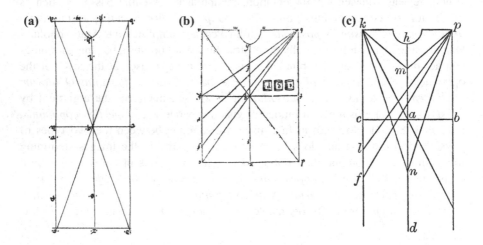

Fig. 6.3 The legacy of De aspectibus, prop. III, 2.27: **a** Alhacen, Paris, BnF lat. 7319, fol. 121v; **b** Witelo, *Perspectiva* (ed. Frederic Risner, *Opticae Thesaurus*, p. 165); **c** Roger Bacon, *Perspectiva* (ed. John H. Bridges, *The 'Opus majus' of Roger Bacon*, II, p. 96). (Permission of the Bibliothèque nationale de France; Witelo, *Optica*, IV, 108, ed. Risner, p. 164; Bacon, *Opus maius*, V, II, II, 3, ed. Bridges, p. 96.)

It appears that *De aspectibus* achieved such undisputed status that it could not be ignored by any scholar embarking on the study of binocular vision, and many authors continued to copy or paraphrase the same passages. One finds lengthy extracts in the *Commentario terzo* of Lorenzo Ghiberti[14] and in many of the notes of Leonardo da Vinci.[15] During the Cinquecento binocular vision was discussed by Girolamo Cardano in *Problematum medicorum*[16] and by Vignola in *Due Regole della prospettiva pratica*. An edition of the latter with commentary was later published by Egnatio Danti.[17] On 23 December 1591 Christoph Grienbergen delivered a lecture at the prestigious Collegio Romano[18] in which he broached the question of binocular vision beginning with a presentation of the theories of Alhacen and Witelo. Alhacen was also cited by the eminent scholar Giambattista della Porta in 1593.[19]

This collection of texts—from translations of *Kitāb al-manāẓir* into Latin and thence into Italian, to Latin commentaries and supercommentaries—provides a picture of the impact of the theories of Ibn al-Haytham on the conceptions of binocular vision in the Latin world. With regard to the accessibility of his texts, it can be taken for granted that the perspectivists had the opportunity to borrow elements from the *Optics* of Ibn al-Haytham in either the Latin or Italian versions. But did they do so? Before addressing this question, let us examine the compatibility of his theory with the problem of representations in perspective.

6.2 Obstacles to the Binocular Theory of Vision

When the theory of binocular vision was applied to perspective it met with objections that we will now pass in review.

---

[14]Lorenzo Ghiberti reproduced Alhacen's chapter on binocular vision almost in its entirety (that is, props. 2.1–2.8, 2.8–2.13, 2.19–2.47, 2.51–2.65 and 2.71–2.86, as they appear in A. Mark Smith's edition), *I Commentari*, ed. Ottavio Morisani, Naples, 1947, pp. 133–136 and 139–149; K. Bergdolt, *Der dritte Kommentar Lorenzo Ghibertis*, pp. 294–304 and 314–368. Experiments performed using Alhacen's ruler are described on pp. 330–348.

[15]For example, *Libro di pittura*, §§ 115 "Perche la pittura non può mai parere spicata come le cose naturali," and 811 "Di prospettiva," ed. A. Borzello, Lanciano, 1924, pp. 76, 401.

[16]Girolamo Cardano, "Problematum medicorum. Sectio Secunda," *Hieronymi Cardani Mediolanensis... Operum tomus secundus*, Lugduni, 1663, pp. 636–642.

[17]Egnatio Danti, *Le Due Regole della prospettiva pratica di M. Iacomo Barozzi da Vignola*, Roma, 1583, pp. 53–55; *Les Deux Règles de la perspective pratique*, French transl. by P. Dubourg Glatigny, Paris, 2003, pp. 222–226.

[18]*Fieri posse... in aliquam mensa lumine*. This text has been edited by Michael J. Gorman, "Mathematics and modesty in the Society of Jesus: the problems of Christoph Grienberger," Mordechai Feingold, ed., *The New Science and Jesuit Science. Seventeenth Century Perspective*, Dordrecht, 2003, pp. 41–48.

[19]Giovanni Battista della Porta, *De refractione optices parte libri novem*, Naples, 1593, Book VI, "Cur binis oculis rem unam cernamus," pp. 139–148.

In the literature one finds various misinterpretations of the functioning of binocular vision, which tend to hierarchize fusion and diplopia, considering simple vision to be the normal mode of seeing and diplopia a marginal condition that might occasionally develop. Since fusion provided a more intuitively convincing explanation of perspective than diplopia, the postulate of monocular vision appeared to be justified. Closer analysis shows that this hierarchy, overtly affirmed in eighteenth-century treatises on optics,[20] is not consonant with the theories of Ibn al-Haytham for it rests on anachronistic notions and simplifications that must be understood if we are to reconstruct the context in which Ibn al-Haytham and his Latin successors conducted their research.

As it is now known that Ibn al-Haytham's theory of binocular vision circulated widely in two versions—Latin and Italian—we will henceforth cite passages from both as we review the development of optics in the Latin West.

6.2.1 Physiological and Pathological Diplopia

Since Book III of Ibn al-Haytham's *Optics* was devoted primarily to the study of optical illusions (*deceptiones visus*), one might be drawn into believing that the theory of binocular vision could be reduced to a theory of pathological diplopia. But Ibn al-Haytham's treatment of this theory is limited to Chap. 2 of Book III, which forms an introduction to the study of optical illusions presented in Chaps. 3–7. The summary to Book III stipulates this clearly:

> The second [concerns certain] things that need to be set forth for the analysis of visual illusions/Secundum de eis que debent proponi sermoni in deceptionibus visus/El sicondo de quelle cose che se debano propore sicondo tuto in le deceptione del uiso.[21]

The closing paragraph of the chapter (prop. 2.86) is no less explicit:

> And now that these points have been explained, it is time to begin the discussion of visual illusions and to describe their causes and their kinds/His autem declaratis, incipiendum est de sermone de deceptione visus et declarare causes et species earum/Dechiarato questo e da guadare in lo sermone dela deceptione del uiso e dechiarare de cagione e le spetie de esse.[22]

As a consequence, it was actually within a general framework that Ibn al-Haytham undertook his study of the conditions for the fusion of quasi-images. This point has not always been clearly grasped by historians of science, who have tended to present the conclusions drawn by Ptolemy and Alhacen from their binocular experiments as follows:

---

[20]See in particular the discussion of Alhacen's theories in William Porterfield, *A Treatise on the Eye*, Edinburgh, 1759, Vol. 2, Book V, pp. 279–328.

[21]Smith, *Alhacen's Theory*, vol. 1, p. 245; MS. Vat. lat 4595, fol. 57rb.

[22]Ibid., p. 286; MS. Vat. lat 4595, fol. 66ra.

In strabismus, most often one of the eyes plays a dominant role that for all practical purposes overrides the contribution of the other; however, the paralysis of an oculomotor muscle, or simply pressure on one of the eye globes, allows the formation of two independent retinal images, thus causing double vision or diplopia.[23]

This reading does not distinguish between physiological and pathological diplopia, and treats all cases of double vision as abnormal. In contrast, the examples of diplopia discussed in chapter III, 2 represent illustrations of normal physiological diplopia and have no particular relationship to strabismus, pathological diplopia, or the phenomenon of optical illusions.[24]

6.2.2 Diplopia Is not an Unusual Phenomenon

The fact that chapter III, 2 opens with a discussion that takes simple vision as the standard case allows one to deduce that diplopia occupied a secondary position in Ibn al-Haytham's theory of optics. For example, his first proposition poses the question: "Since it is so, we must determine how a single, distinct object is seen simultaneously by the two eyes as one image, *most of the time and in most cases*, and how it happens that a unique object can be identically situated in relation to the two eyes, *most of the time and in most cases*."[25] The Latin and Italian versions of his text read:

Prop. 2.1: So we need to explain how a single visible object is generally perceived as single by both eyes in many [different] situations, as well as how the situation of a single visible object will generally be equivalent with respect to both eyes under various conditions/Unde oportet nos declarare quomodo unum visum comprehenditur a duobus visibus unum in *maiori parte temporis* et in pluribus positionibus, et quomodo positio unius visi ab ambobus oculis *in maiore parte temporis* et in pluribus erit consimilis/Doue fia di bisogno noi dechiarare como uno uiso zioe como una cosa ueduta si comprendre da dui uisi como el uno *in magiore parte di tempi* e in piu dispositione e chomo la positio de uno seno (?) da amedui lochij *in magiore parte del tempo* e in piu sera consimille.[26]

All the same, the phenomenon of diplopia is not rare. Ibn al-Haytham acknowledges this, for example, in prop. 2.9, which introduces the two cases of homonymous and crossed diplopia, and by prop. 2.44 and prop. 2.45, which describe how the lines on his ruler are perceived by the two eyes:

---

[23]Gérard Simon, *Le Regard, l'être et l'apparence dans l'optique de l'Antiquité*, Paris, 1988, pp. 131–132.

[24]Further confirmation is provided by three passages in *De aspectibus* which interpret mistaken perceptions of size as a consequence of monocular vision; see prop. 2.51, 2.52 and 2.53.

[25]*Optics*, vol. 1, pp. 228–229 (italics mine).

[26]Smith, *Alhacen's Theory*, vol. 1, p. 247. "*In maiori parte temporis*" should be translated as "most of the time" since "generally" already translates *universaliter* (prop. 2.3); MS. Vat. lat 4595, fol. 57vb. See also prop. 2.49.

Prop. 2.9: Furthermore, the axes of both eyes *often* intersect on some visible object while the two eyes perceive another visible object that is not correspondingly situated with respect to the eyes in terms of direction.../Et *multotiens coniunguntur* duo axes amborum visuum in aliquo viso, et cum hoc duo visus comprehendent aliam rei visam cuius positio in respectu duorum visuum erit diversa in parte.../E *molte uolte se congiungeno* le axe de tuti dui li uisi in alchuna cosa uisa e cun queste dui uisi comprendeno comprenderano laltra cosa uisa dela quale la positione in rispecto de amedui i uisi sera diuersa in la parte....[27]

Prop. 2.44: Yet this line, and everything that lies on it, except for the peg that is placed in the center, *invariably* appears double if the two [visual] axes intersect at the peg placed in the center/Et ista linea et omnia posita super ipsam preter individuum positum in medio *semper videntur duo* cum duo axes concurrerint in individuo posito in medio/E questa linea e tute quelle cose che sono posite sopra quella ultra lindividuo posito in lo megio *sempre pareno due* quando due asse serano concurse in lindividuo posito in lo megio.[28]

Prop. 2.45: On this basis it has been therefore been shown that a visible object that lies on different sides of the two [visual] axes *always* appears double/Declaratum est igitur ex hac dispositione quod visum cuius positio in respectu duorum axium est diversa in parte *semper videtur duo*.../Dechiarato e adunche da quella dispositione che quello che si uede quando la positione de esso in rispecto de due asse e diuersa in parte *sempre pareno due*....[29]

The adverbs "often" (*multotiens*) (2.9) and "always" (*semper*) (2.44, 2.45) preclude limiting diplopia to exceptional cases of abnormal vision. Therefore, the meaning of passages in which Alhacen describes cases of simple vision as "frequent" must not be misinterpreted. Objects are seen as single images if they lie in corresponding or almost corresponding directions. In all other cases they will be seen as double images.

6.2.3   Images Are not Blurred in Diplopia

Another interpretation that permits the hierarchization of simple and double vision, with simple vision being considered as normal and double vision as the exception, is based on the argument that diplopic images would be out of focus. In fact, all retinal images tend to be blurred around the edges because the conditions of stigmatism are no longer met and the cones, which are more sensitive to light, are less numerous toward the periphery. This argument cannot be found in *Kitāb al-manāẓir* because the anatomical knowledge on which it is based was discovered much later. Here, for example, is what the Latin and Italian versions of Alhacen's text have to say about those cases in which quasi-images are disparate:

Prop. 2.19: Nevertheless, its form will be indefinite rather than definite/Sed tamen forma eius non erit verificata sed dubitabilis/Ma niente meno la forma de esso non sera certificata seno dubitabile.[30]

---

[27]Smith, *Alhacen's Theory*, vol. 1, p. 252; MS. Vat. lat 4595, fol. 59ra.

[28]Ibid., p. 268; MS. Vat. lat 4595, fol. 62va.

[29]Ibid., p. 269; MS. Vat. lat 4595, fol. 62va.

[30]Ibid., p. 259; MS. Vat. lat 4595, fol. 60va.

Prop. 2.20: So the form of its extremities will be indefinite rather than definite/Quapropter forma extremorum erit dubitabilis, non certificata/Per la quale cosa la forma degli estremi essi dubitabili o uoi sera dubitabile, non certificata.[31]

Prop. 2.21: And so the form of such visible objects will be indefinite under all circumstances... /Et sic forma huiusmodi visibilium erit dubitabilis in omnibus positionibus.../ Cosi la forma de questi uisibili sera dubitabile in tute le dispositione....[32]

Although the fusion of corresponding points leads to "definite" vision in the sense that it is certified by the organ of sight (*certificata, verificata*), diplopia does not produce a blurred image. The entire argument here rests on the word *dubitabilis* —"dubious, what may be doubted"—which brings one back to the duality of the quasi-images produced by the two eyes. However imperfectly superimposed, these quasi-images do not in the end merge in a synthesis that "glues together" their salient traits. It is precisely because they are clearly distinct that the eye hovers in doubt as to what it is seeing; that is, it does not know which of the two images it should choose.

6.2.4   Different Theories Regarding the Unification of Binocular Images

The theory of the *suppression* of one of the two images in binocular vision and the critique of this theory, which will be examined further below, can only be understood within the framework of the notions that were conceived in order to account for the unification of the visual sensations received by the two eyes.[33] These theories can be divided into three groups: the theory of permanent fusion, the theory of conditional fusion, and the theory of the absence of fusion.

(1) *Permanent fusion*

According to the earliest of all the theories, that of permanent fusion, the images received by the two eyes are always associated and joined together. This was proposed most notably by Pseudo-Aristotle (*Problemata* XXXI, 4, 7), who retained that the functioning of the two eyes could not be disassociated, either in terms of motor function or of perception. Is it not, as he asked, "[...] because the eyes, although two in number, depend on one and the same principle?"[34] This theory would be adopted by John Pecham who, considerably simplifying the text of Alhacen on this point, wrote: *"The duality of the eyes must be reduced to unity/Oculorum dualitatem necesse est reduci ad unitatem."*[35]

---

[31] Ibid., p. 261; MS. Vat. lat 4595, fol. 60vb.

[32] Ibid., p. 262; MS. Vat. lat 4595, fol. 61ra.

[33] On this point the best account is that of Nicholas J. Wade, "Descriptions of visual phenomena from Aristotle to Wheatstone," *Perception* 25 (1996): 1137–75; *Idem* and Hiroshi Ono, "Early studies of binocular and stereoscopic vision," *Japanese Psychological Research* 54 (2012): 54–70.

[34] Aristote, *Problèmes*, tome III, ed. P. Louis, Paris, 1994, p. 50.

[35] Lindberg, *John Pecham and the Science of Optics*, p. 116.

The considerations of Galen (*De usu partium* X, 14) and Pseudo-Aristotle (*Problemata* XXXI) survive in a question in *Problematum medicorum* by Girolamo Cardano. The precise date of this text is not known because it was never published during the author's lifetime;[36] all that can be affirmed is that the work was written before the death of Cardano in 1576. Question 3 reads:

> Why does the myopic see better with both eyes than with either one of the two eyes, while those who have good vision see as much with a single eye?—the proof is that they close one eye when they want to aim [their vision] in a straight line. But rather than aiming, would it not be better just to see? Is it that, on closer inspection, the visual species of a single eye is more concentrated in the cone? I also state that the myopic only sees a portion [of the visual field] with each eye, the other part is not seen, and so the two [eyes] are needed to see everything. The remaining [visual power] helps the operation of the other eye. Suppose that K sees *ABCD* straight ahead and *BDEF* on the other side; that L sees *BDEF* straight ahead and *ABCD* on the other side. Nevertheless K and L will perfectly see all parts, and all the better the whole that K [will see] his half *ABCD*, and L his own *BDEF*. But for perfect vision, the [visual] power of the closed eye combines with that of the other [eye], and as a result, he sees clearly; and rather better when he looks with one eye rather than two [eyes]. Why, when the organ is injured and corrupted, the strong (for example, that having strong fire from a little water) is easily strengthened? Is it because he gathers his power, or is this not always true? Where the light is lacking, as in old men, light added to the other eye allows them to see better, while those who are filled with [light] are prevented from doing so. This is why the same [old men], because they see better at a distance, as far as they are brought by nature, see quite well with both eyes."[37] (Fig. 6.4).

This text is surprisingly anachronistic. Cardano revives the theory of extramission, which was abandoned by most medieval scholars after their exposure to the work of Alhacen. It also illustrates the difficulty that Cardano had in choosing between the conflicting theses of Pseudo-Aristotle; visual fixation is better using one eye, but visual perception is clearer using two eyes (*Problemata* XXXI, 4 and 10). Cardano attempts to reconcile the two notions by proposing a less than

---

[36]Jean-Pierre Nicéron, *Mémoires pour servir à l'histoire des hommes illustres dans la République des Lettres*, Paris, 1731, vol. 14, p. 267.

[37]"Cur lusciosi melius cernunt ambobus oculis simul quam alterutro oculo quovis qui autem acie valida sunt perspiciendum uno tantum; indicio est quod ubi collimare velint alterum oculorum claudunt? An quod collimare melius, non sit melius videre? at hoc sit ut rectius inspiciant cum in unum oculum species visibilis in conum magis coangustetur? Vel dico quod lusciosi parte una oculi vident alia non vident & ita utroque totum. Et reliqua adiuvant operationem alterius oculi, velut K, inspiciat *ABCD* recte perfecte & *BDEF* contra: L, recte & perfecte *BDEF* contra *ABCD*: ita K, L, utranque partem bene atque perfecte videbunt & aliquanto melius etiam totum quam K, suam medietatem *ABCD*, & L, suam *BDEF*. At in oculo perfecto uno occluso tota vis & spiritus in unum coeunt & ita optime videt, ac melius nec solum collimat quam ambobus. An quod membrum laesum a quovis incommodo diversitur & offenditur; robustum (Exemplo praevalidi ignis ab aqua pauca) corroboratur modo leve sit: ut quia suas vires colligat, vel quod non semper verum sit, sed ubi lumen deficit, ut in senibus maxime, lumine addito ex altero oculo melius videant, contra illi a superfluo impediuntur. Ad hoc videmus in iisdem qui ob hoc procul melius vident quamvis hoc magis naturae aduersetur procul, quam duobus oculis bene videre," Girolamo Cardano, *Opera Omnia, tomus secundus*, Lyon, 1663, pp. 640–641.

Fig. 6.4 The theory of
binocular vision according to
Girolamo Cardano, *Opera
Omnia, tomus secundus*,
p. 640

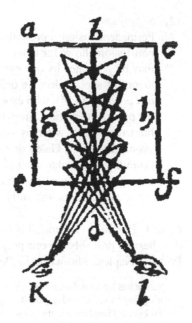

satisfactory compromise—allowing for fusion in the case of myopathy and suppression in the case of normal vision.

The theory of permanent fusion would then be taken up by Egnatio Danti who, citing Risner's edition of the works of Alhacen and Witelo,[38] kept nothing of the theory of correspondence that formed the core of the analysis of binocular vision. The doctrine of permanent fusion would reappear in Descartes' *Dioptrique*.[39] Published in Leiden in 1636, *Dioptrique* ignores the teachings of Aguilonius and Scheiner which were published in 1613 and 1619. According to Descartes, the unification of visual sensations is controlled by a small gland which he denominated the *glandula pinealis* or *conarium*. In reality this gland is responsible for the synthesis of melatonin, but Descartes believed it to be "the seat of common sense." In the sixth Discourse he reiterates his thesis by drawing the famous parallel:

> And as the blind man does not judge a body to be double although he touches it with his two hands, so too, when both our eyes are disposed in the manner required to direct our attention to one and the same place, they need only make us see a single object there, even though a picture of it is formed in each of our eyes.[40]

---

[38]Danti, *Le Due Regole della prospettiva*, pp. 12, 33. Danti's text is discussed below.

[39]René Descartes, *Le Discours de la méthode suivi d'extraits de la Dioptrique*, etc., Paris, 1966, pp. 139–140.

[40]"Et comme cet aveugle ne juge point qu'un corps soit double, encore qu'il le touche de ses deux mains, ainsi, lorsque nos yeux sont tous deux disposés en la façon qui est requise pour porter notre attention vers un même lieu, ils ne nous y doivent faire voir qu'un seul objet, nonobstant qu'il s'en forme en chacun d'eux une peinture," Descartes, *La Dioptrique, Sixième Discours*, p. 149.

(2) *Suppression*

Permanent fusion represents a single theory of vision, whereas the thesis of non-fusion allows for two distinct types of vision in which a monocular image may be generated by one eye or by the two eyes alternately. The sources of the theory of suppression are quite ancient; they can be traced back to Galen (*De usu partium* X, 14), who espoused the theory of extramission and believed that the pneuma descended from the encephalon and was shared by the two eyes in binocular vision, whereas it was transmitted integrally to one eye in monocular vision. From this Galen drew the conclusion that monocular vision was more acute than binocular vision.[41] Some of his successors took the argument to its extreme and claimed that, the two eyes being intact, the processing of images was conducted monocularly.

Case 1 The image is furnished by one of the two eyes. The best known version of this thesis is probably the one proposed by the Jesuit scholar Giovanni Battista della Porta of Naples, who wrote in *De Refractione*:

> Nature has bestowed on us two eyes, one on the right, the other on the left, so that when we are going to see something on the right, we see with the right eye, and on the left with the left eye… Therefore we always see with a single eye… Hence the two eyes are not able to see the same thing at the same time.[42]

In 1679 Sébastien Le Clerc, a French engraver who had taken up the study of geometry, wrote a short treatise in support of the thesis of suppression.[43] Mistaken ideas are often slow to be abandoned; Le Clerc drew his principal argument not from the works of Della Porta, but from the lines of investigation pursued by Ibn al-Haytham and others. He based his thesis of suppression on the conundrum posed by the disparity between images, finding himself unable to explain how they were unified, even though Aguilonius had already presented a convincing geometric analysis of the problem in 1613.

In the eighteenth century Porterfield[44] would formulate, on anatomical grounds, a critique of the ancient idea that the unification of visual sensations takes place at the level of the chiasma. The role assigned by Descartes to the pineal gland (which the optic nerves do not actually reach) is ascribed by Porterfield to this crossing point between the two optic nerves. Being aware of the latest discoveries regarding

---

[41]Galien, *Oeuvres anatomiques, physiologiques et médicales*, ed. Ch. Daremberg, Paris, 1854, vol. 1, p. 648.

[42]"Oculos binos natura largita est nobis a dextris unum, a sinistris alterum, ut si a dextris aliquid visuri sumus, dextro utamur, at si a sinistris sinistro, unde semper uno oculo videmus… Unde non simul videre possunt rem eandem," Della Porta, *De refractione optices parte libri novem*, Naples, 1593, pp. 142–143.

[43]Sébastien Le Clerc, *Discours touchant le point de veuë, dans lequel il est prouvé que les choses qu'on voit distinctement, ne sont veuës que d'un oeil*, Paris, 1679. His ideas are discussed below.

[44]William Porterfield, *A Treatise on the Eye. The Manner and Phænomena of Vision*, 2 vols., Edinburgh, 1759, pp. 308–328.

the horopter, Porterfield acknowledges that when images are disparate they are seen as double. Even so, he sought to minimize the consequences of this by hypothesizing that the soul possesses a 'faculty of learning' that allows it to reunify double images.[45] The following year Du Tour[46] conducted experiments on binocular vision and established that if two images are disparate they are not fused, but rather are seen sequentially. This paved the way for the modern theory of retinal rivalry, according to which two images are perceived alternately if there is a large disparity between them in terms of form or color.

Case 2 The image is furnished by only one eye, generally the dominant one. This thesis was defended by Francis Bacon in a short treatise published in 1627.[47] One encounters it again in the work of Giovanni Alfonso Borelli who, based on his personal observations, concludes that the left eye is always dominant.[48] These notions would be criticized in *La Vision parfaite* by Chérubin d'Orleans, who challenges the thesis of suppression—both alternate (case 1) and constant (case 2)[49]—and then in the *Physical Essay on the Senses* by Le Cat, who observes that vision can vary from individual to individual, with either one eye or the other being dominant, and sometimes with no dominance at all (case 2).[50]

In French it became the custom to refer to these theories by the term "suppression" in cases where inhibition involves only a part of the image seen in one eye, and the term "neutralization" when this inhibition involves the entire image

---

[45]"That we have here given the true Account of this *Phænomenon*, will be further evident to any one who considers that, when the Mind does not mistake the Situation of the Eye, as in those who by Custom have, from their Infancy, contracted a Habit of moving their Eyes differently, all Objects appear single as to other Men; and this likewise is the Reason why, in the Case before us, all Things come in Time to be seen single; for, by repeated Experiences, the Mind becomes wiser, and by Degrees learns to form a right Judgment concerning the Direction of the Eye," Porterfield, *A Treatise on the Eye*, pp. 314–315. See also p. 328.

[46]Étienne-François Du Tour, "Discussion d'une question d'Optique" with an "Addition," *Mémoires de mathématique et de physique présentés à l'Académie royale des Sciences par divers savans* 3 (1760): 514–530; 4 (1763): 499–511.

[47]"We see more exquisitely with one eye shut, than with both open. The cause is, for that the spirits visual unite themselves more, and so become the stronger," *Sylva sylvarum* (1627), ed. by J. Spedding et al., *The Collected Works of Francis Bacon*, London, 1857, 2, p. 628; quoted by Nicholas Wade, "Early studies of eye dominances," *Laterality* 3 (1998): 104.

[48]Giovanni Alfonso Borelli, "Observations touchant la force inégale des deux yeux," *Journal des Sçavans* 3 (1672): 295–298.

[49]"The man who has two healthy eyes naturally well-formed and jointly open, necessarily sees with both eyes at the same time, and not with just one eye at a time or with the two eyes alternately/L'homme qui a les deux yeux sains naturellement bien conformez, & conjointement ouverts, voit necessairement des deux yeux au mesme temps, & non pas seulement d'un oeil à la fois ou des deux yeux alternativement," Chérubin d'Orleans, *La Vision parfaite ou Le concours des deux axes de la vision en un seul point de l'objet*, Paris, Chez Sebastien Mabre-Cramoisy, 1677, Epitre au Roy.

[50]Claude-Nicolas Le Cat, *Traité des Sens*, Amsterdam, 1744, pp. 204–208.

seen in one eye, whereas the English refer to *suppression theory*, which assumes the partial or total inhibition of the summing of binocular images.[51]

(3) *Conditional fusion*

In comparison to the preceding theses of unconditional fusion and suppression, the theories in the third group appear quite sophisticated. Here again one notes the presence of two types of phenomena: fusion takes place if the direction of the two eyes corresponds exactly or if they correspond approximately.

Case 1 On the condition that the direction of the two eyes corresponds exactly. It is in this group that one finds once again the classic theories of Ptolemy and Ibn al-Haytham which were examined in Chap. 5, and of their Latin successors mentioned at the beginning of this chapter. Geometry remaining the preferred paradigm in ancient and medieval optics, the notion of correspondence was always interpreted in the strict sense; that is, the images must appear in similar positions: *fī nuqṭatayni mutashābihatay al-waḍ', in duobus punctis... consimilis positionis, in two points correspondingly situated.* Ibn al-Haytham may have ventured slightly beyond this limit by studying the admissible tolerances in the fusion of images, but his observations were not, as we have already said, sufficient to challenge the empirical definition of the fusional area. Therefore, while medieval Latin authors (such as Bacon, Witelo and Pecham) may have been familiar with these problems, they made no real contribution to the study of optics, which had to await the intuitive discovery of the circular form of the horopter by Aguilonius and its systematic definition by Vieth and Müller.[52]

Case 2 On the condition that the direction of the two eyes corresponds, but only approximately. The development of experimental physiological optics led to the adjustment of the theory of the horopteric circle without calling into question the validity of the concept of "correspondence." The first steps were taken by Charles Wheatstone in 1838;[53] with the aid of a stereoscope he showed experimentally that the unification of visual sensations took place as long as the direction of the two eyes diverged slightly, and that this slight discrepancy contributed to the perception of relief or stereopsis (Fig. 6.5).

Wheatstone did not specify the fusion intervals and it was not until the work of Panum[54] that these conditions were defined. This would result in the abandonment of the thesis of the Vieth-Müller circle and its replacement by Panum's fusional area.

---

[51]The terminology has been clarified, especially by David Stidwill and Robert Fletcher, *Normal Binocular Vision*, London, 2011, pp. 61–62.

[52]See Chap. 5.

[53]Charles Wheatstone, "Contributions to the physiology of vision.—Part the First. On some remarkable, and hitherto unobserved, phænomena of binocular vision," *Philosophical Transactions of the Royal Society of London* 128 (1838): 371–394.

[54]Peter Ludwig Panum, *Physiologische Untersuchungen über das Sehen mit zwei Augen*, Kiel, 1858.

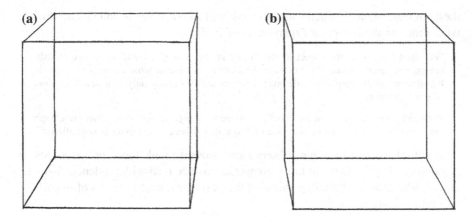

Fig. 6.5 Binocular disparity according to Charles Wheatstone, "Contributions to the physiology of vision," p. 372, Plate XI, Fig. 15

With regard to this detailed classification of past optical theories, which we have simplified in order to be able to regroup them, modern physiological optics (most of whose theories fall into the third group) distinguishes between three fundamental situations relative to the integration of visual sensations:

(1) an exact correspondence between the image points on the two retinas (the theoretical horopter) results in the simple *fusion* of quasi-images;
(2) a slight disparity (of less than 1°) in the image points on the two retinas (Panum's area) triggers *stereopsis*, i.e., the fusion of the two images with the perception of relief;
(3) a marked disparity triggers *physiological diplopia*, a complex phenomenology that includes homotopic diplopia (the superpositioning of two distinct and disparate images), retinal rivalry (disparate images are seen alternately at a rhythm of 2 to 3 cycles per second), suppression (the partial effacement of one quasi-image in favor of the other), and neutralization (the complete suppression of one of the two images).[55]

6.2.5   Neutralization Is not Constant

If we bring together all the historical texts dealing with binocular vision and compare what they have to say on the unification of visual sensations, we find that: (1) the cases of normal and pathological diplopia were regularly confused; (2) the predominant explanation regarding how quasi-images become unified was the

---

[55]Henry Saraux and Bertrand Biais, *Physiologie oculaire*, Paris, 1983, pp. 393–394.

thesis of neutralization. Such was the opinion of Albert Lejeune and Gérard Simon, who wrote on the binocular experiments of Ptolemy:

> We always focus on the object to which our attention is given and we unconsciously neutralize everything that could damage the clear perception of what we wish to see. It is true that the observation of double images is very difficult and is only achieved at the cost of much practice.[56]

> Such redoublings, which should greatly interfere with peripheral vision, generally escape our notice; we often perceive only one of two unfused images, the other is neutralized.[57]

It suffices to consider their observations together with those of specialists in physiological optics to grasp that their works must be read with prudence. Yves Le Grand, who examined the hypothesis of the permanent neutralization of one of the two retinal images, wrote:

> One may have the neutralization of one of the retinal images, which would prevent one from seeing double. This was the opinion of Porterfield (1759), who considered the perception of double images as abnormal or at the very least artificial... Yet, with a little practice, one can perceive very well these double images, as we will show shortly, and the hypothesis of constant neutralization is therefore inadmissible.[58]

This thesis, which was probably drawn from past sources—including not only Porterfield, but also Della Porta, Gassendi, Tacquet and Le Clerc—does not suffice to explain the functioning of binocular vision for two principal reasons.

Primo, the thesis of constant neutralization would negate the very real benefits associated with binocular vision. Specialists have listed at least ten positive effects, which can be classified in two groups based on the contributions of binocular vision to vision in general and to the perception of three-dimensional space in particular.[59]

Among its generic properties, it is known that binocular vision: (i) lowers the threshold for the detection of light; (ii) shortens the reaction time to visual stimuli; (iii) increases visual acuity; (iv) increases the threshold of sensitivity to contrast; and (v) increases the sensitivity to spatial contrasts.

Among the properties directly involved in the perception of space, it is known that binocular vision: (vi) facilitates the perception of relief by detecting the curvature of objects in the frontal plane; (vii) allows for the localization of an object in space by concentrating attention on the fixation point; (viii) is responsible for the spatial arrangement of objects through depth perception (Euclid was aware that two eyes could perceive a larger portion of a sphere than one eye alone);[60] (ix) allows one to infer the distance of an object from the fixation point based on the degree of

---

[56]Albert Lejeune, "Les recherches de Ptolémée," p. 79.

[57]Gérard Simon, *Le Regard, l'être et l'apparence dans l'optique de l'Antiquité*, Paris, 1988, pp. 131–132.

[58]Yves Le Grand, *Optique physiologique*, vol. 3, p. 209.

[59]These properties are expounded in all textbooks on physiological optics. See, for example, Stidwill and Fletcher, *Normal Binocular Vision*, pp. 29–30.

[60]Euclid, *Optica*, prop. 28–30; Theisen, *Liber de visu*, pp. 78–80; Kheirandish, *The Arabic Version of Euclid's Optics*, pp. 88–99.

binocular convergence; and (x) generates a mental image beginning with the quasi-images produced by each eye, the most complex phenomenon discussed in Chap. 5. And yet by making the assumption that we see through just one eye at a time, the thesis of constant neutralization denies the existence of all of these properties and in particular, by annulling binocular parallax, suppresses any possibility of the perception of relief and depth.

Secundo, if one accepts the notion that the phenomena of neutralization, suppression and retinal rivalry can be merged, "neutralization" only manifests itself in well-defined situations, as has been illustrated by experiments with black discs or vertical and horizontal grids. One may then question the validity of an extrapolation of the thesis of neutralization beyond these cases to the reduction of double images to single ones.

What was Ibn al-Haytham's position on this point? According to him, simple images are seen when they are situated exactly or approximately in the same frontal plane (the horopter), which corresponds to the case of the fusion of quasi-images. Otherwise what one has is physiological diplopia, that is to say, stereopsis or true diplopia. In the case of fusion, the form of the object is certified (*certificata*); otherwise, the two images are superimposed (*due forme erunt se penetrantes*) and the form of the object is open to doubt (*forma eius non erit certificata sed dubitabilis*). In the follow-up to this analysis, Ibn al-Haytham is led to specify the conditions for the perception of these double images. According to him, we do not notice the doubling of an image if the object is of a single color or texture. Here is the relevant text in three languages:

> Prop. 2.19: Rather, the form of every point that lies far from the point of intersection will be impressed on two points of the two eyes that correspond... And if the visible object is of one color, then the effect of doubling will hardly be noticed because of the correspondence in color and the sameness of the form. If, however, what is seen is multicolored, or if there is some design, or depiction, or [if there are] subtle features in it, then the effect of doubling will be noticeable.../Sed forma cuiuslibet puncti remoti a puncto concursus figetur in duobus punctis amborum visuum... Et si visum fuerit unius coloris, tunc illud fere nichil operabitur in ipsum propter consimilitudinem coloris et ydemptitatem forme. Si autem visum habuerit diversos colores, aut fuerit in eo lineatio, aut pictura, aut subtiles intentiones, tunc illud operatur in ipsum.../Ma la forma de ziascheduno punto rimoto dal punto del concorso se fichara in dui punti de amedui i visi... E sel viso fosse de uno colore alora quello apena operarebe in esso e quasi nulla operarebe per la similtudine <dela forma e dela identi> del colore e dela identita dela forma. Ma se el viso havesse havuto diversi colori o pictura o intentione subtile alora questo opera in esso ...[61]

The doubling of images is perceptible then, if at least one of the above conditions is met: i.e., the object must be multicolored, or present a complex outline or texture. In all other cases the doubling will pass unnoticed. Ibn al-Haytham's solution offers the advantage of not relying on the theory of constant neutralization and therefore constitutes an important milestone in the development of modern optical theory.

---

[61]Smith, *Alhacen's Theory*, vol. 1, pp. 260–261; MS. Vat. lat 4595, fol. 60vb.

6.3 Conclusion

A perusal of the literature on optics shows that the theory of Ibn al-Haytham formed the subject of learned commentary from the Middle Ages to the classical period. It can be concluded therefore that his work was generally accessible to the scientific community in Latin Europe. Physiological diplopia was not regarded at the time as an abnormal phenomenon.

Ibn al-Haytham explains the conditions under which objects will be seen as simple images in binocular vision: they must lie exactly in the same frontal plane or nearly so. Otherwise there will be a disparity, which may go unnoticed if the object is a single color or its texture or pattern are uniform. By bringing a rigorously constructed solution to bear on the problem of depth perception—the basis of the theory being the distinction between the case of direct diplopia (the doubling of images of objects located behind the fixation point) and the case of crossed diplopia (the doubling of images of objects located in front of the fixation point)—could this medieval theory of binocular vision have provided a valid foundation for the development of perspective by offering a solution to the problem of how objects located at different distances from the spectator could be depicted?

Ibn al-Haytham's theories would have been accessible to scholars through the many texts and commentaries then in circulation. Since the problems he was studying corresponded to those that artists were seeking to resolve, his work could have found direct applications in the representation of perspective.

Chapter 7
The Rejection of the Two-Point Perspective System

Abstract During the classical period many theoreticians (Vignola, Danti, Bassi, Huret and Le Clerc) published detailed critiques rejecting the perspective system based on two vanishing points located on the same horizon. A study of their texts establishes unambiguously that the system, by then judged by the main theorists to be unorthodox, was closely linked to the principles of binocular vision. The very fact that these discussions were being carried on in the academies and in circles close to them, in France as well as in Italy, shows that the postulate of monocular vision, which was a prerequisite for linear perspective, met with considerable resistance from painters and architects during the classical period. The validity of drawing a perspective using just one eye was still being debated in 1679, two and a half centuries after Brunelleschi.

There is ample documentation showing that the notion of "two-point perspective" explicitly linked to binocular vision was rejected by the principal theoreticians of perspective during the classical period. Various texts written between the middle of the sixteenth and the end of the seventeenth centuries inform us as to the debates that were being conducted between painters, architects and mathematicians who adopted opposing positions on this topic. All of them drew a link between the heterodox construction of two-point perspective and the assumption among its practitioners that the system of monocular perspective attributed to Brunelleschi and Alberti could not be used because man has two eyes rather than one. Hence the experience of binocular vision was a permanent and non-contextual condition of visual perception. This explains why one finds compositions based on the two-point system of perspective so frequently in paintings from the classical period.

© Springer International Publishing Switzerland 2016   115
D. Raynaud, *Studies on Binocular Vision*,
Archimedes 47, DOI 10.1007/978-3-319-42721-8_7

With the exception of the notes of Lorenzo Ghiberti[1] and Leonardo da Vinci[2] which were discussed above, there are very few documents that present an application of the principles of binocular vision. One finds instead a series of treatises criticizing the perspective system based on two vanishing points situated on the same horizon. These texts were written in the sixteenth century by scholars associated with the first Italian art academies (Giacomo Barozzi da Vignola, Egnatio Danti and Martino Bassi) and in the following century by theorists close to the circles of the Académie royale de Peinture de Paris, such as Grégoire Huret and Sébastien Le Clerc.[3] We will now attempt to identify the practitioners who utilized the two-point construction that was so uncompromisingly condemned by theorists.

7.1 In the Earliest Italian Art Academies

7.1.1 *Vignola and Danti* (1559–1583)

The first work testifying to this debate is Jacopo Barozzi da Vignola's *Due Regole della prospettiva pratica*, which includes a chapter entitled "*That all things terminate at a single point/Che tutte le cose vengano à terminare in un solo punto*". In it he refutes the perspective system based on two principal vanishing points:

> By the shared opinion of those who drew in perspective, it was concluded that all things visible at the sight should tend at a single point. *However some have thought that, the man having two eyes, [Perspective] should end in two points...* Who has studied the anatomy of the head can have seen that the two optic nerves join together. Similarly, the thing seen, though it enters both eyes, ends in a single point to common sense... Being unified, we see only one view. So far as I have been trained in this art, I do not think we can operate rationally by more than one point/*Che tutto le cose vengano à terminare in un solo punto. Per il commune parere di tutti coloro, che hanno disegnato di Prospettiua, hanno concluso, che tutte le cose apparenti alla vista vadiano a terminare in un sol punto: ma per tanto si sono trovati alcuni, che hanno havuto parere, che havendo l'huomo due occhi, si deve terminare in due punti... & chi ha veduto l'annotomia della testa, puo insieme haver*

---

[1]Ghiberti discusses binocular vision on two occasions, Bergdolt, *Der dritte Kommentar Lorenzo Ghibertis*, pp. 294–304, 314–368. One also finds vestigial references in other texts, for example when Ghiberti, going back to Roger Bacon, writes, "in due occhi si fanno due diuersi iudicij," p. 118.

[2]Paris, Institut de France, ms. A, fol. 1b, 70r, ms. D, fol. 5r, 9r; Windsor, *Disegni anatomici*, fol. 146r, 198r; British Museum, fol. 115a; *Libro di pittura*, §§ 115, 482, 811. See *The Notebooks of Leonardo da Vinci*, ed. Jean-Paul Richter, New York, 1972, pp. 53, 77, 129.

[3]We will not consider here a passage by Andrea Pozzo which begins "Immaginatevi dunque un'Uomo con due occhi...," because it presents a simplified hypothesis in which the two eyes could assume any position; Andrea Pozzo, *Perspectiva pictorum et architectorum*, vol. 2, Roma, 1698, Fig. 4.

Fig. 7.1 Diagram of the optic tracts, after Egnatio Danti, *Le Due Regole della prospettiva pratica*, p. 54

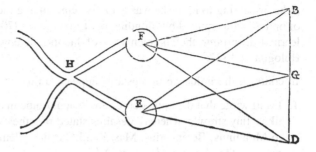

veduto, che li due nervi de gli occhi vanno ad unirsi insieme, & parimente la cosa vista, benche entri per due occhi, va a terminare in un sol punto nel senso commune… & stando la vista unita non se ne vede se non una. Ma sia come si voglia, per quanto io mi sia travagliato in tal' Arte, non so trovare che per piu d'un punto si possa con ragione operare.[4]

In his commentary on Vignola's text, the mathematician Egnatio Danti supported the position that two-point perspective should be excluded using an argument based on the physiology of vision. Since visual images fuse at the chiasma (Fig. 7.1), only one-point perspective is possible:

[The optic nerves] join together at point *H*, where the species coming to the common sense, blend together… It follows that, with two eyes, we see only one thing and that in perspective there is only one point where you draw what you see at a glance without changing position. It is not possible to work in this art *with two horizontal points placed on the same plane*, and you must operate with only one point which all major parallel lines meet/[I nervi della vista] si congiungono insieme nel punto *H*, dove le specie, che da gli spiriti visuali sono portate al senso commune, si mescolano insieme… ne segue, che con due occhi si vegga une cosa sola, & che nella Prospettiua sia un punto solo, disegnandoci ella quel che si vede in un' occhiata, senza muoversi punto: & che non sia possibile operare in quest' arte *con due punti orizontali posti nel medesimo piano*… che non si possa operare se non con un punto principale, al quale vanno tutte le linee parallele principali.[5]

This shift from a geometry-based theory of optics towards one based on the physiology of vision reflects the fact that the author's brother, Vincenzo Danti, was a practicing anatomist and Egnatio himself had carried out numerous dissections in Florence and Bologna.[6] All the same, his assertion that his writings were based on experiments and authoritative sources: *"Et questa è la descritione dell'occhio tratta da' libri dell'annotomia di Vincentio Danti… Vessallio, e altri… Valverde…"*[7] has

---

[4]*Le Due Regole della prospettiva pratica*, Roma, 1583, p. 53, French translation by P. Dubourg Glatigny, Paris, 2003, p. 222 (italics mine).

[5]*Le Due Regole della prospettiva pratica*, p. 54, French p. 225 (italics mine).

[6]Pascal Dubourg Glatigny, "La merveilleuse fabrique de l'oeil," *Roma moderna e contemporanea*, 7 (1999), p. 374.

[7]Danti, *Le Due regole*, pp. 2–3.

a rhetorical ring to it, for he was actually reproducing the Arab–Galenic conception of optics as transmitted by Mondino dei Liuzzi (ca. 1270–1326), which he probably learned of through the commentaries of Jacopo Berengario da Carpi published in Bologna in 1522.[8]

In contrast, the influence of Vesalius appears unlikely:

1. Danti wrote that the optic nerves are "vacui come una picciola cannucia [hollow like a tiny straw]," whereas Vesalius stated that they were solid and—in contrast to Mondinus, Berengarius, Massa and Curtius—denied that they crossed at the point of the chiasma (*decussatio*).[9]
2. Danti supported the thesis of intromission, which Vesalius rejected.[10]
3. According to Vesalius the crystalline lens is located in the center of the eye, whereas Danti—in agreement with Realdo Colombo—placed it immediately behind the pupil, at a point corresponding to one-fifth of the diameter of the eye.[11]
4. With regard to the fusion of images in binocular vision, Danti's argument was exactly the same as that of Ḥunayn ibn Isḥāq as transmitted by Mondinus; images are carried by the optic nerves as far as the chiasma, the point at which they "return to unity".[12]

Danti's critique poses a historical problem. The passage "*However some have had the opinion that, since the man has two eyes, the [perspective] must terminated at two points/Ma per tanto si sono trovati alcuni, che hanno havuto parere, che havendo l'huomo due occhi, si deve terminare [la prospettiva] in due punti*" indicates that two-point perspective was in fact accepted and was being used at the time. Otherwise neither Vignola nor Danti would have devoted an entire chapter in their texts on optics to a refutation of the system. Their rejection of two-point

---

[8]David C. Lindberg, *Theories of Vision from Al-Kindi to Kepler*, Chicago, 1976, pp. 33–44.

[9]Vesalius, *De humani corporis fabrica*, Basileae, 1543, pp. 324–325: "However they do not cross, but the right tends to the right eye, and the left tends to its left eye/Tamen non incruciantur, sed dexter ad dextrum oculum, sinister ad suum sinistrum tendit," Ruben Eriksson, *Andreas Vesalius' first public anatomy at Bologna*, Uppsala, 1959, p. 220.

[10]"Then, consequently, the other couples of nerves appear to be bent to the eyes, by which the visual spirit pass to the eyes/Deinde consequenter, ceteras coniugationes nervorum ostendebat tendentes ad oculos, per quos spiritus visiui ad oculos transeant," *Vesalius' first public anatomy*, p. 220, *De fabrica*, p. 324.

[11]"In his section on the eye, Vesalius discusses the point... and completely errs by considering that the crystalline humor is precisely situated in the center of the eye/Vessalij in historia de oculo nullo negotio deprehendes... et tota errat via, existimans cristallinum humorem in centro oculi exquisite situm esse," Realdo Colombo, *De re anatomica libri XV*, Venezia, 1559, p. 220.

[12]Max Meyerhof, *The Book on the Ten Treatises on the Eye ascribed to Hunain ibn Is-hâq*, Cairo, 1928; Ernest Wickersheimer, *Anatomies de Mondino dei Luzzi et de Guido da Vigevano*, Paris, 1926. The "spirito visivo" (Danti) as well as its Latin model in Mondinus derive from the Arabic *al-rūḥ al-bāṣir* (Ḥunayn).

perspective had the same authority as Alberti's critique of the *superbipartiens* method[13] *(De pictura, I, 19)* and the firmness with which they presented their arguments suggests that their condemnation was no mere rhetorical artifice. It was aimed at discrediting heterodox practices then in use that have been completely forgotten today.[14]

All that is known of this binocular perspective is that it was based on two vanishing points located on the horizon (*two horizontal points placed on the same plane/due punti orizontali posti nel medesimo piano*). The observation by Vignola *"Several have found/Si sono trovati alcuni"* refers to traditional practices that considerably pre-date the first edition of the text, which appeared in 1559.[15] Can the *alcuni* who were using this heterodox system of perspective well into the Cinquecento be identified? Danti's biography and his *Carteggio*[16] provide the names of painters with whom he was in contact during his lifetime. If one analyzes the corpus of works by these artists at least two examples of two-point perspective emerge. One is the *Holy Virgin in Glory* painted by Bartolomeo Passerotti in 1565 for the Church of San Giacomo Maggiore in Bologna (Fig. 7.2) and the other is the *Communion of Saint Jerome* by Agostino Caracci (ca. 1592).

In these two paintings one can immediately recognize Vignola's description of two points (*F* and *F'*) lying on the same horizon (*due punti orizontali posti nel medesimo piano*) and their chronology is extremely significant. *Due Regole* was published in 1583 and Passerotti's *Virgin* was produced two decades earlier, while Caracci's *Saint Jerome* appeared one decade later. Egnatio Danti dedicated his commentary on Vignola's work to Passerotti, who had collaborated with Vignola in the past on various commissions.[17] The finding of a direct link between such

---

[13]"Here some would draw a transverse line parallel to the base line of the quadrangle. The distance which is now between the two lines they would divide into three parts... and they would add the remaining lines, so that the space between the antecedent lines and the following lines would always be, as the mathematicians say, *superbipartiens*. I can say those who would do thus, even though they follow the good way of painting in other things, would err.../Hic essent nonnulli qui unam ab divisa aequidistantem lineam intra quadragulum ducerent, spatiumque, quod inter utrasque lineas adsit, in tres partes dividerent... ac deinceps reliquas lineas adderent ut semper sequens inter lineas esset spatium ad antecedens, ut verbo mathematicorum loquar, superbipartiens. Itaque sic illi quidem facerent, quos etsi optimam quandam pingendi viam sequi affirment, eosdem tamen non parum errare censeo..." Alberti, *De pictura*, ed. J.L. Schefer, 1992, p. 116.

[14]The *superbipartiens* method is well documented historically: "Some painters of the time of Alberti could have completed the method described... Finally we know that Pisanello was one of those painters/Alcuni pittori dei tempi di Alberti avrebbero completato il metodo descritto... Finalmente sappiamo che Pisanello era uno di quei pittori," Pietro Roccasecca, "Punti di vista non punto di fuga," p. 42.

[15]Christoph Thoenes and Pietro Roccasecca, "Per una storia del testo de 'Le due regole della prospettiva pratica'," R.J. Tuttle et al., eds., *Jacopo Barozzi da Vignola*, Milan, 2002, p. 367.

[16]Pascal Dubourg Glatigny, *Il Disegno naturale del mondo*, Milan, 2011.

[17]Danti, *Le Due Regole*, p. 97, French transl. p. 310.

heterodox artists and the theorists Vignola and Danti sheds a completely different
light on the meaning of Danti's dedication. Caracci had studied under Passerotti and
used the two-point construction in 1592 without taking into account Vignola's
refutation of it in *Due Regole*. These examples underline the limited impact that
theoretical rules had on well-established practices.

7.1.2 Martino Bassi (1572)

The second important comment on two-point perspective can be found in *Dispareri in materia d'architettura* by Martino Bassi, who faults the painter Pellegrino Tibaldi of Bologna with using multiple vanishing points. Although there is no record of Tibaldi's defense of his approach, it can be inferred from Bassi's critique that the painter used the argument of binocular vision. What is more surprising is that, to support his condemnation of the practice of two-point perspective, Bassi evoked traditional optical theory but in a selective and biased manner. Completely disregarding the theories of binocular vision, he claimed that optical theory in antiquity and the Middle Ages was based on the functioning of one eye at a time:

> If you think my discourse inappropriate, check what Euclid and Witelo say in their demonstrations, i.e., *ponatur oculus* [the eye is placed] and never *ponantur oculi* [the eyes are placed]; *conum esse figuram quem habet verticem in oculo* [the cone is the figure that has its vertex in the eye], and not *in oculis* [in the eyes]; *ponatur, radios ab oculo emissos in rectam lineam ferri* [the radius sent out from the eye is to be carried in straight line] and not *ab oculis* [from the eyes]..."[18]

Bassi's attack explicitly focused on a bas relief sculpted in Milan by Tibaldi between 1561 and 1572. Tibaldi does not seem to have taken any notice of this and continued to employ two vanishing points long afterwards, as for example when he was summoned to the court of Philip II of Spain. In the library at the Escorial is a fresco entitled *The School of Athens* painted by Tibaldi between 1590 and 1592[19] (Fig. 7.3).

The perspective lines marked by the receding rows of pedestals conform to the two vanishing points *F* and *F'* as defined by Vignola (*due punti orizontali posti nel medesimo piano*). Thus, even though he had completed his apprenticeship in Bologna and Rome where artists were fully conversant with the latest theories on perspective, Tibaldi seems to have preferred to adhere to the older, now heterodox tradition.

---

[18]"Ma se questo mio parlare non vi pare à proposito, notate in Euclide, & in Vitellione ciò, che essi dicono nelle loro demostrationi, cioè, *ponatur oculus*, & non mai ponantur oculi, & *conum esse figuram, quem habet verticem in oculo*, & non in oculis, & *ponatur, radios ab oculo emissos in rectam lineam ferri*, & non ab oculis..." Martino Bassi, *Dispareri in materia d'architettura, et perspettiva*, Brescia, 1572, p. 16 (italics mine).

[19]Michael Scholz-Hansel, "Las obras de Pellegrino Tibaldi en el Escorial," *Imafronte* 8/9 (1992/3), p. 391.

Fig. 7.3 Pellegrino Tibaldi, *The School of Athens (Schola Atheniensium)*, wall painting, ca. 1592 (El Escorial, Library of the Monastery of San Lorenzo), author's reconstruction after a photography of themathematicaltourist.wordpress.com

7.2 In the Circles of the Académie Royale de Peinture

7.2.1 Grégoire Huret (1670)

Some years after his admission to the Académie royale de Peinture et de Sculpture, on 7 August 1663[20] the draughtsman and engraver Grégoire Huret published *Optique de portraiture et de peinture*, the third important text on two-point perspective to come down to us. A supporter of the theory of the unconditional fusion of visual sensations (see Chap. 6), Huret carried the argument against binocular perspective even further. Here are some of the most pertinent passages from his text:

> *Secret: why architectures and other reduced topics, according to the geometrical rules of perspective, can only be properly seen by one eye, and by a pinhole placed at the point where the picture should be seen.* Since every natural subject sends its appearance to each of our eyes, and that such appearances form each a kind of cone, by the radiations of vision... it follows that, since the Picture or transparency is interposed between the subject and the viewer, it will be cut by each of the two cones, as is seen in projection, *Fig. 33*, to which the only front line *BC* sends the rays *BbO* and *CdO* to the right eye *O* of the viewer, which rays cut the transparency *YFEZ* at points *b* and *d*, moreover it sends to the left eye *A* the rays *BEA* and *CDA* that will cut the same transparency to the said points *E* and *D*. This shows that the said line *BC* sends to the two eyes, two appearances *bd* and *ED* on the Picture... And likewise the other magnitude *GH* will send its appearance to the eye *O* by means of the rays *HLO* and *GIO*, and to the eye *A* by means of the rays *HKA* and *GFA*, which also receives a double interlaced appearance *IL* and *FK*. And likewise the

---

[20]For biographical information on this artist, who also took a deep interest in the theoretical problem of perspective, see: Emmanuelle Brugerolles and David Guillet, "Grégoire Huret, dessinateur et graveur," *Revue de l'art* 117 (1997): 9–35.

Fig. 7.4 Binocular
perspective, Grégoire Huret,
*Optique de portraiture et de
peinture*, Plate V, Fig. 33

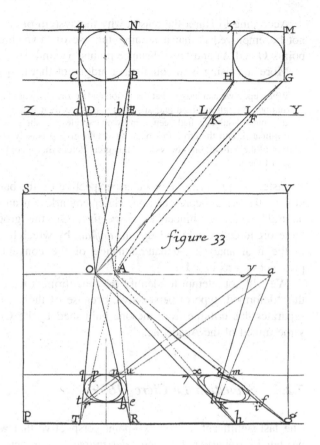

squares *CBN4* and *HGM5*, and the circles which are inscribed to them, will each send
similarly, for example on Picture *gVSP*, a double appearance... But here is the last point of
the secret why Geometry allows the Picture to be seen with one eye only: if it were seen by
both eyes together, it would have to subordinate its double cones of appearances, and thus
establish the direct two points *O* and *A*, with each his point of distance, such as *y* and *a*, as
you can see to the said perspective, *Fig. 33.*[21] (Fig. 7.4).

---

[21]*"Secret pourquoy les architectures & autres sujets reduits, suivant les regles geometriques de la
Perspective, ne peuvent estre bien vûes que par un seul oeil, & par un pertuis posé au point d'où le
Tableau doit estre veu.* Puisque chaque sujet naturel envoye son apparence en chacun de nos yeux
& que lesdites apparences forment chacunes suivant le rayonnement de la vûë côme une espece de
cone... il s'ensuit que puisque le Tableau ou transparence est interposé entre le sujet & le
regardant, il se trouvera coupé par chacun desdits deux cones, ainsi qu'il se voit au geometral
Fig. 33. auquel la seule ligne de front *BC* envoye les rayons *BbO* & *CdO* à l'oeil droit *O*, du
regardant, lesquels rayons coupent la transparence *YFEZ* aux points *b* & *d*, plus elle envoye encore
à l'oeil gauche *A*, les rayons *BEA* & *CDA* qui couperont la mesme transparence ausdits points *E* &
D, ce qui fait voir que ladite *BC* envoye pour venir aux deux yeux, deux apparences *bd* & *ED* sur
le Tableau... Et il en sera de mesme de l'autre grandeur *GH*, qui envoyera son apparence à l'oeil
O suivant les rayons *HLO* & *GIO*, & à l'oeil *A* suivant les rayons *HKA* & *GFA* qui porte aussi une
double apparence entre-lassée *IL* & *FK*. Il en sera encore de mesme des quarrez *CBN4* & *HGM5* &

According to Huret the reason why the system of two-point perspective should not be employed is that it requires the use of "two direct points" (the vanishing points O and A) and "two distance points" (y and a).

More surprising is the qualitative judgment that he passes on this perspective:

> Whereupon we can judge what confusion and deformity would affect a painting done in this manner, though artificial and absolutely accurate. But if during this wonderful event the said natural subjects had been seen by one single eye, the said appearances do not appear duplicated, and therefore this painting would give precisely the same visual sensation as that of the natural, but always with this great defect that it could be seen only with one eye at a time ...[22]

Instead of justifying monocular perspective on the basis of the conventions that govern the art of representation, Huret embarks on an evaluation of the "supernatural" nature of binocular perspective. On the grounds of symmetry he is therefore forced to acknowledge the *deffaut*, by which he means the arbitrary and— in the final analysis—unnatural nature of the convention on which monocular perspective was based.

We will not attempt to identify the practitioners who were accused of applying this 'deformed' type of perspective, because of the very short period of time that separates this critique from the one published by Le Clerc, who belonged to the same milieu of the Académie.

7.2.2 Sébastien Le Clerc (1679)

The last significant text on two-point perspective that will be discussed here was written by Sébastien Le Clerc, *dessinateur et graveur ordinaire du Roy* and a protégé of Charles Le Brun, who sponsored his admission on 16 August 1672 to the Académie royale de Peinture. Le Clerc would later be appointed professor of geometry and perspective at the Académie. His condemnation of two-point

---

(Footnote 21 continued)

des cercles qui leurs sont inscrits, qui envoyeront semblablement, par exemple sur le Tableau *gVSP*, chacun une double apparence... Or voicy le dernier point du secret, pourquoy la Geometrie ne permet la vûë du Tableau qu'à un seul oeil, qui est que s'il devoit estre vû par les deux ensemble, il faudroit qu'elle fist travailler sur la sujetion de ses doubles cônes d'apparences, & partant qu'elle établist les deux points directs *O & A*, ayant chacun son point de distance, comme *y & a*, ainsi que vous voyez audit perspectif, Fig. 33," Grégoire Huret, *Optique de portraiture et de peinture*, Paris, 1670, pp. 59–60.

[22]"Surquoy on peut juger quelle confusion & quelle difformité auroit un Tableau fait de cette maniere, quoy que surnaturelle & absolument precise, mais si lors de ce merveilleux évenement lesdits sujets naturels n'avoient esté veus que par un seul oeil lesdites apparences ne se trouveroient pas doublées, & par consequent ce Tableau donneroit précisément la mesme sensation visuelle que celle du naturel, mais toûjours avec ce grand deffaut qu'il ne pourroit estre bien veu que d'un seul oeil à la fois..." Huret, *Optique de portraiture et de peinture*, p. 60.

perspective appears in a short and somewhat eccentric tract on the relationship between perspective and mathematics:

> While I was in a company where the art of painting was the subject of conversation, and where several questions about this art were agitated, this, among others, was put forward: whether the Perspective, which is part of Painting, must have some rank in Mathematics. There were some who said that there was not doubt, and the reason they alleged was that Geometry produces its rules. Others said, however, that we should not attribute to Geometry, which is the source of all mathematical truths, rules that were based only on false principles, and therefore could only produce errors. And supporting this argument with warmth, they argued a number of things, which mainly tended to show that Painters could not *by means of a single viewpoint* reach their goal, which is to imitate nature. The Painter, they said, *considers the objects with his two eyes*, and he professes to paint as he sees them; thus *the art of Perspective admitting only one point to represent the two eyes*, its rules can not meet the intention of the Painter... A very enlightened person who was there spoke, and said that Mr. Huret, the author who most considered the secrets of Perspective, had decided this question when he proved the necessity for the Painters to admit a single point of view... and that it was a mockery to ask anyone to close one eye and deprive himself from half the view to accommodate the failure of this art... and thereupon he asked me to say what I thought of it.[23]

Le Clerc attempted to defend monocular perspective on physiological grounds, arguing as follows. There are two opposing theses: either retinal images are fused or they are not. The disparity of images has been established on experimental grounds (Fig. 7.5) and therefore visual images cannot be fused. Adopting a somewhat simplistic solution based on the thesis of suppression (in opposition to the conclusions of scholars from Ibn al-Haytham to Aguilonius), Le Clerc declares: "The images in both eyes are dissimilar, so they cannot agree together; and if they cannot

---

[23]"M'estant trouvé dans une compagnie où l'art de peindre faisoit le sujet de la conversation, & où plusieurs questions qui le concernent estoient agitées; celle-cy entre autres y fut proposée: sçavoir, si la Perspective, qui est une partie de la Peinture, devoit avoir quelque rang dans les Mathematiques. Il y en eut qui dirent qu'on n'en devoit pas douter, & la raison qu'ils alleguerent estoit que la Geometrie en produisoit les reigles. D'autres dirent au contraire, qu'on ne devoit pas attribuer à la Geometrie qui est la source de toutes les veritez Mathematiques des reigles qui n'estoient fondées que sur de faux principes, & qui par consequent ne pouvoient produire que des erreurs, & soûtenant cette proposition avec chaleur, ils avancerent plusieurs choses, qui principalement tendoient à faire voir que les Peintres ne pouvoient *par le moyen d'un seul point de veuë* arriver à leur but, qui est d'imiter la nature. Le Peintre, disoient-ils, *considere les objets de ses deux yeux*, & il fait profession de les peindre comme il les voit; ainsi *l'art de la Perspective n'admettant qu'un point pour representer les deux yeux*, ses reigles ne peuvent responde à l'intention du Peintre... Une personne, qui se trouvait là, fort éclairée prit la parole, & dit que Monsieur Huret qui estoit l'Auteur qui avoit le plus examiné les secrets de la Perspective, avoit décidé cette question lors qu'il avoit prouvé *la nécessité où estoient les Peintres de n'admettre qu'un point de veuë*... & que c'estoit une raillerie de vouloir que l'on fermast un oeil & que l'on se privast de la moitié de la veuë pour s'accommoder au defaut de cet art... & là-dessus il me pria de vouloir dire ce que j'en pensois," Sébastien Le Clerc, *Discours touchant le point de veuë, dans lequel il est prouvé que les choses qu'on voit distinctement, ne sont veuës que d'un oeil*, Paris, 1679, pp. 1–4 (italics mine).

Fig. 7.5 Binocular disparity,
Sébastien Le Clerc, *Discours
touchant le point de veuë*,
p. 47

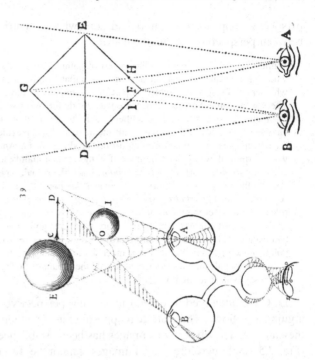

agree together, much less can they be merged... The soul necessarily perceives only
one of these two images."[24]

But there is more: the nature of the subject and the terminology used might lead
one to conclude that Le Clerc wrote his booklet in order to counter the views
expressed by Chérubin d'Orleans in *La Vision parfaite* published just two years
earlier, in 1677. Essentially dedicated to "Oculaire binocle" ('the binoculars'),
Chérubin's treatise begins with a discussion of binocular vision in general and a
radical critique of the theory of suppression. Taking as his departure point the
axiom that we see better with two eyes than with one, Chérubin poses the problem
of the fusion of quasi-images with great clarity: "The challenge ... is to see ... the
composition of the homonymous parts of the two paintings of the object formed by
the optical beams ... and how ... the union of similar parts of these paintings is
done, so that the visual power ... receives only a very simple and perfectly unique
idea of them." However, he stopped short of drawing any conclusions: "So in any
place and in any way this composition, or meeting, of similar parts of the two

---

[24]"Les images des deux yeux sont dissemblables, donc elles ne peuvent convenir ensemble; & si
elles ne peuvent convenir ensemble, à plus forte raison ne se peuvent-elles réunir... Il faut que
l'ame n'ait que la perception d'une de ces deux images," Le Clerc, *Discours touchant le point de
veuë*, pp. 50 et 53. The arguments of Le Clerc—most of which are fallacious—are included in his
later *Système de la vision fondé sur de nouveaux principes*, Paris, 1712.

paintings of the object takes place in the brain; it can only be conjectured, not firmly determined by Philosophy. As I am dealing with Optics here, I ignore it, because for me it is enough to have shown its necessity."[25] It would be peremptory to ascribe Le Clerc's reversion to the theory of suppression to shortcomings in Chérubin's theory, but one thing is certain: the theory of the fusion of quasi-images casts doubt on the unicity of the vanishing point in central perspectives, a concept that Le Clerc was determined to defend.

In any event, Le Clerc's argumentation shows that the validity of the monocular and binocular postulates was still being debated at the end of the seventeenth century and there was strong resistance to the rules of perspective in some quarters (*others ... supporting this argument with heat/d'autres ... soutenant cette proposition avec chaleur*). Whatever the validity of Le Clerc's argument, he notes that there were members of his entourage who continued to challenge the underlying principles of a system that had been established two and a half centuries earlier.

Who were these individuals, not named by Le Clerc but described by him as belonging to "*a company where the art of painting was the subject of conversation/une compagnie où l'art de peintre faisoit le sujet de la conversation*"? Knowing that the term *compagnie* was used to refer to the members of the Académie, it seems likely that this point was debated during a session of the academy sometime between 6 August 1672, when Le Clerc was admitted as an academician, and 12 August 1679, when *Discours touchant le point de veuë* was published. The author presented a copy of his work to the Académie on August 26th of the same year. By studying the *procès-verbaux* of the Académie for this period,[26] it can be determined who was present at the sessions in which Le Clerc

---

[25]"La difficulté... consiste à faire voir... le concours des parties homonymes des deux peintures de l'objet qui sont formées par les pinceaux optiques... & comment... la réunion des parties semblables de ces deux peintures se fait, en sorte que... la puissance visive n'en reçoive qu'une idée tres-simple, & parfaitement unique"; "Donc en quelque lieu, & en quelque maniére que ce concours, ou réünion des parties semblables des deux peintures de l'objet, se fasse dans le cerveau; cela ne pouvant estre que simplement conjecturé, & non pas solidement déterminé par la Philosophie, traitant icy de l'Optique, j'en fais abstraction, me suffisant d'en avoir démontré la necessité," Chérubin d'Orleans, *La Vision parfaite ou Le concours des deux axes de la vision en un seul point de l'objet*, Paris, Chez Sebastien Mabre-Cramoisy, 1677, pp. 20–21.

[26]Anatole de Montaiglon, *Procès-verbaux de l'Académie royale de Peinture et de Sculpture, 1648–1793*, tome I: *1648–1672*, Paris, 1875, p. 398 (August 6, 1672); tome II: *1673–1688*, Paris, 1878, p. 5 (April 29, 1673), pp. 12–14 (October 6, 1673), p. 15 (October 10, 1673), pp. 59–60 (November 16, 1675), p. 67 (January 25, 1676), pp. 68–70 (February 1st, 1676), pp. 74–77 (March 31, 1676), pp. 121–123 (December 22, 1677), pp. 124–125 (January 4, 1678), pp. 126–127 (January 29, 1678), pp. 140–141 (December 31, 1678), p. 153 (August 26, 1679). The discussions are not recorded in the minutes of the meetings of the Academy.

Fig. 7.6 Le Clerc, *Renouvellement d'Alliance entre la France et les Suisses*, etching, 1680, Paris, Bibliothèque nationale de France, département Estampes et photographies, Réserve FOL-QB-201 (46)

participated. Scrutiny of the perspective methods of the academicians makes it clear that Le Clerc's book was in fact directed at some of his closest colleagues.

In 1680 a *Histoire du Roy* was published (in installments 'chez Jeaurat' between 1665 and 1680), which contains an etching that was signed by Charles Le Brun, Pierre de Sève, Jean-Baptiste Nolin and Sébastien Le Clerc.[27] With so many artists involved it is somewhat problematic to determine which hand was responsible for the perspective drawing, but the fact remains that the geometry of *Renouvellement d'Alliance entre la France et les Suisses fait dans l'église de Nostre Dame de Paris par le Roy Loüis XIV, et les Ambassadeurs des XIII cantons et de leurs alliez le XVIII novembre M.DC.LXIII* was faulty, being based on the two-point system of perspective. Moreover, the vanishing points *F* and *F'* coincide with the candle rings of the two chandeliers, which is no accident (Fig. 7.6).

Thus Sébastien Le Clerc, professor of geometry and perspective at the Académie royale de Peinture, collaborated on an engraving in which it appears probable that either Charles Le Brun or Pierre de Sève applied a traditional two-point construction that fell outside the rules of correct perspective. The fact that such an

---

[27]We can read in the lower left-hand corner of the border: *Car. Le Brun inuen. Pet. Seue pinxit.*; in the lower right-hand corner of the scene: *Io. Nolin sculpsit.*; and in the lower right-hand corner of the border: *Sim Le Clerc sculps. 1680.* In another version without a border (anterior or posterior), the lower left-hand corner bears the words: *Car. Le Brun inuen.* and the lower right-hand corner: *S. Le Clerc sculps.* These inscriptions attribute the preparatory drawing to Le Brun and the engraving of the plate sometimes to Jean-Baptiste Nolin and sometimes to Sébastien Le Clerc.

anomalous work could be produced by prestigious members of the Académie at the end of the seventeenth century shows that the debate regarding binocular versus monocular vision was still current and, as Le Clerc himself stated, provided the impetus for *Discours touchant le point de veuë*.[28]

7.3   Conclusion

What is the common thread that runs through these three texts written between 1559 and 1679? Each of them attempts to justify the practice of monocular linear perspective by offering arguments against the two-point perspective, describing the latter as a heterodox system lying outside the limits of sound perspective practice. Why would this type of construction have formed the subject of such determined attacks if it was not still being used at quite a late date, when the rules of perspective were not only well known but had been taught for many years? What is perhaps most striking is its survival in the very circles where the modern rules of perspective were being applied and, on at least two occasions, in close proximity to the theorists who had expressed these condemnations—Passerotti, to whom Danti had dedicated his work *Due Regole*, and Le Brun and de Sève, who were close collaborators with Le Clerc.

The aim of the chapters that follow will be to resurrect these systems of perspective and to reveal the mechanisms on which they were based, after explaining the methodology that was used to reconstruct as closely as possible the geometric lines of these works.

---

[28]"And thereupon he asked me to say what I thought of it... But because I have undertaken to gradually meet many difficulties that have been proposed to me, I made a switch from a small project to a larger one/Et là-dessus il me pria de vouloir dire ce que j'en pensois... Mais parce qu'insensiblement je me suis engagé à respondre à beaucoup de difficultez qui m'ont esté proposées, j'ay passé d'un petit projet à un plus grand," Le Clerc, *Discours touchant le point de veuë*, pp. 4–5.

Part III
Sifting the Hypotheses

Chapter 8
The Properties of Two-Point Perspective

Abstract Thirty works painted between the end of the Duecento and the middle of the Quattrocento using two-point perspective are reconstructed here using the process of error analysis described in Appendix A. In terms of the classification of different systems of representation, these constructions do not correspond to any type of perspective recognized today. Notably, we show that they do not constitute either orthogonal projections or linear perspectives. Nor do they meet the criteria of the "bifocal perspective" that Parronchi believed could be identified in such representations. We then examine these works in relation to the principles of binocular vision laid out by Ibn al-Haytham and his Latin successors, and show that the reconstruction of their plans and elevations based on the perspective view produces coherent spaces that are in conformity with the architectural models of the period. In addition, a qualitative relationship linking the distance between the vanishing points and the position of the fixation point is brought to light.

The analysis of pre-perspectives—that is, the modes of representing three-dimensional space in use before the introduction of linear perspective—poses specific problems. The first stems from the great variety of perspective approaches that characterize paintings produced before the Cinquecento. The second arises out of the largely conjectural nature of the interpretations that have been proposed. Let us examine these two points in detail.

1. Before the principles of linear perspective were adopted in the Renaissance, painters and architects experimented with different systems of representation. Some works from the Middle Ages were composed using an oblique perspective (a form of parallel perspective that in reality has no connection with linear perspective apart from its name) or different axonometric techniques. To cite just one example, Giotto di Bondone did not hesitate to use more than one system at a time. For the frescoes in the upper church of the Basilica of St. Francis in Assisi, painted between 1296 and 1305,[1] he and his assistants

---

[1] Giuseppe Basile, *Giotto. Le storie francescane*, Milan, 1996.

© Springer International Publishing Switzerland 2016   133
D. Raynaud, *Studies on Binocular Vision*,
Archimedes 47, DOI 10.1007/978-3-319-42721-8_8

employed an oblique perspective for *Extasis*, an exploded view for *The Death of the Knight of Celano* and a two-fold exploded view for *The Recovery of the Wounded Man of Lerida*, but also a dimetric perspective for *The Founding of the Order of Saint Clare*, a central perspective for *The Apparition to Gregorius IX*, and a two-point perspective for *St. Francis Preaching Before Pope Honorius III*. This spectrum of approaches employed by the same atelier over the period of a decade illustrates the broad range of techniques that were in use before a consensus was reached and the preference for linear perspective took root. In any case, it ought to come as no surprise to find a nascent artistic movement accompanied by uncoordinated individual initiatives. In this context the history of pre-perspective techniques was anything but linear.

Therefore, to study the pre-perspective compositions of the period we must divide them into groups and focus on a homogenous 'family' of representations, even if this family may have been quite *heterodox* in relation to the emerging canons of perspective representation.

2. Another difficulty lies in the interpretation of the geometric constructions underlying these perspectives. While it may be a relatively simple matter to carry out a reconstruction ex post facto, we have almost no idea what meaning was accorded in the Middle Ages to the lines and points used to construct perspectives. Artists and architects relied on the mechanical arts and learned mainly by example while working in the ateliers of master painters, a process that remains poorly documented. We therefore know little about the theoretical notions which were applied when composing a perspective and a fortiori the exact signification that was assigned to them.

We must try as far as possible to avoid the mistake of projecting modern theories of perspective onto the art of the period.[2] Notions such as "horizon line," "orthogonals," "vanishing point," "transversals," the "reduction of intervals," and so on did not acquire a mathematical meaning until the sixteenth and seventeenth centuries with the work of Danti and Guidobaldo del Monte. It would therefore be preferable to employ neologisms. At the same time, the need to clearly describe the geometric construction underlying a work recommends that we use modern terminology. In this book we have sought to maintain a clear distinction between the different functions of the concepts referred to. As a consequence, when we employ modern

---

[2]The notion of the vanishing point not having yet been clearly established in the Duecento and Trecento, some have argued that painters in the Middle Ages did not use the vanishing point at all; see Andrés de Mesa Gisbert, "El 'fantasma' del punto de fuga en los estudios sobre la sistematización geométrica de la pintura del siglo XIV," *D'Art* 15 (1989): 29–50. This ingenious proposition, however useful in that it does not tempt one into the over-interpretation of early examples of perspective, runs into various difficulties that will be examined in Chap. 11. An analogous problem has long influenced our interpretation of the Renaissance concept of perspective; we now know that the term *costruzione legittima* used to designate Alberti's method was introduced quite recently, in the nineteenth century.

terms, it should be understood that they are devoid of any *semantic function* and are only being used to *index the figures*. This approach will hold for Chaps. 8–10.

The objective of the present chapter is: (i) to identify a corpus of works that conform to the definition of two-point perspective laid out in Chap. 7; (ii) to demonstrate that these representations do not correspond to any known type of perspective; and (iii) to show that they are the result of a qualitative application of the principles of binocular vision.

8.1   Typical Constructions

After identifying about fifty works that appeared to conform to the principle of two-point perspective, a group of them was selected for study based on four criteria. First, we eliminated those paintings whose geographic and cultural context differed from the majority of the works.[3] Second, we ruled out works that were not produced during the period in which this system of representation was most widely used.[4] Third, we excluded paintings whose architectural framework was not sufficiently detailed to allow us to determine how the perspective was constructed.[5] Finally, paintings that contained serious errors of perspective were eliminated.[6] At the end of this selection process we found ourselves left with thirty works produced in central Italy during the period 1295–1450 whose architectural framework

---

[3]Works that were excluded from our analysis based on the criterion of geography: Jan van Eyck, *Portrait of Arnolfini and his Wife* (1434); Jan van Eyck, *Dresden Triptych* (1437); Enguerrand Quarton, *Requin Altarpiece* (ca. 1450); and Jos Amman von Ravensburg, *Annunciation to the Virgin* (1451).

[4]Works that were excluded based on the historical criterion: Benozzo Gozzoli, *St. Francis' Death and Ascent into Heaven* (1452); Giovanni Bellini, *The Coronation of the Virgin* (1471–1474); Benozzo Gozzoli, *Joachim Driven from the Temple* (1491); and Leonardo da Vinci, *The Last Supper* (1495–1497). It seemed preferable to limit our corpus to a clearly defined geo-historic context, allowing us to follow the diffusion of this system of representation (something that would have been impossible if the corpus had involved a broader geographic distribution and time frame). It is known, for example, that Giotto worked on a commission for the Palazzo della Ragione of Padua some years previously to Giusto de' Menabuoi, who also utilized a two-point construction (Appendix B, Nos. 19 and 21).

[5]Works that were eliminated: Master of the Rebel Angels, *Fall of the Rebel Angels* (ca. 1340); Bartolo di Fredi, *Presentation to the Temple* (ca. 1365); Giusto de' Menabuoi, *Annunciation* (1374–1378); Giusto de' Menabuoi, *The Seated Madonna* (ca. 1380); Gentile da Fabriano, *Madonna with Child* (ca. 1420); Fra' Angelico, *Birth of Saint Nicholas* (1437); Paolo Uccello, *Scenes from the Life of Noah: The Flood* (1446); and Fra' Angelico, *Ordination of St. Lawrence by Saint Sixtus* (1447–1449).

[6]Works that were eliminated: Giotto, *The Approval of the Franciscan Rule* (1301–1302); Giotto, *Marriage at Cana* (1304–1306); Duccio di Buoninsegna, *Christ Taking Leave of the Apostles* (1308–1311); Ambrogio Lorenzetti, *Saint Nicholas* (1327–1332); Maso di Banco, *The Miracle of the Bull* (1340), Altichiero; *Saint George Baptizing the King* (1373–1379); Giusto de' Menabuoi, *Dragon Hunted* (ca. 1375); and Donatello, *The Flagellation of Christ* (1425).

allowed us to reconstruct the geometric projections of their perspective views. Since the intention in this chapter is to analyze how the perspectives were drawn, we will not enter into questions of dating or attribution, instead endorsing—for the information of the reader—the analyses and conclusions of experts (see Appendix B).

With regard to geographic provenance, we find that more than two-thirds of the thirty works came from an area extending from Padua to Assisi and down to Florence. As a comparison of their dates shows, artistic output during this period of one and a half centuries was uneven. After a highly productive period between 1295 and 1320, a fall-off in activity can be detected beginning around 1350 and probably attributable to the economic and social consequences of the Black Death (1346–1353). A second period of reduced productivity around 1450 seems to have been due to the introduction of new perspective techniques and the relative reluctance of artists to adopt the procedures laid out by Alberti in his 1435 treatise.

Our decision to concentrate on works produced between 1295 and 1450 springs from a simple observation: during this period an equivalent number of works that adhered to the rules of linear perspective did not exist. Piero della Francesca's *Flagellation* was one irreproachable example, but other works—from Masaccio's *Trinity* and Donatello's *Feast of Herod* to Ghiberti's panels depicting the biblical stories of *Isaac* and *Joseph* and Uccello's *Profanation of the Host*—present false perspectives, notably because the shortening of the parallel frontal lines resulted in a distance point that was far removed from the horizon line. Since linear perspective was not yet the dominant system, the two-point approach cannot be dismissed as a minor and unimportant chapter in the history of perspective. It was used with such regularity that it may be considered typical of the entire period from 1295 to 1450, a fact that demands an explanation.

8.2   Perspectives in Which the Spectator Is not Placed at Infinity

Let us begin by introducing a classification system for the different modes of representation that emerge from an analysis of ten key paintings (Fig. 8.1).[7]

All parallel projections, in particular the axonometric projections (isometry, dimetry, trimetry) and by extension the cavalier and military perspectives, place the spectator at an infinite distance from the painting. This type of projection therefore has an important property: two parallel object lines are represented by two parallel object images.

An examination of the works in our corpus (Appendix B, Nos. 1–30; Appendix E, Plates E.1-E.14) shows that this property is not respected. The orthogonals of the architectural framework are always drawn as sets of lines receding toward two

---

[7]Jean-Claude Ludi, *La Perspective "pas à pas". Manuel de construction graphique de l'espace et tracé des ombres*, Paris, 1989, pp. 138–139.

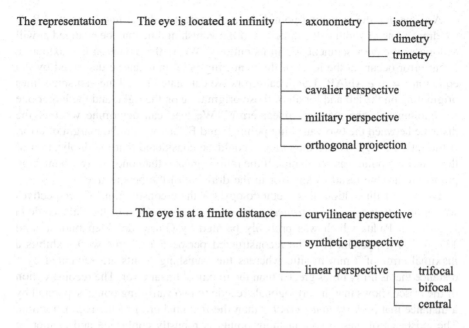

Fig. 8.1 Classification of the different systems of representation, adapted from Jean-Claude Ludi, *La Perspective pas à pas*, p. 139

vanishing points. Since the property of parallelism is not conserved, these works presuppose a finite distance between the spectator and the painting, which means that they must be placed in the second class of representations.

8.3 Perspectives that Differ from Linear Perspective

Once we accept the kinship between the two-point construction and linear perspective, we must clarify the relationship between the two-point system and the three known types of linear perspective: central, bifocal and trifocal.

8.3.1 Central Perspective

In a central linear perspective, all of the vanishing lines meet at one point located in the center of the painting. A simple hypothesis that might allow us to place our group of works within the framework of central perspective would be to suppose that the two vanishing points are the result of an *accidental* duplication of the single vanishing point. Indeed, such errors in construction can be calculated using the method described in Appendix A.

Assume that the vanishing point F is in alignment with the line segment AB. If we draw a line of width *e*, the direction of the vanishing line may be changed at will as long as it covers segment AB in its entirety. Where the deviation is maximal, a linear error occurs at the level of the vanishing point that can be described by the equation $\epsilon = \pm\ e\ AF/AB$. The linear errors are calculated for all the vanishing lines originating on the left and for all of those originating on the right, and the larger one is denominated "the maximal linear error." We then can determine whether the distance between the two vanishing points F and F' falls within the margin of error. If the ratio FF'/ϵ_{max} is less than one, it could be considered that the duplication of the vanishing point was accidental. If the ratio is greater than one, the two vanishing points are not the result of an error in the drawing of the perspective.

By way of illustration, this methodology for the reconstruction of perspectives was applied to the *Saint Enthroned*, a fresco on the north wall of the Palazzo della Ragione in Padua which was probably painted by Giusto de' Menabuoi around 1370–1380 (Appendix A). The reconstructed perspective of this work exhibits a maximal error of 7 mm in situ, whereas the vanishing points are separated by a distance that is thirty times greater than the maximal linear error. The reconstruction of the fresco shows that the orthogonals recede to two vanishing points separated by a distance that is *thirty times greater* than the maximal error of the reconstruction. The existence of these two vanishing points is robustly confirmed and cannot be contested on the grounds of errors in the perspective drawing. The two vanishing points are therefore *not* accidental, a conclusion that is supported by two other observations.

1. If one positions the two points in situ, they coincide with the centers of the rosettes sculpted on the back of the cathedra; the center point of the right rosette is perfectly visible (Appendix A, Fig. 4b). This observation casts into doubt the hypothesis advanced by Andrés de Mesa Gisbert[8] that artists in the Trecento could have constructed their receding lines based on ratios of proportionality (which would have made the use of the vanishing point unnecessary). But if this was so, why would the vanishing points coincide with points so precisely defined within the architectural framework?
2. The application of the construction based on two vanishing points was anything but sporadic. If we return to the Palazzo della Ragione in Padua and analyze other frescoes painted in the same period (ca. 1370–1380), we discover that the same schema was used no less than twelve times:

North wall (5)
Saint Justine, lower register, panel LI
Seated saint, lower register, panel LXI

---

[8]Andrés de Mesa Gisbert, "El 'fantasma' del punto de fuga en los estudios sobre la sistematización geométrica de la pintura del siglo XIV," *D'Art* 15 (1989): 29–50. This question will be discussed in Chap. 11.

Saint enthroned, lower register, panel LXII
Seated saint, lower register, panel LXXIII
Seated saint, lower register, panel LXXIV

East wall (1)
The Coronation of the Virgin, upper register, panel 293

South wall (1)
Saint Antonio, lower register, panel VIII

West wall (5)
Saint Mark the Evangelist, upper register, panel 149
Bishop and saint, lower register, panel XXXIV
Seated saint, lower register, panel XXXVI
Seated saint, lower register, panel XLI
The trial of Pietro d'Abano, lower register, panel XLIII

The use of this method sheds new light on the chronology of perspective representation. Every operator is free to adjust his reconstructed lines in such a way as to achieve a priori a given perspective and this may be the main reason why so many "correct" perspectives have been identified in Western art beginning in the Renaissance. If no error calculation is carried out, the operator risks forcing his reconstruction to fit a preconceived model.

Let us return to our group of thirty paintings based on the two-point perspective (Appendix B). If the maximal linear error for each of the works is calculated, it becomes clear that the two vanishing points cannot be interpreted as an error in the drawing of the perspective either. The distance between the vanishing points F and F' is always greater—on average 48 times greater—than the value of the maximal linear error. In none of the works does the polygon of error encompass the two vanishing points. Therefore, these works are clearly not based on a central linear perspective. Furthermore, since it is known that, far from following a codified procedure, painters in the Quattrocento[9] adopted a variety of approaches in constructing their perspectives, these observations would tend to eliminate the descriptive value of the classical division between the periods 'Middle Ages' and 'Renaissance,' at least when it comes to the question of perspective.

---

[9]Raynaud, *L'Hypothèse d'Oxford*, pp. 72–120.

8.3.2   Bifocal Perspective

Let us instead suppose that our paintings represent examples of a bifocal linear perspective. In the case of central linear perspective, the object being depicted in perspective is laid out parallel to the picture plane; in a bifocal (or angular) perspective, there is a single vertical edge of the object pointing towards the picture plane (Fig. 8.2).

A simple hypothesis that would explain two-point perspective within the framework of linear perspective consists in assuming that it derives from the geometry of the space represented. Take a city square lined by two rectilinear buildings whose facades are not parallel but *converge in the opposite direction to the observer*. It is to be expected that the right and left vanishing lines would cross and meet at two distinct points. One example is San Marco Square in Venice with its Napoleonic wings; each of the two lateral facades of the square has its own vanishing point (Fig. 8.3).

Under these conditions, could the two-point perspective that we find in so many paintings represent spaces whose sides are not parallel? Various examples can be cited that weaken this hypothesis.

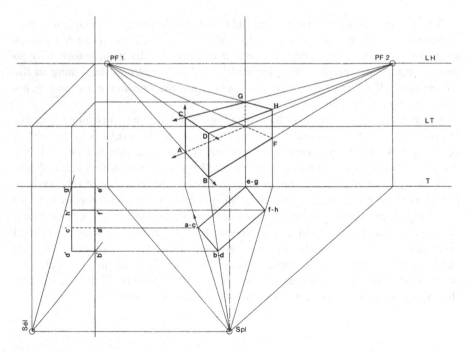

Fig. 8.2 The construction of a bifocal perspective. Jean-Claude Ludi, *La Perspective pas à pas*, p. 65

Fig. 8.3 The lines of the piazza S. Marco in Venice. Author's reconstruction after a plate by Carlo Naya, ca. 1850

Fig. 8.4 The lines of the Piazza S. Marco in Venice. Author's reconstruction after François de Nomé, *A View of Venice*, oil on canvas, ca. 1620 (private collection)

1. The two-point construction was also used by artists to depict spaces whose sides *converge toward the observer*. For example, in 1620 François de Nomé (an artist from Metz who settled in Italy, where he was known as Monsù Desiderio) painted San Marco Square as viewed across the piazza from the side along the lagoon. A map of Venice shows unambiguously that the two lateral sides of the square converge towards the sea; therefore, the lines projecting towards points F and F' in this painting should not have crossed (Fig. 8.4).

Fig. 8.5 François de Nomé, *S. Maria delle Grazie*, oil of canvas, 1619. Author's reconstruction (Courtesy of the Galerie Jean-François Heim, Basel)

2. We discover that François de Nomé used the same construction to depict conventional rectangular spaces whose opposite sides were parallel in perspective paintings such as *The Tomb of Solomon* and *S. Maria delle Grazie* (Figs. 8.5). If the same construction could be used to represent the sides of a space that are parallel, divergent or convergent, we are forced to conclude that artists regarded the technique as applicable independently of the objective structure of the space they were representing.
3. The "scenographic hypothesis," according to which parallel walls were represented spaced apart to better show the figures or objects arranged in them, poses another problem. We find a large number of works in which this space is nearly empty. There are only three figures standing in the nave of François de Nomé's *S. Maria delle Grazie*, three in the nave of his *Tomb of Solomon*, and just one in the foreground of the painting *Saint Enthroned* by Giusto de' Menabuoi (analyzed in Appendix A).

Let us now consider the possibility that the two vanishing points constitute the lateral points of a bifocal perspective. We then observe that the attempt to reconstruct the plans and elevations underlying the perspective views leads to aberrant

(a)   (b)

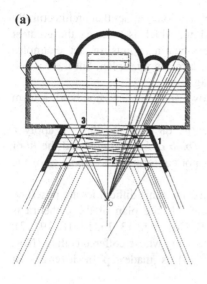
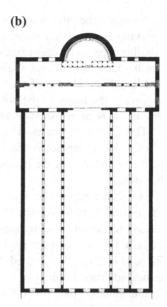

Fig. 8.6 **a** Ground plan of the church of St. Nicholas in the painting by Gentile da Fabriano, reconstructed in accordance with the principles of bifocal perspective and **b** Ground plan of the Basilica di San Paolo fuori le Mura, Roma, from *European Architecture*, eds. George H. Chase and al., Boston, University Prints, 1916, Plate G124

results.[10] If we examine the space depicted in *The Crippled and the Sick Cured at the Tomb of St. Nicholas* by Gentile da Fabriano (Appendix B, No. 27), whose architectural frame echoes the typology of a basilica such as Saint Paul Outside the Walls in Rome, we find that the plan reconstructed from the perspective view does not meet a proper basilical plan (Fig. 8.6).

Primo, since the walls running down the nave are not parallel, the span of the transverse ribs quickly becomes much longer than anything the medieval engineers would have been able to achieve. There are only two elevations that might correspond to such a perspective. In one the shape of the transverse ribs would have to gradually change along the axis of the nave, the artist eventually resorting to basket-handle arches, of which there are no examples in medieval architecture. Alternatively, the shape of the transverse ribs may remain unchanged while the ceiling increases in height as the walls recede from the choir, a solution that was equally unknown in the Middle Ages.

---

[10]Under certain conditions one can apply the rules of perspective in reverse to reconstruct the plans and elevations of a space beginning with the perspective view. This method was used to study the three-dimensional space in certain works by Leonardo da Vinci; see Giovanni degl'Innocenti, "Restitutions perspectives: hypothèses et vérifications méthodologiques," in Carlo Pedretti, *Léonard de Vinci architecte*, Milan/ Paris, 1983, pp. 274–289.

Secundo, the groin vaults are wider than they are long, a fact that architecturally speaking runs counter to the purpose of a ceiling, which is to cover the greatest possible area using the smallest number of stress-bearing points. In fact, examples can be cited of medieval vaults constructed on a rectangular plan, but their width and length are always of the same order of magnitude.

Tertio, the bases of the columns running down the nave are rhomboid, a form that is not seen in medieval architecture.

These inconsistencies demonstrate that Gentile da Fabriano did not apply a bifocal, linear perspective in *The Infirm at the Tomb of St. Nicholas* and the floor plan associated with the perspective view does not conform to the typology of the basilica.

In the majority of the works under examination the same difficulties are posed by the bifocal hypothesis, the most frequent being: (1) the plan of the cathedra is always rectangular, not trapezoidal (Appendix B, Nos. 5, 6, 13, 15, 17, 18, 19, 22); and (2) we know of no example of a medieval building whose coffered ceiling (Nos. 7, 9, 12, 23, 29) or tiled floor (Nos. 10, 14) is made up of lozenge- or trapezoid-shaped units.

Our findings therefore do not support Alessandro Parronchi's thesis that there is a connection between the bifocal and the binocular perspective constructions. In the wake of Panofsky's research on the various systems of representation devised by artists, Parronchi studied solutions that did not coincide with the theory of perspective invented by Brunelleschi. He mistakenly saw in the sinopias of Paolo Uccello's *Nativity* "an attempt to find a solution to the problem of binocular vision" (Fig. 8.7).[11]

This preparatory sketch has three vanishing points on the same horizon line: a "principal vanishing point" in the center of the composition towards which the vanishing lines of the tiles in the pavement converge, and two lateral "distance points" towards which the respective sets of diagonals meet. We can identify in this drawing the superimposition of a central and a bifocal perspective that does not in any way infringe on the rules of a *monocular* linear perspective.

Parronchi resorted to some less than convincing arguments to support his thesis that "the two distance points are based on a model of binocular vision"[12] and we present the principal ones here. *Primo*, he believed that according to medieval optics, as argued notably by Roger Bacon, the visual axis was a kind of diaphragm separating the visual field into two hemifields,[13] an interpretation that renders null and void the experiments on binocular vision conducted by Ibn al-Haytham and his Latin successors. *Secundo*, Parronchi believed that the distance point perspective

---

[11]Alessandro Parronchi, *Studi sulla dolce prospettiva*, Milan, 1964, pp. 326–327.

[12]. Parronchi, "Prospettiva e pittura in Leon Battista Alberti," *Convegno internazionale indetto nel V Centenario di Leon Battista Alberti*, Rome, 1974, p. 215.

[13]Parronchi, "Prospettiva e pittura," p. 215.

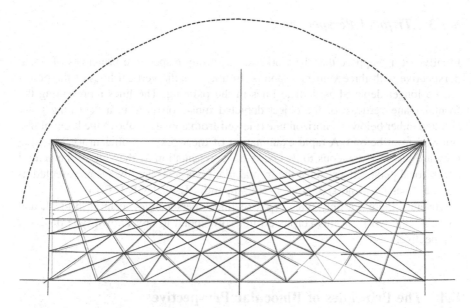

Fig. 8.7 Paolo Uccello, sinopia for the *Nativity*, 140 × 215 cm, ca. 1436–1437 (Florence, Galleria degli Uffizi, formerly in the cloister of San Martino alla Scala). Author's drawing after Maria Clelia Galassi, *Il disegno svelato*, Nuoro, 1998, p. 24, Fig. 6 (*thick lines* show the incisions; *thin lines* show the geometrically true construction)

stemmed from the merging of the figure as seen in binocular vision with its "reflection in a plane mirror,"[14] whereas medieval treatises on optics treated these two problems separately. *Tertio*, according to Parronchi binocular vision could be summed up as the ability of the eyes "to look in two diverging directions,"[15] whereas in reality this is impossible because in normal vision the direction of the two eyes is coupled together, a fact that was already known in antiquity (Pseudo-Aristotle, *Problemata* XXXI, 7). *Quarto*, Parronchi retained that the "horizontal movement of the eyes plays an integral role in binocular vision,"[16] whereas the mobility of the eyes exists independently of binocularity and diplopia occurs in the absence of any movement of the eyes.

These points of confusion prevent us from accepting Parronchi's hypothesis that there is a link between bifocal perspective and binocular vision (he cites the *Nativity* by Uccello as an example). There is absolutely no connection between binocular vision and the two distance points of a bifocal perspective. The conditions of binocular vision produce diplopia, that is to say, two non-corresponding retinal images and consequently two "principal vanishing points," which have no relationship whatsoever with two "distance points."

---

[14]Ibid., p. 215.
[15]Ibid., p. 216.
[16]Ibid., p. 214.

8.3.3 Trifocal Perspective

Finally, let us suppose that the works in our group respect the principles of linear perspective with three vanishing points. In this case the vertical lines of the object are no longer depicted as vertical lines in the painting. The lines representing the frontal plane verticals of the object depicted must converge in a vanishing point situated either below the horizon line (viewed from above) or above the horizon line (viewed from below). A rapid examination of the works is sufficient to show that they are just as extraneous to this class of representation as they are to the central and bifocal perspectives, since the frontal plane verticals are always represented as vertical lines (Appendix B, Nos. 1–30; Appendix E, Plates E.1–E.14).

Because there are only three types of linear perspective—central, bifocal, and trifocal—the works in this corpus do not correspond to any known type of linear perspective.

8.4   The Principles of Binocular Perspective

The use of two-point perspective in our group of paintings therefore still awaits explanation. Here I will lay out the hypothesis that they stem directly from propositions to be found in the optical treatises of the period under consideration.

8.4.1 The Vantage Point and the Vanishing Point

Contrary to the texts of the Middle Ages, which generally reasoned within the framework of binocular vision and only rarely mentioned experiments involving just one eye,[17] Renaissance studies on perspective systematically adopted the postulate of monocular vision.

In his account of the perspective experiment that Brunelleschi was reputed to have carried out in Florence, Antonio di Tuccio Manetti observed: "It is necessary that the painter postulate beforehand a single point from which his painting must be viewed."[18] Brunelleschi's panel has been lost, but the fact that he pierced a single hole through which the observer was supposed to gaze with one eye indicates that his experiment was based on the assumption of monocular vision. The same

---

[17]Only a few counter-examples can be cited. Pecham used the formula "If a one-eyed man looks/Si monoculus aspiciat" once when describing the size of objects as a function of distance, *Perspectiva communis*, I, 74, ed. Lindberg, *John Pecham and the Science of Optics*, pp. 146–147.

[18]"El dipintore bisognia che presuponga uno luogo solo donde sa a uedere la sua dipintura...," Antonio di Tuccio Manetti, *Vita di Filippo di Ser Brunelleschi*, fol. 207v.

principle would be taken up by Alberti and Piero della Francesca. As the latter wrote regarding the construction of a quadrilateral, "So let there be constructed in proper form a square surface which is BCDE, then mark the point A which will be the eye."[19] A linear perspective is therefore always constructed using one eye, and should be viewed with one eye.

Each time that we encountered an example of two-point perspective we took the liberty of calling the points F and F′ either "points of concurrence" or "vanishing points." The second term is inappropriate because the concept itself did not exist before the Cinquecento, as can be seen from Alberti's treatise: "Then I establish a point in the rectangle [i.e. the picture] wherever I wish; and as it occupies the place where the centric ray strikes, and I shall call it the centric point."[20] What is this centric ray? Alberti describes it as follows: "Among these visual rays there is one which is called the centric ray... because it meets the [visible] surface in such a way that it makes equal angles on all sides."[21] The centric ray corresponds to the optical axis when the visible surface lies in the frontal plane. According to Alberti this ray is *"the most keen and vigorous/acerrimum et vivacissimum/galliardissimo et vivacissimo,"*[22] a quality that many authors, from al-Kindī to Bacon, have ascribed to the axis of the visual pyramid.[23] The point of concurrence of a set of orthogonals is not conceived as a vanishing point in the modern sense of the term. In medieval and Renaissance optics the "central vanishing point" corresponds rather to the point of intersection between the axis of the visual pyramid and the picture plane.

8.4.2 Depth Gives Rise to Disparate Images

The treatises on optics of Bacon, Witelo and Pecham closely follow Ibn Al-Haytham's *De aspectibus* (III, 2), presenting the same commentary and figure with few variations.

---

[19]"Adunqua facise in propia forma una superficie quadrata la quale sia ·BCDE· poi se punga il puncto ·A· il quale sia l'occhio," Piero della Francesca, *De prospectiva pingendi*, III, 1, ed. Nicco-Fasola, p. 130.

[20]"Post haec unicum punctum quo sit visum loco intra quadrangulum constituo, qui mihi punctus cum locum occupet ipsum ad quem radius centricus applicetur, idcirco centricus punctus dicatur/ Poi dentro questo quadrangolo, fermo uno punto il quale occupi quello luogo doue il razzo centrico ferisce et per questo il chiamo punto centrico," Alberti, *De Pictura*, I, 19; *Della pittura*, I, fol. 124v. The translation is that of Cecil Grayson in Leon Battista Alberti, *On Painting*, London, Penguin Books, 1991, p. 55.

[21]"Est quoque ex radiis mediis quidam... dicatur centricus, quod in superficie ita perstet ut circa se aequales utrinque angulos reddat/Ecci fra i razzi uisiui uno detto centrico. Questo, quando giugnie alla superficie, fa di qua et di qua torno ad sé gli angoli retti et equali," Alberti, *De Pictura*, I, 5; *Della pittura*, I, fol. 121v; *On Painting*, p. 40.

[22]Alberti, *De Pictura*, I, 8; *Della pittura*, I, fol. 122r; *On Painting*, p. 43.

[23]Bacon, *Opus maius*, IV, III, 3, ed. Bridges, p. 125–127.

Fig. 8.8 Witelo, *Optica*, III, 37, ed. Frederic Risner, *Opticae Thesaurus*, Basel, 1572, p. 102

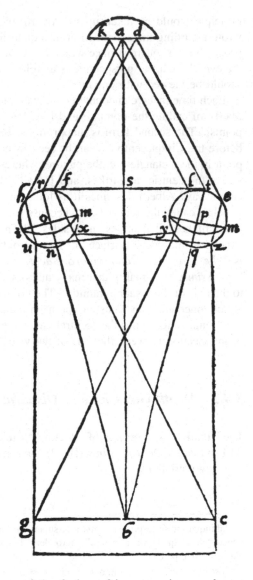

Witelo (Fig. 8.8) studied the process of the fusion of images using a schema drawn from *De aspectibus*, III, 2.15. As he wrote: "The two forms, which enter two corresponding points on the surface of the two eyes, arrive at the same point of concavity of the common nerve, and are superimposed at that point to be reduced to one form."[24] The re-composition of the quasi-images produced by the two eyes

[24]"Due forme que infinguntur in duobus punctis consimilis positionis apud superficies duorum uisuum perueniunt ad eundem punctum concauitatis nerui communis et superponuntur sibi in illo puncto et erunt una forma," Witelo, *Optica*, III, 37, ed. Risner, p. 102–103.

occurs at the chiasma, the crossing point of the optic nerves. Fusion therefore takes place internally and is a product of the nervous system.

One finds a similar treatment of these topics by Bacon and Pecham. Let us take, for example, Pecham's *Perspectiva communis*, prop. I, 32: "The duality of the eyes must be reduced to unity"[25] and prop. I, 80: "An object appears double when it has a sensibly different position relative to the two [visual] axes."[26] His division of the two propositions reflects the distinction that he made between the mental image, which results from the combination of sensations at the point of the chiasma (prop. I, 32), and the sensory stimuli of images received by the two eyes (prop. I, 80). Like his predecessors, Pecham distinguished between external quasi-images and the internal image resulting from their fusion, and this had immediate consequences in the area of visual representation. A scene viewed in perspective is an *external* image and is therefore subject to the principles of binocular vision and the composition of quasi-images. It remains to determine in what form.

8.4.3   The Determination of the Fixation Point

Let us return to Ibn al-Haytham's experimental ruler—a strip of wood one cubit long and four fingerbreadths wide, with a slight depression (*MN*) at the observer's end to serve as a nose rest, and the axes *HZ* and *KT* and the diameters *AD* and *BC* painted in different colors.[27] In his third experiment, which focused on diplopia, al-Haytham

---

[25]"Oculorum dualitatem necesse est reduci ad unitatem," Lindberg, *John Pecham and the Science of Optics*, p. 116.

[26]"Ex variato sensibiliter situ visibilis respectu duorum axium ipsum duo apparere," Lindberg, *John Pecham and the Science of Optics*, p. 150.

[27]"Take a smooth wooden plaque that is one cubit long and four digits wide, and let it be perfectly flat, even, and smooth. Let the edges along its length, as well as those along its width, be parallel, and let there be two diagonals intersecting one another at a point through which a straight line is drawn parallel to the edges along the length. Then, through that [same] intersection-point let a straight line be drawn perpendicular to the first line, passing through [the plaque's] middle, and let [each of] these [two perpendicular] lines be painted a different color, both colors being bright so that they are readily visible, but let the two diagonals be painted the same color. Then, in the middle of the bottom edge of the plaque, between the [endpoints of the] two diagonals, let a rounded notch be cut, but one that narrows inward so that, when the plaque is brought up to it, the bridge of the nose can fit into it in such a way that the two corners of the plaque almost touch, but do not actually touch, the two midpoints of the surfaces of the two eyes/Accipiatur tabula lenis ligni, cuius longitudo sit unius cubiti et cuius latitudo sit quattuor digitorum, et sit bene plana et equalis, et lenis. Et sint fines sue longitudinis equidistantes, et sue latitudines equidistantes, et sint in ipsa duo dyametri se secantes a quorum loco sectionis extrahatur linea recta equidistans duobus finibus longitudinis. Et extrahatur a loco sectionis etiam linea recta perpendicularis super lineam primam positam in medio, et intingantur iste linee tincturis lucidis diuersorum colorum ut bene appareant, sed tamen duo dyametri sint unius coloris. Et concauetur in medio latitudinis tabule apud extremum linee recte posite in medio, et inter duos dyametros, concauitate rotunda, et cum hoc quasi piramidali tantum quantum poterit intrare cornu nasi quando tabula superponetur illi quousque tangent duo anguli tabule fere duo media superficierum duorum uisuum, tamen non

placed three wax columns at points *L*, *Q* and *S*. When the eyes (*A* and *B*) focus on the wax column *Q*, the other two columns *L* and *S* are seen as double images. It follows that "the object on which the two visual axes intersect always appears single... and the object seen by rays differently located [with respect to the visual axes] appears double."[28] The double image of *L* seen on the nearer side of the fixation point was the result of *heteronymous (crossed) diplopia*, while the double image of *S* beyond the fixation point was due to *homonymous (direct) diplopia*[29] in modern terminology.

Given the fact that physiological diplopia operates under normal viewing conditions at moderate distances, Ibn al-Haytham's experiments can be applied to the problem of representing what one sees. But what does one see? The scenes that correspond to the case of two-point perspective are always composed of a central figure placed in the *foreground* and an architectural frame, most of which appears in the *background*. This spatial organization favors double vision, which is a regular occurrence at distances of between 0 and 30 m. Under these conditions the two eyes must select the object on which they will focus. Since the fixation point is generally a conspicuous figure standing in the foreground, the architectural frame produces two disparate images, in conformity with the case of homonymous diplopia. Because the frame is made up of rectilinear edges, these lines are doubled and meet at two lateral vanishing points. This can be shown through a simple experiment: if one holds up a finger while gazing down a straight section of track, the rails will appear to converge on two distinct vanishing points.[30]

(Footnote 27 continued)

tangent," *De aspectibus*, III, 2.26, A. Mark Smith, *Alhacen's Theory of Visual Perception*, Philadelphia, 2001, pp. 263–264. Roger Bacon ascribed scant importance to the length of the ruler, which might measure from four to six palms in length (about 30–46 cm): "An experimenter can prove this by taking a board of a palm's width, four or five or six palms in length, and with a smooth surface/Nam hoc potest experimentator probare, accipiendo unum asserem latitudinis unius palme et longitudinis quatuor, uel quinque, uel sex, et sit superficies eius leuis," *Perspectiva* II, II, 2, ed. Bridges, p. 95, ed. Lindberg, p. 186.

[28]"Ex hac igitur experimentatione et expositione declaratur bene quod uisum in quo currunt duo axes semper uidetur unum... et quod uisum quod comprehenditur per radios diuerse positionis in parte uidetur duo...," *De aspectibus*, III, 2.48, ed. Smith, *Alhacen's Theory of Visual Perception*, pp. 269–270. Bacon's exposition was more succinct: "For if the axes of the eyes A and B are fixed with an attentive gaze on part O of visible object MON, visible point K inside the intersection of the axes and visible point H outside this intersection will both necessarily appear double/Nam si oculorum A et B axes figantur diligenti intuitione in O partem uisibilis MON, tunc K uisibile infra concursum axium apparebit duo, et H uisibile ultra concursum similiter uidebitur duo necessario," *Perspectiva*, II, II, 2, ed. Bridges, p. 95; ed. Lindberg, pp. 184–186.

[29]Henry Saraux and Bertrand Biais, *Physiologie oculaire*, Paris, 1983, p. 391.

[30]According to Ibn al-Haytham, if when looking at an object the two eyes are focused in exactly or approximately the same direction, a single image is seen; otherwise it is double, Sabra, *Optics*, vol. 1, p. 239. These results are striking in their clarity and in their consistency with the conceptions of modern physiological optics, which state that the two images are amalgamated only when the object lies in Panum's fusional area, that is to say, if the retinal points correspond precisely or approximately (creating a more or less focused image). Diplopia takes place when this condition is not met.

Like the Arab texts on which they were based, Latin treatises drew the reader's attention to the arrangement of quasi-images and how certain of their properties could be studied quite easily through direct observation. Roger Bacon's *Perspectiva* provides an example in this regard. After reiterating the definitions of homonymous and crossed diplopia[31] and presenting the results of his experiments using Ibn al-Haytham's ruler (the phenomenon of diplopia, the reduction of the horopter to the frontal plane, the tolerance of fusion), Roger Bacon then passes without any transition from experimentation to observation, because: "Even without such a board, an experimenter can test many things relevant to these matters. For at night he can place his finger between himself and a candle; if he then fixes the axes of his eyes on the candle, that finger will appear double."[32] This is an example of crossed diplopia.

An observation corresponding to homonymous diplopia is described shortly afterwards: when one focuses on distant objects such as the stars, which lie beyond the fixation point, they appear to be doubled.[33] One discovers here an idea that has a direct bearing on the problem of perspective. Suppose an artist wishes to represent a scene based on the principles of binocular vision. He cannot simultaneously fix his eyes on the figures in the foreground and the architectural framework in the background. It is natural that his gaze will focus on the most important element in the scene, which will usually be a figure standing in the foreground (for example, Saint Francis in Giotto's *Approval of the Rule*). This produces a case of homonymous diplopia. The doubling will involve all of the objects situated beyond the fixation point, including the architectural frame that provides the orthogonals. Being doubled, these will necessarily meet at two vanishing points.

8.4.4 The Crossing of the Orthogonals on the Axis Communis

The fact that in the perspectives under examination here the vanishing lines originating on the right side converge at one point on the left and those originating on the left side meet at one point on the right can be understood by referring once again to Bacon's *Opus majus*. Accompanying the passage cited above is a figure illustrating the phenomenon of crossed diplopia, where *M* designates the finger of the experimenter, *A* the candle, and *dexter* and *sinister* the positions of the two eyes (Fig. 8.9).

---

[31]Bacon, *Opus maius*, V, II, II, 2, ed. Bridges, p. 95; ed. Lindberg, p. 184.

[32]"Et experimentator potest sine tabula experiri multa in hac parte. Nam potest de nocte eleuare digitum inter ipsum et candelam; si igitur figat axes super candelam, uidebitur unus digitus duo," Bacon, *Opus maius*, V, II, II, 3, ed. Bridges, p. 96; ed. Lindberg, p. 188.

[33]Bacon, *Opus maius*, V, II, II, 3, ed. Bridges, p. 97; ed. Lindberg, p. 188.

Fig. 8.9 Homonymous and
crossed diplopia, after Roger
Bacon, *Perspectiva*, II, II, 3,
ed. John H. Bridges, *The
'Opus majus' of Roger Bacon*,
London, 1897, II, p. 97

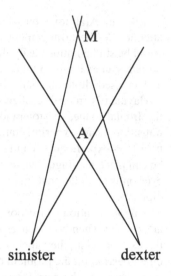

sinister   dexter

Thus it is certain that when the axes of the eyes are fixed on visible object A, the visual power nevertheless extends to visible object M, but the species coming from the right proceeds to the left if it passes beyond [this point], and by the same token the species of the left eye proceeds to the right. For these species intersect at point M, and [afterwards] they separate so that the right-hand species crosses to the left side and the left-hand species to the right side, as is evident to sense. Now since M appears double, the image corresponding to the right eye extends to that side.[34]

In homonymous diplopia the outcome is the reverse. If the eye fixes on the finger M, a double image of the candle A is produced.[35] The quasi-image produced by the left eye A_L appears to the left of M, while the quasi-image produced by the right eye A_R appears to the right. The principal difference between objects in the background as seen in these experiments and as depicted in perspective is that the candle flame can be regarded as a single point whereas the architectural framework extends backwards in space. Consequently it is not a limited image that is doubled, but a

---

[34]"Certum igitur est, quod axibus oculi fixis super A uisibile, nichilominus tamen uirtus uisiua tendit in M uisibile, sed species ueniens ab oculo dextro tendit in sinistrum si procedatur ultra, et similiter species oculi sinistri tendit ad dextrum, nam hee species intersecant se in M puncto, et separantur ita ut dextra transeat ad sinistram partem et sinistra ad dextram, ut patet ad sensum," Bacon, *Opus maius*, V, II, II, 3, ed. Bridges, p. 97; ed. Lindberg, pp. 188–190.

[35]"Yet the right-hand image does not always disappear when the left eye is closed, nor the left image with the closing of the right eye; rather it can easily happen that when the right eye is closed the right-hand image disappears, and when the left eye is closed the left-hand image disappears/Et tamen non semper disparet ymago dextra oculo sinistro clauso, nec sinistra ad clausionem dextri oculi. Sed bene accidit quod clauso dextro oculo ymago dextra dispareat, et clauso sinistro ymago sinistra disparebit," Bacon, *Opus maius*, V, II, II, 3, ed. Bridges, p. 97; ed. Lindberg, p. 190.

large area of the visual field. Following on from this is the question of determining which of the two images imposes itself in each of the visual hemispheres.

Unlike modern optics, which offers no all-encompassing hypothesis, medieval optics favored a form of cross-neutralization, because it was assumed that the rays in a visual pyramid are not of equal power. The argument, drawn from Ptolemy, was taken up by Alkindi and Alhacen, and then appeared in the Latin texts of Bacon, Witelo, Pecham, Henry of Langenstein, and finally Alberti:

Ptolemy: "The objects that lie directly in front of, and at right angles to the rays are seen more clearly than those that do not. For everything that falls orthogonally strikes its subjects more intensely than whatever falls obliquely."[36]

Alkindi: "A more powerful vision makes a complete transformation. It therefore produces a perfect ray, that is to say, strong... As a result, an object on which falls a stronger ray will be seen more clearly... Thus I say that the strongest ray falls on the center of vision."[37]

Alhacen: "Furthermore, the effect of light arriving along perpendiculars is stronger than the effect of light arriving along oblique lines... From this experiment it will therefore be clear that vision [taking place] through the center of the eye, along the [visual] axis as defined by us, is clearer than vision at the edge of the eye, along the lines surrounding the [visual] axis. It has therefore been shown that vision [taking place] along the axis of the visual cone will be clearer than the vision [taking place] along the other radial lines..."[38]

Bacon: "And Jacob [Alkindi] finds the cause of this beyond what has already been said... And the reason for this is that the strength of the pyramid derives [especially] from this ray, for it is shorter that all others... Thus it will have more of the virtue produced by the depth of the agent, and therefore the axis of the pyramid is the product of greater virtue than any other line"/"And therefore it is perpendicular to the body, and is the axis of the pyramid, and therefore is stronger and has more force."[39]

---

[36]"[...] Ea quorum situs directus est super radios ad angulos rectos, uidentur magis quam que non ita se habent. Omnia enim quorum casus fit secundum perpendiculares lineas, habent incubitum super subiecta magis quam ea quorum casus fit obliquus," Ptolemy, *Optica* II, 19, ed. Lejeune, pp. 19–20, ed Smith, pp. 76–77.

[37]"Potentior ergo uisus conuersionem efficit perfectam. Ipse igitur efficit radium perfectum, scilicet fortem... Corpus ergo, super quod cadit radius fortior, comprehenditur manifestius... Dico ergo quod super centrum uisus cadit radius fortior," al-Kindī, *De aspectibus* 12, ed. Rashed, p. 471.

[38]"Et operatio lucis uenientis super perpendiculares est fortior operatione lucis uenientis super lineas inclinatas... Manifestabitur ergo ex hac experimentatione quod uisio per medium uisus et per axem quem distinximus est manifestior uisione per extremitates uisus et per lineas continentes axem. Declaratum est ergo quod uisio per axem piramidis radialis manifestior quam uisio per omnes lineas radiales," *De aspectibus* I, 6.24 et II, 2.30, ed. Smith, *Alhacen's Theory of Visual Perception*, pp. 33 and 97.

[39]This property is stated in two different places: "Et huius causam preter dicta dat Iacobus [Alkindi]... Et causa huius est quia fortitudo pyramidis est ab hoc radio, nam hic radius est brevior omnibus... quare axis pyramidis a maiori uirtute causabitur quam alie linee," *De multiplication specierum* II, 9, ed. Lindberg, *Roger Bacon's Philosophy of Nature*, pp. 164–166; "Et ideo est perpendicularis super corpus, et axis pyramidis, et ideo est fortior, et plus habet de virtute," Bacon, *Opus maius*, IV, III, 3, ed. Bridges, p. 127.

Witelo: "Distinct vision occurs only along perpendicular lines from a point of the thing seen to the surface of the eye... It is thus clear that vision is made only by perpendicular lines."[40]

Pecham: "Therefore only that perpendicular called the axis, which is not refracted, manifests the object efficaciously, and other rays are correspondingly stronger and better able to manifest [the object] as they are closer to the axis."[41]

Langenstein: "The question is whether the perpendicular ray is the strongest."[42]

Alberti: "The central ray is that single one which alone strikes the quantity directly, and about which every angle is equal. This ray [is] the most active and the strongest of all the rays..."[43]

Thus, according to Roger Bacon the centric ray of the visual pyramid "... *is stronger, and has more force/et ideo fortior, et plus habet de uirtute*" than all the others, which is to say that those objects located along the axis of the visual pyramid are seen more distinctly and have greater visual force. The further the object lies from the axis, the less visible it will be.[44] In binocular vision the axes of the two pyramids pass through the fixation point *M*. Beyond the *axis communis*, the ray *dexter-M* extends to the left hemifield, which is seen more powerfully by the right eye, and the ray *sinister-M* extends to the right hemifield, which is seen more powerfully by the left eye. Medieval optics posited a form of cross-neutralization of the images from the right and left hemifields. These images being of the architectural frame, the edges of the frame will give rise to rectilinear vanishing lines. Those that appear in the right hemifield will have their vanishing point on the left, while those that appear in the left hemifield will have their vanishing point on the right. Herein lies the explanation for the crossing of the vanishing lines on the *axis communis*.

To summarize, in homonymous diplopia the orthogonals projecting from the right cross over to the left and the orthogonals projecting from the left cross over to

---

[40]"Visio distincta fit solum secundum perpendiculares lineas a punctis rei uise ad oculi superficiem productas... Patet quod secundum solas perpendiculares lineas fit uisio," Witelo, *Optica* III,17, ed. Risner, p. 92.

[41]"Unde sola perpendicularis illa que axis dicitur, que non frangitur, rem efficaciter representat, et alii etiam radii quo ei sunt propinquiores eo fortiores et potentiores in representando," Pecham, *Perspectiva communis* I, 38, ed. Lindberg, p. 120.

[42]"Queritur utrum radius perpendicularis sit fortissimus," Henri de Langenstein, *Questiones super perspectivam*, I, q. 6a (ca. 1397), Valencia, 1503.

[43]"Centricum radium dicimus eum qui solus ita quantitatem feriat ut utrinque anguli angulis sibi cohaerentibus respondeant. Equidem et quod ad hunc centricum radium attinet uerissimum est hunc esse omnium radiorum acerrimum et uiuacissimum," Alberti, *De pictura* I,8, ed. Grayson, p. 90.

[44]Modern physiological optics arrived at a similar conclusion by other paths, arguing that the quality of central vision is due to the high concentration in the fovea of the cones that are responsible for visual acuity. The fact remains that in the case of pathological diplopia (e.g., strabismus) the problem created by the disparity in images generally leads to the suppression of the central zone of the image perceived by the skewed eye (neutralization scotoma), whereas the uniform perception of the visual field would in theory lead to the neutralization of the entire image in the skewed eye.

the right. The "binocular perspective" defined by this rule corresponds exactly to the perspective drawing in the works under consideration here.

8.4.5   Examination of the Plans and Elevations

The painters of the Duecento and Trecento are not likely to have applied exact methods of projection to represent three-dimensional space. It is nevertheless useful to retrace the vertical and horizontal projections that correspond to the perspective views in order to test the coherence of the spaces represented. Such reconstructions are generally based on the postulate of monocular vision, but can be extended to the case of binocular vision. The procedure adopted here is the inverse of the usual one that consists in constructing a perspective beginning with the plans and elevations. The only difficulty lies in determining the dimensions of the frontal lines in relation to those of the orthogonal lines, which requires that one identify an object whose width to depth ratio is known (e.g., the homothetic figures in the foreground or background, the tiles on the floor, the sunken panels in a coffered ceiling, or some other standard architectural element). Once the ratio has been established, the position of the eyes O and O' can be determined. In this way the perspective view is associated with its corresponding plan and elevations (Appendix E, Plate E.1). One then calculates the scale of the geometric projection as a function of the position and size of a figure, and measures the distance between the eyes O and O'.

The application of this method to our set of paintings led to two main conclusions:

1. The distance between the eyes is always overestimated, varying between 12 and 440 cm, with an average of 128 cm (Appendix D, Nos. 21 and 30). This observation suggests that the painters of the Duecento and Trecento sketched their perspective drawings directly without relying on preparatory plans and elevations.[45]
2. Reconstructing the plans and elevations based on binocular vision produces coherent spaces that are consonant with the architectural models of the period (Appendix E). This shows that the hypothesis being advanced is plausible.

8.4.6   An Exploration of the Relationship FF' ∝ XP/HX

We can now use Ibn al-Haytham's experimental ruler to examine the case of homonymous diplopia corresponding to the situation of the paintings under study.

---

[45]Plans and elevations were nevertheless used in this period, as is attested to by Giotto's *Drawing for the Campanile of the Cathedral of Florence,* dated 1334 (Siena, Museo dell'Opera).

By introducing variations in the initial conditions, it can be shown that the degree of disparity in the quasi-images is a function of the distance separating the fixation point from the background and the distance separating the spectator from the fixation point.

Let us begin with two wax columns positioned at Q (the fixation point) and Z (the background). The perceived distance between the quasi-images will be equal to CD. If we reduce the distance between the two wax columns by moving the first column from point Q to point S, the perceived separation between the quasi-images will then be equal to PR, which is less than CD (Fig. 8.10). This demonstrates that the disparity between quasi-images decreases as the distance between the fixation point and the background (measured along the *axis communis*) decreases.

Let us once again place two wax columns at points Q and Z. The perceived gap between the quasi-images remains equal to CD. If we move these columns to points L and S, maintaining the distance between them constant, the separation between the two quasi-images will be equal to UV, which is greater than CD (Fig. 8.11). In this case the disparity between the quasi-images increases as the distance between the spectator and the fixation point (measured along the *axis communis*) diminishes.

Since the vanishing point of a central perspective is located along the *axis communis*, it is subject to the same phenomenon of disparity as the wax column Z in the experiments just described. This explains why there are two vanishing points in all of the paintings in our corpus. If the artists who painted them were aware of the *qualitative relationships* between the disparity of the images and the respective distances between the spectator, the fixation point and the background, there should exist a valid statistical relationship between the parameters used to construct their perspectives.

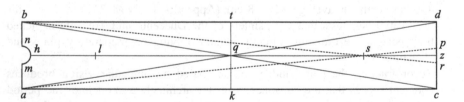

Fig. 8.10 The first relationship of disparity between the quasi-images. Author's drawing

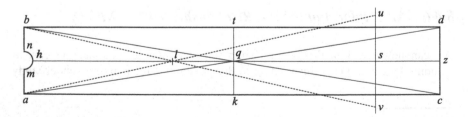

Fig. 8.11 The second relationship of disparity between the quasi-images. Author's drawing

Let us label the two eyes O and O' and the two vanishing points F and F'. Along the *axis communis HP*, one finds midway between the two eyes the cyclopean point (H), the fixation point (X), and the point in the background (P) (Fig. 8.12).

This relationship can be expressed as $\frac{FF'}{OO'} = \frac{XP}{XH}$ or, if the interpupillary distance OO' (a fixed datum) is disregarded:

$$FF' \propto \frac{XP}{XH}$$

One can then compare the *observed distance* between the vanishing points (FF'_{obs} measured on the geometric projection) and the *theoretical distance* between the vanishing points (FF'_{theo}) calculated from the relationship given above. These values, determined for all the works in our corpus (Appendix D), exhibited no systematic correspondences. Therefore the painters did not fix the two vanishing points based on a mathematical knowledge of this relationship. Nevertheless, if one

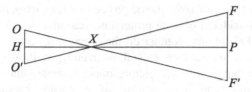

Fig. 8.12 The relationship FF' to XP/HX. Author's drawing

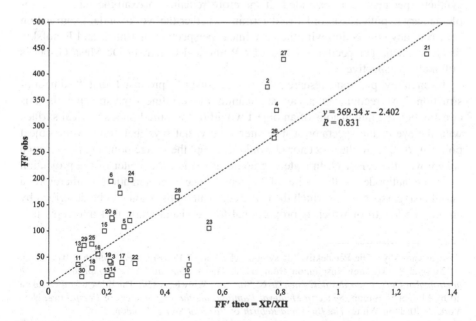

Fig. 8.13 A graph FF'obs ⊥ XP/HX. Author's drawing

plots $FF'_{theo} = XP/XH$ (abscissa) against FF'_{obs} (ordinate), the graph reveals a statistical relationship between the variables: $R = 0.831$ (Fig. 8.13).

This correspondence could signify that the painters and architects of the period, who did not yet have a projection method capable of fixing the vanishing points with precision, applied a *purely qualitative relationship* of the type: "The closer the subject of the scene is to the observer and the further it is from the background, the greater the distance between the two vanishing points." Knowledge of this relationship, which was accessible to them through Alhacen's binocular ruler, provides a fresh argument in support of the hypothesis that two-point perspectives derived from the medieval theory of binocular vision.

8.5   Conclusion

Let us return to the question of the classification of the different types of representation described at the beginning of this chapter. As we have shown, two-point perspectives are not parallel axonometric projections (isometry, dimetry or trimetry) nor can they be considered oblique projections (cavalier or military) in which the spectator is placed at infinity. Neither do they represent central, bifocal or trifocal linear perspectives. Insofar as there exist three main types of projection in which the distance of the spectator from the picture plane is finite—linear, curvilinear and synthetic perspective—the interpretation of two-point perspectives based on binocular vision cannot be entertained unless any association with curvilinear or synthetic perspective is excluded. It therefore remains to examine various complementary hypotheses in order to determine whether the works under examination here bear any connection with the curvilinear perspective of Hauck and Panofsky, or the synthetic perspective proposed by White and Carter, or De Mesa Gisbert's arithmetic perspective.

Inspired by previous research, Erwin Panofsky[46] proposed that "axial constructions" be recognized as an application of curvilinear perspective. In this construction the dimensions of an object would be measured on a spherical surface with the eye of the spectator at its center. The visual rays that connect the object points $A, B, C \ldots$ to the spectator O would intercept the image points $a, b, c \ldots$ on a sphere with the center O. In a stereographic projection the center of the projection O is the antipode of the point of tangency. The representation is therefore a curvilinear perspective in which the entire segment on the right can be described by an arc, the length of which is proportional to the visual angle that intercepts it.[47]

---

[46]Erwin Panofsky, "Die Perspektive als symbolische Form," *Vorträge der Bibliothek Warburg* 4 (1924/5): 258–331, drew inspiration from Guido Hauck, *Die subjektive Perspektive und die horizontalen Curvaturen des dorischen Styls*, Stuttgart, Wittwer, 1879. This idea was then taken up by Miriam S. Bunim, *Space in Medieval Painting and the Forerunners of Perspective*, New York, 1940; John White, *The Birth and Rebirth of Pictorial Space*, London, 1967.

[47]André Flocon and André Barre, *La Perspective curviligne*, Paris, 1968.

The same property would hold for an orthographic projection, with the center of projection being extended to infinity. All perspective drawings obtained on a plane tangent to the sphere—except for the gnomonic projection which is equivalent to linear perspective—satisfy the same property: the lines are rendered in the form of *curvilinear* arcs. This condition is not met by any of the works in our corpus (Appendix B), and they therefore cannot be classified as examples of curvilinear perspective.[48] Nor can any connection be drawn between them and Fouquet's attempts to construct curvilinear perspectives, as in *The Arrival of Emperor Charles IV at the Basilica of Saint Denis* or *The Banquet Held for the Emperor Charles IV* (Paris, BnF, ms. fr. 6465).[49]

Panofsky was employing the term "curvilinear perspective" in a broader sense that lay midway between linear and curvilinear perspective. The measurements made on the projection circle being transferred onto a picture plane, the object lines are rendered by image lines. This case demands separate study.

John White and Bernard Carter[50] provided a more specific formulation of Panofsky's concept, calling it "synthetic perspective" by which they meant a combination of "artificial" and "natural" perspective. The dimensions of objects, measured on a projection circle, are in this case transferred onto the picture plane by means of parallel lines. White and Carter's hypothesis rests on a mathematical property that cannot be found in any of the thirty works under study. Since it is known that receding lines converge in well-defined vanishing points, the property which states that in a synthetic perspective three orthogonals do not allow for a single vanishing point excludes the possibility that our works can be interpreted as examples of such a construction.

A final, and more radical, objection was advanced by Andrés de Mesa Gisbert, according to whom the pre-perspectives of the Duecento and Trecento were not constructed using geometric methods but rather on the basis of ratios of proportionality. As a consequence the historian would not even be able to trace the lines of the perspective, and these lines could not be extended up to the vanishing points.[51]

The next chapters will be devoted to a critical examination of these hypotheses: Panofsky's "curvilinear perspective" in Chap. 9, White's "synthetic perspective" in Chap. 10, and De Mesa's "arithmetic construction" in Chap. 11.

---

[48]With the exception of a few obscure and contentious mentions, serious study of curvilinear perspective only began during the nineteenth century, William Herdman, *A Treatise on the Curvilinear Perspective of Nature*, London, 1853.

[49]John White, "Developments in Renaissance Perspective: I," *Journal of the Warburg and Courtauld Insitutes* 12 (1949): 58–79, Plates 23B and 24B.

[50]White, *The Birth and Rebirth of Pictorial Space*, op. cit.; Bernard A.R. Carter, "Perspective," *The Oxford Companion to Western Art*, ed. by H. Osborne, Oxford, 1987, pp. 840–861.

[51]Andrés de Mesa Gisbert, "El 'fantasma' del punto de fuga en los estudios sobre la sistematización geométrica de la pintura del siglo XIV," *D'Art* 15 (1989): 29–50.

Chapter 9
The Hauck–Panofsky Conjecture Regarding Curvilinear Perspective

Abstract In the wake of Guido Hauck's work on "subjective perspective," Erwin Panofsky concluded that perspectives based on what he referred to as a "vanishing axis" could be interpreted as a form of curvilinear perspective. The present chapter aims to refute this conjecture by demonstrating that such constructions do not correspond either to the linear or the curvilinear perspective. They cannot be classed as linear perspectives because the distance between their two vanishing points is always greater than the maximum reconstruction error. Nor can they be considered curvilinear perspectives, in which the vanishing lines always converge in two well-defined vanishing points, because lines based on the mapping of projection circles on the picture plane do not end in points of concurrence. The refutation of the Hauck–Panofsky conjecture provides support for the interpretation set out in Chap. 8. Panofsky's erroneous reading stems from his failure to recognize that if the vanishing lines taken from the same side of the axis are extended beyond this axis, they will converge on a defined vanishing point.

Before adopting the principles of linear perspective, painters and architects utilized a wide range of systems to represent perspective: axonometry, oblique perspective, splayed views, etc. Among these systems was an axial form of construction that dates back to antiquity and is still described today as "the fishbone perspective." Guido Kern,[1] who identified it, initiated a line of research that during the course of the last century explored the possible existence of a curvilinear alternative for the representation of perspective. According to Kern, painters gradually advanced from "an axial construction with parallel vanishing lines" to an "an axial construction with convergent vanishing lines," before finally conceiving an approach in which the lines meet in a single vanishing point (central linear perspective).

---

[1] Guido J. Kern, "Die Anfänge der zentralperspektivischen Konstruktion in der italianischen Malerei des 14. Jahrunderts," *Mitteilungen des Kunsthistorischen Instituts in Florenz* 2 (1913): 39–65.

© Springer International Publishing Switzerland 2016
D. Raynaud, *Studies on Binocular Vision*,
Archimedes 47, DOI 10.1007/978-3-319-42721-8_9

Erwin Panofsky[2] sought to extend the empirical basis of perspective painting by adding the principles of the axial construction, and proposed a hypothesis that he believed would take into account this construction. He claimed that a similar system was already used by the artists of Pompeii working in the third and fourth styles, but agreed with Kern that the construction was characteristic above all of the Middle Ages. For example, in his analysis of a work by Duccio di Buoninsegna, Panofsky wrote: "The orthogonals of the lateral sections of the ceiling at first run entirely parallel with the brackets dividing the ceiling, *thus in a pure vanishing-axis construction.*"[3] He based his reconstruction on a brief, enigmatic passage in Vitruvius' *De Architectura*: "Scenography is the illusionistic reproduction of the facade and the sides, and the correspondence of all lines with respect to the center of the circle [actually the 'compass point']/Item scaenographia est frontis et laterum abscedentium adumbratio ad circinique centrum omnium linearum responsus."[4]

Influenced by Hauck's notion that the curvilinear construction was a "natural" form of perspective,[5] Panofsky deduced that painters during antiquity and the Middle Ages measured the objects they were depicting on a curved surface, which he denominated the "projection circle" and whose center corresponded to the eye of the viewer. He applied this hypothesis to the representation of a 'space box' (*Raumkasten*). The steps involved in the construction of the *Raumkasten* will be described below.

While the distinction between linear and curvilinear perspective is clear today, such was not always the case with the ancient texts, in which the curvilinear alternative was sometimes acknowledged and sometimes not.

Maurice Pirenne[6] was the first to challenge the notion formulated by Hauck and then adopted by Panofsky that curvilinear perspective was more suited to the subjective representation of space than linear perspective, because the former was "natural" whereas the latter was "artificial." Examining the alternatives:

---

[2]Erwin Panofsky, "Die Perspektive als symbolische Form," *Vorträge der Bibliothek Warburg* 4 (1924/5): 258–331; *Perspective as Symbolic Form*, New York, 1991.

[3]Panofsky, *Perspective as Symbolic Form*, p. 121 (italics mine).

[4]From Panofsky's own translation, *Perspective as Symbolic Form*, p. 100. Granger, who translates *circini centrum* as "vanishing point" has instead interpreted the passage from Vitruvius as follows: "Scenography (perspective) also is the shading of the front and the retreating sides, and the correspondence of all lines to the vanishing point, which is the center of the circle," F. Granger, ed., *De Architectura*, I, I, Cambridge, 1956, vol. 1, p. 26.

[5]Guido Hauck, *Die subjektive Perspektive und die horizontalen Curvaturen des dorischen Styls*, Stuttgart, 1879.

[6]Maurice Pirenne, "The scientific basis of Leonardo da Vinci's theory of perspective," *British Journal for the Philosophy of Science* 3 (1952): 165–185; idem, *Optics, Painting and Photography*, Oxford, 1970.

A. *A natural system of perspective exists that corresponds to what we see*

   Aa. *This natural system coincides with the perspective system developed in the Renaissance*
   Ab. *This natural system differs from the Renaissance system of perspective*

B. *This natural system does not exist; instead many different systems of perspective are admissible*

Pirenne demonstrates that linear perspective is just as well adapted to the perception of space as the curvilinear perspective. His principal argument is based on physiological optics—if, as Panofsky assumed, a rectilinear object is always perceived as a curvilinear form, there is no need to employ a curvilinear perspective for the object to conform to our perception of it, because the drawing itself is an object. On the contrary, the use of the curvilinear convention would deform the object, leading to representations that do not correspond to what one is attempting to depict. "Renaissance perspective remains a fairly good approximation, and, more important, it is probably the best possible approximation."[7]

Other authors, including Decio Gioseffi,[8] have attempted to refute on geometric and physiological grounds Panofsky's relativist thesis regarding the multiplicity and the essential arbitrariness of the perspective systems available to artists. According to Gioseffi, the equivalence between perspective representations in the art of Pompeii and the *perspectiva artificialis* of the Renaissance remains to be proven.

James Elkins[9] extended the critique of Panofsky's theses by questioning the notion that early intimations of a Renaissance form of curvilinear perspective are to be found in Leonardo da Vinci's notebooks. He argues that Leonardo's comments regarding the axiom of angles referred to natural vision rather than to perspective representations.

Richard Tobin showed that Panofsky was only able to compare and contrast the "axiom of angles" (curvilinear perspective) with the "axiom of distances" (linear perspective) based on a misreading of theorem 8 in Euclid's *Optics*.[10] It might be added that, far from sustaining the "axiom of angles" everywhere in his treatise, Euclid sometimes resorted explicitly to the "axiom of distances," e.g. in propositions 19–22.[11] This shows to what degree research on the precursors to curvilinear perspective was based on a subjective evaluation of the question.

---

[7]Pirenne, "The scientific basis of Leonardo da Vinci's theory," pp. 181, 183.

[8]Decio Gioseffi, *Perspectiva artificialis*, Trieste, 1957.

[9]James Elkins, "Did Leonardo develop a theory of curvilinear perspective?" *Journal of the Warburg and Courtauld Institutes* 51 (1988): 190–196. See also *The Poetics of Perspective*, Ithaca, 1994.

[10]Richard Tobin, "Ancient Perspective and Euclid's Optics," *Journal of the Warburg and Courtauld Institutes* 53 (1990): 14–41. Different versions of theorem 8 (prop. 9) appear in *Euclidis Optica*, ed. Heiberg, Leipzig, 1895, pp. 14–17; Elaheh Kheirandish, *The Arabic Version of Euclid's Optics*, New York, 1999, pp. 26–29; Wilfred R. Theisen, "Liber de visu," *Mediaeval Studies* 41 (1979), p. 67.

[11]Kheirandish, *The Arabic Version of Euclid's Optics*, pp. 56–69; Theisen, "Liber de visu," pp. 72–74.

This chapter will begin with a critical review of the passages that Panofsky devoted to the system of perspective based on the vanishing axis, but it will follow a different path from that of Pirenne, Elkins and Tobin, who sought to undermine the foundations of Panofsky's chain of reasoning rather than the conjecture itself that axial compositions constitute examples of curvilinear perspective. It will then remain to determine whether the paintings analyzed here conform to the system imagined by Panofsky.

The two limitations to the present study are the following:

1. The question of true and false perspectives will not be explored. It appears in effect that the evaluations one makes of modes of representation are in part determined by aesthetic biases of a complex nature. It seems more useful to limit the analysis to the operations involved in the construction of a perspective, identified ex post facto beginning with the architectural framework of the composition and the texts that render them intelligible.

2. The investigation will be limited to the question of vanishing lines, which is certainly just one aspect of perspective constructions. This reduction in the scope of the problem is justified from an analytical point of view, because two distinct operations are involved in the construction of a perspective. The first consists in mapping the *vanishing lines* (running parallel to the axis of the viewer's gaze) and the second in restoring the *frontal lines* (lying in a plane perpendicular to the axis of the viewer's gaze). The marking of a point anywhere in the composition is equivalent to the construction of a vanishing line and a frontal line. Since the notion of the vanishing axis is connected to the tracing of the *vanishing lines*, it is possible to analyze the latter without taking into account operations involving a reduction in the depth intervals.

9.1   Compositions Based on a Vanishing Axis

What is a vanishing axis? It is, according to Panofsky, the ancestor of the "vanishing point" in modern linear perspectives. This notion is based on the observation that the perspective lines in certain works do not converge towards a single point but meet along a vertical axis. Panofsky also uses the term "fishbone construction" to designate the arrangement of the various elements resulting from the use of a vanishing axis in a perspective view. He wrote: "The extensions of the orthogonals do not merge at a single point, but rather only weakly converge, and thus meet in pairs at several points along a common axis... This creates a 'fishbone effect'."[12] This description does indeed apply to many works produced during the Middle Ages.

---

[12]Erwin Panofsky, "Die Perspektive als symbolische Form", *Vorträge der Bibliothek Warburg* 4 (1924/5): 258–331, *Perspective as Symbolic Form*, p. 38.

The relationship perceived by Panofsky between a curvilinear perspective and a perspective with a vanishing axis is based on a short passage from Vitruvius' treatise on architecture that we will examine below. I will first review the construction hypothesis of "projection circles" and then present the arguments that can be raised against the Hauck–Panofsky conjecture.

9.1.1 The Basis of Panofsky's Interpretation

If one follows Panofsky's reasoning, the representation of the vanishing axis is based on a combination of the curvilinear and linear perspectives. The idea stems from a passage in which Vitruvius states: "Scenography [that is, perspective] also is the shading of the front and the retreating sides, and the correspondence of all lines to the center of the circle."[13] From this Panofsky derives the notions that: (a) in antiquity a "projection circle" centering on the eye of the observer (*circini centrum*) was used, and (b) compass measurements along this projection circle could then be transferred to the picture plane. The error regarding the convergence of the vanishing lines might result from this complex relationship, effectuated point by point using the method outlined below.

Panofsky examines the following case: suppose that one wishes to depict "the inside of a box" (*Raumkasten*). To points *I*, *II* and *I'*, *II'* which define the anterior and posterior faces of the box seen in elevation, correspond points *1*, *2* and *1'*, *2'* on the vertical projection circle whose center is *O*. Corresponding to points *A*, *B*, *C* and *A'*, *B'*, *C'* ... in the plane of the box are points *a*, *b*, *c* and *a'*, *b'*, *c'* ... on the horizontal projection circle with its center *O*. There are two possible outcomes in this case:

- the measurements made on the projection circles are transferred by the artist to the picture plane by means of *transfer lines*, such that points *11'*, *22'*, *aa'*, *bb'*, *cc'* ... are situated along parallel lines. This construction, which provided the basis for the White–Carter conjecture, will be examined in detail in Chap. 10.
- the measurements made on the projection circles are transferred to the picture plane after the *development* of projection circles on a picture plane. It is this construction, denominated the "Hauck–Panofsky conjecture," that will be examined in the present chapter.

Assume that the vertical and horizontal projection circles are developed on the planes $\mathscr{V}$ and $\mathscr{H}$. The points are transferred to the picture plane with point *I* corresponding to point *i*, point *A* corresponding to point *a*, point *B* corresponding

---

[13]Vitruvius, *De architectura*, I, 2, 2: "Scenography [that is, perspective] also is the shading of the front and the retreating sides, and the correspondence of all lines to the center of the circle [literally, the center of the compass]/Item scaenographia est frontis et laterum abscedentium adumbratio ad circinique centrum omnium linearum responsus", Vitruvius, *On Architecture*, ed. Granger, London, 1956, p. 26.

to point *b*, and so on. In the geometric figure obtained, the lower left corner of the box is formed from the intersection of lines *2'* and *a*; the upper left corner by the intersection of lines *1'* and *a*; and so on. One can then note that the transfer of measurements by means of the development of projection circles results in a perspective in which the vanishing lines corresponding to lines *AA'*, *BB'*, *CC'* … do not converge towards a single vanishing point as in a central perspective, but towards a vanishing axis *AF* (Fig. 9.1).

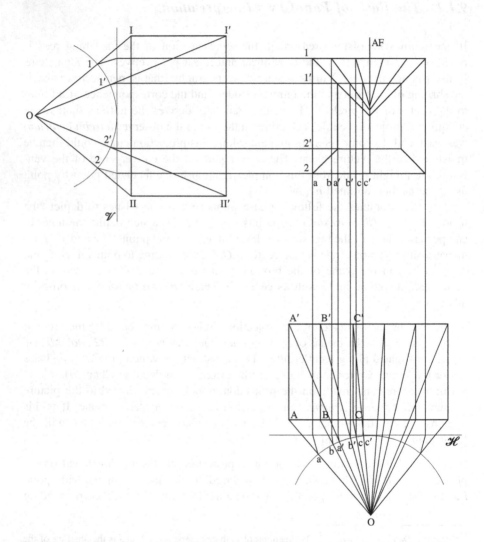

Fig. 9.1 The construction of a *Raumkasten*. Author's drawing

9.1.2 The Limitations of Panofsky's Interpretation

One can raise various arguments against such an ingenious hypothesis. The first concerns the many lacunae and insufficiencies in Vitruvius' *De architectura* with regard to this perspective construction. Panofsky admits that it is a somewhat audacious interpretation on his part and even acknowledges: "It is not clear that such an interpretation of the passage from Vitruvius can be sustained."[14] For example, while Vitruvius refers to a *circini centrum* there is nothing to indicate that its center corresponds to the eye of the observer (even if this free interpretation might correspond to some of Euclid's propositions on optics). Nowhere in *De architectura* is it suggested that one should take compass measurements along this hypothetical "projection circle." Nothing indicates that these measurements should be transferred to the picture plane. And there is no evidence either that medieval artists followed Vitruvius' *De architectura* in their practice of perspective. This speaks to the conjectural nature of Panofsky's interpretation. If one adds that the construction is quite complicated with regard to the operations that must be carried out in the case of a linear perspective, it is difficult to imagine practitioners who were already reluctant to follow the earliest canons of perspective spontaneously choosing to adopt a geometric construction that was even more complex and tedious.

9.2 Our Methodological Approach

The Hauck–Panofsky conjecture raises a sufficient number of doubts and difficulties to justify more thorough analysis. Can perspective representations from the Middle Ages be identified that meet the conditions of this construction? Let us focus on thirty works produced between 1295 and 1450 that approach the model of the vanishing axis (see Chap. 8 and Appendices B–E). By carrying out a rigorous reconstruction of the system of vanishing lines in these works, it can be determined whether they corroborate Panofsky's hypothesis or if instead compositions based on a vanishing axis can be reduced to cases of the more well-known model of linear perspective.

9.2.1 Errors in Panofsky's Reconstruction

The perspective lines of all the works in our corpus were analyzed based on ex post facto reconstructions. This methodology is subject to two broad types of error, one linked to the vantage point of the viewer and the other to the graphic reconstruction of the perspective.

---

[14]Panofsky, *Perspective as Symbolic Form*, p. 39.

The analyses were conducted on photographs of the original works (including certain frescoes that are more than 3 m in width), a measure which inevitably results in a loss of information. Every system of optical representation will introduce some geometric aberrations and it is important to verify that the perspective representation is not modified as a consequence. Three forms of distortion—spherical aberration, coma, and astigmatism—do not lead to a noticeable transformation in the perspective image. In contrast, negative or positive distortions (referred to as "barrel" and "pincushion" distortions) transform the tangential lines into curves. A maximum curvature of 1 % in the outermost line of the perspective image is admissible. In short, rarely will the point of view correspond to the central axis of a painting. If the picture plane is rectangular, it will often be depicted in the form of a more or less irregular quadrangle. This deformation does not invalidate the conclusions, because the quadrilateral $A'B'C'D'$ is a linear transformation of the rectangle $ABCD$.

Let us now examine the errors linked to the reconstruction of the vanishing lines. Let AB represent the visible portion of a vanishing line in the painting, which should continue as far as the eventual vanishing point C. If I draw a line of width e, the entire vanishing line covering the visible segment AB of length d can be retraced. The maximum deviation appears when A and B are located on opposite sides of this line. The angular error $\alpha = \pm \arctan(\epsilon/d)$ and the metric error $\epsilon = d' \tan(\alpha)$ can then be calculated. Working from small-scale images with a reduction coefficient of about $K = 10$ in relation to the original work, the metric error in situ is generally about ± 10 mm. The metric error for each work was calculated following the methodology described in Appendix A.

9.2.2   The Criteria Used to Test Panofsky's Hypothesis

Granting these conditions, one can test three concurrent hypotheses regarding curvilinear perspective:

Hypothesis 1. The works examined corroborate Panofsky's thesis of "projection circles."
Hypothesis 2. These constructions can be classified as examples of linear perspective.
Hypothesis 3. These perspectives invalidate the Hauck–Panofsky conjecture, but without showing any ties to the principles of linear perspective.

The best way to begin this analysis is to decide between the three hypotheses based on geometric criteria.

Hypothesis 1. Geometric analysis of the "vanishing axis" perspective constructed from Panofsky's projection spheres shows that the vanishing lines lying on the same side of the axis do not meet in a single point.

The vanishing lines of the perspective in Panofsky's *Raumkasten* have been extended in Fig. 9.2. One observes that the vanishing lines *ef*, *gh* and *ij* do not concur in a single point. The two vanishing lines *ef* and *gh* meet at point *r*, whereas *gh* and *ij* meet at point *s*, and *ef* and *ij* meet at point *t*.

The absence of a point of concurrence for the vanishing lines *ef*, *gh* and *ij* is a consequence of the geometric properties inherent in the Hauck–Panofsky hypothesis.

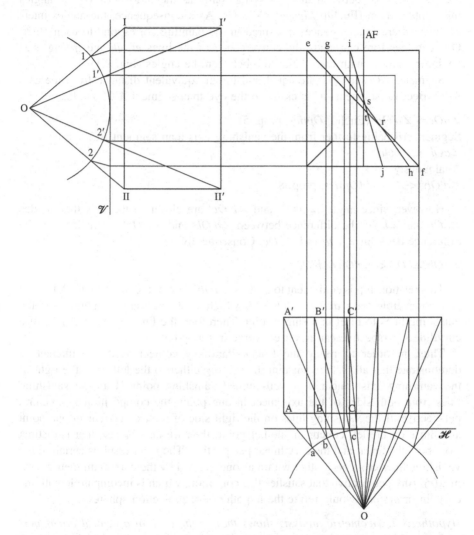

Fig. 9.2 Geometric diagram showing that the vanishing lines in Panofsky's *Raumkasten* do not end in a single point of concurrence. Author's drawing

The existence of a unique point of concurrence in the vanishing lines (Fig. 9.1) would be equivalent to the condition:

$$ab/bc = a'b'/b'c'$$

On the picture plane, the distance between two points issuing from the horizontal and vertical circles, which we will call ab, is equal to the length of the homonymous arc $\overset{\frown}{ab}$. The ratio between the arcs is the same as the ratio between the angles intercepting them (Euclid, *Elements* VI, 33). As a consequence, the angles intercepting the arcs and distances measured in the painting correspond to each other. One can therefore can study the convergence of the lines *ef*, *gh* and *ij* (Fig. 9.2) beginning simply with the relationship between the angles (Fig. 9.1).

Segments *AC* and *A'C'* are equal and lie an equivalent distance from the axis *AFO*. Because segment *AC* is closer to the eye than segment *A'C'* one has:

$\angle aOc > \angle a'Oc'$ (Euclid, *Optics*, prop. 5)
Segment *AB* being further from the vanishing axis than segment *BC*,
$\angle aOb < \angle bOc$
Analogously
$\angle a'Ob' < \angle b'Oc'$ (*Optics*, prop. 8)

However, since angles $\angle a'Ob'$ and $\angle b'Oc'$ are closer to the axis than angles $\angle aOb$ and $\angle bOc$, the difference between $\angle a'Ob'$ and $\angle b'Oc'$ is smaller than the difference between $\angle aOb$ and $\angle bOc$. Consequently

$\angle aOb/\angle bOc < \angle a'Ob'/\angle b'Oc'$

This relationship is equivalent to $ab/bc < a'b'/b'c'$ by virtue of *Elements* VI, 33. It is incompatible with $ab/bc = a'b'/b'c'$, which is a necessary condition for the vanishing lines to meet in a single point. Therefore, the lines *ef*, *gh* and *ij* … of a curvilinear perspective can never converge in one point.

Thus, in order to refute the Hauck–Panofsky conjecture it is sufficient to demonstrate that all of the vanishing lines lying either to the left or to the right of the vanishing axis concur in a well-defined vanishing point. If all the vanishing lines from both sides of the axis meet in one point, the composition is a central perspective; if the vanishing lines on the right side of the axis concur in one point and those on the left concur in another point, then we have a case that is distinct from both the linear and the curvilinear perspective. The latter result is certain if the vanishing lines cross each other within a zone defined by the maximum metric error $\max(\epsilon)$. Any perspective that satisfies this condition, which is incompatible with the curvilinear system, would refute the hypothesis of projection spheres.

Hypothesis 2. Geometric analysis shows that in the case of a central linear perspective all the vanishing lines must meet in a single point, which has been denominated the vanishing point (referred to by Alberti as the punctum centricus*).*

To disprove this hypothesis, it must be established that the vanishing lines of the painting do not converge in a single vanishing point (allowing for a slight margin of error), but instead either fail to converge within a zone defined by the maximum metric error or converge in more than one vanishing point separated by distances considerably greater than $\max(\epsilon)$.

Hypothesis 3. Any perspective can be assigned to one of the three categories defined in the initial hypotheses. This hypothesis, being residual to hypotheses 1 and 2, is heterogeneous and does not require further qualification.

Such is manifestly the case with the perspective in the *Saint Enthroned* by Giusto de' Menabuoi (Appendix A). It does not meet the criteria of hypothesis 1 because all of the vanishing lines issuing from the same side of the axis concur in a single point. It also fails to meet the criteria of hypothesis 2 as the ensemble of vanishing lines on the right and left sides of the axis concur in two well-defined points. It remains to be determined whether this property holds in general for all the works in our corpus.

9.3   Analysis of a Corpus of Works

An analysis of paintings selected according to these test criteria identified thirty works with a perspective construction that did not conform to Panofsky's thesis of projection circles nor to the principles of central linear perspective (Appendices B–E). Artists working between 1295 and 1450, from Giotto, Martini, and Lorenzetti to Gentile da Fabriano, Donatello and Uccello, all utilized this type of construction. The traditional division of the broad arc of time under consideration into two distinct periods—the Middle Ages and the Renaissance—is in fact inappropriate when describing the construction of pictorial representations, because although the rules of perspective were sometimes scrupulously applied during the Renaissance (for example, in Piero della Francesca's *Flagellation*), many compositions deviated from the most elementary rules of linear perspective. The existence of constructions with two vanishing points shows that more than one system for the representation of perspective was used during this period.

Around 1447 Paolo Uccello painted *A Scene from the Life of Noah. The Flood* for the *chiostro verde* of the Church of Santa Maria Novella in Florence (Fig. 9.3, 215×510 cm). This work was analyzed based on a reduced-scale version of the original (54×366 mm; reduction coefficient $K = 13.9$). The reduction coefficient, being greater than 10, does not compromise the validity of the results because the visible sections of the vanishing lines here are quite long.

Our findings corroborate the observations made by art historians. John White analyzed the fresco in detail and came to the same conclusion as Parronchi, who

Fig. 9.3 Paolo Uccello, *The Flood and the Receding of the Waters (Diluvio e Recessione delle acque)*, fresco, 215 × 510 cm, ca. 1447 (Florence, Santa Maria Novella). Author's reconstruction after a Wikimedia Commons image

had already noted that Uccello's work did not adhere to the rules of linear perspective.[15] The ex post facto reconstruction of the vanishing lines demonstrates that the horizontals of the side walls of the ark converge on two distinct vanishing points, F_1 and F_2, a fact that in itself disproves the hypothesis of projection circles. The maximal error max(ϵ) of the reconstruction is ±0.33 mm (in situ error ±5 mm). The distance separating the points of concurrence is 26 mm, which is 78 times greater than the max(ϵ), and means that the two points can never meet in a single vanishing point. It has been proposed that the doubling of the vanishing point may be due to a defect in the juxtapositioning of the cartoons, that is to say, to an "accidental error." Such a possibility can never be excluded, but fails to explain why works based on a two-point perspective appeared with such regularity. Indeed, we know of at least one other example of this construction in a work dating to the same period as the fresco by Uccello.

Between 1445 and 1448 Donatello created a series of bronze bas-reliefs on sacred themes, of which the *Confession of the Newborn Child* (Appendix B, No. 29, 57 × 123 cm) was chosen for analysis in the dimensions 94 × 201 mm ($K = 6.12$). One might object that the surface of a bas-relief is too uneven to allow for a geometric analysis, but in this case the section depicting the coffered ceiling is sufficiently flat to permit a reconstruction (only the coffers have been sculpted in shallow *intaglio*). The ceiling appears to conform to the rules of perspective based on a vanishing axis AF, obtained by drawing the orthogonals of the ceiling coffers. If these are extended, two vanishing points F_1 and F_2 symmetrically located on either side of the axis are revealed. An *ad hoc* vanishing point (F_3) can also be observed, which in all likelihood was created by the artist to avoid having the

[15]White, *The Birth and Rebirth of Pictorial Space*, Faber and Faber (London, 1967); Alessandro Parronchi, "Le fonti di Paolo Uccello: I perspettivi passati," *Paragone* 89 (1957): 3–32.

lowest vanishing lines extend below the ground line. The maximal metric error of the reconstruction is ±2.4 mm (in situ error ±15 mm). The distance between the points of concurrence is 12 mm, a gap that is five times greater than the max(ϵ) and allows one to envisage their meeting in a single vanishing point. Clearly, the drawing is composed of two pencils of lines.

Another work chosen for study was a large fresco, *The Funeral of Saint Francis*, dating to 1315–1317, which was painted by Simone Martini (1284–1344) in the Lower Basilica of Saint Francis of Assisi (Appendix B, No. 11, 270 × 230 cm). The image was analyzed in a format measuring 185 × 158 mm ($K = 14.6$). Based on the alignment of the capitals and bases of the columns sustaining the vault, a fishbone construction with a series of points of concurrence along the vanishing axis AF was revealed. Even so, if the vanishing lines are extended, they converge in two points, F_1 and F_2. Thus, as in the preceding examples, the vanishing axis resolves itself in two distinct points in a system that Martini applies in a coherent and systematic manner (with the exception perhaps of the vanishing lines following the base of the posts, which deviate slightly from the perspective). The maximal error of the vanishing lines is ±1.7 mm and the in situ error is ±24 mm. The distance (19 mm) between the points of concurrence is eleven times greater than the maximal error and it is therefore not possible for them to merge in a single point.

In the Lower Basilica of Assisi there is also an earlier fresco, *The Approval of the Franciscan Rule by Innocent III* (Appendix B, No. 2, 270 × 230 cm), thought to have been painted by Giotto between 1296 and 1299, which was analyzed in a reduced format 210 × 179 mm ($K = 12.8$). The lateral walls, sustained by the interplay of the consoles and lintels of the side portals, seem to converge toward a central vanishing axis AF in conformity with Panofsky's hypothesis. However, if these lines are extended they meet in two distinct vanishing points, F_1 and F_2, which are symmetrical to the vanishing axis. The max(ϵ) of the vanishing lines is ±1.6 mm (in situ error = ±21 mm). The fresco also presents an irregularity; the four consoles depicted in an end view by the artist produce construction lines that cross at two new points, F_3 and F_4, lying on either side of the central axis. The max(ϵ) of these vanishing lines is ±0.9 mm (in situ error = ±12 mm). They form two pencils of lines, invalidating the thesis that a projection circle was used to draw the perspective. The distance separating the points of concurrence is 179 mm, a gap that is 112 times greater than the maximal metric error. It is therefore impossible for the two vanishing points to be reduced to a single point and the central perspective in this fresco is neither curvilinear nor linear.

The works just examined all conform to the same perspective scheme:

1. The vanishing axis always resolves itself in two vanishing points lying on the same horizon line. *Axial perspectives may therefore be classified as a form of two-point perspective.* Such a conclusion simultaneously refutes hypothesis 1 (that these axial perspectives represent projection circles), and hypothesis 2 (that they can be interpreted as a form of linear perspective.

2. The two pencils of lines cross along the central axis. The lines issuing from the left side of the painting meet at an apex located on the right, while the lines issuing from the right side of the painting meet at an apex located on the left.

These works exemplify a perspective scheme that was observed in thirty works produced over a period of one and a half centuries in various parts of Italy. Since the perspective has been drawn in precisely the same fashion in all cases, it seems improbable that the final results were due to accidental errors in the perspective drawing. We find ourselves instead in the presence of a system that was typical of the entire period between 1295 and 1450.

9.4   Conclusion

The results of this study raise several questions, the main one being: What led Panofsky to make such an error in judgment as to interpret works based on a vanishing axis to be examples of curvilinear perspective? One can easily arrive at the nature of this error by observing the partial equivalence of the two following diagrams (Fig. 9.4).

Figure 9.4a shows the network of vanishing lines in an axial perspective, in which the vanishing axis is AF. In Fig. 9.4b the same lines meet at points F F'. If—like Panofsky—one does not extend the lines beyond their crossing points along the axis AF, it cannot be conceived that the two pencils of lines will converge and meet in two well-defined points beyond this axis. If instead the vanishing lines are prolonged beyond the vanishing axis, the vanishing points will supplant in importance the vanishing axis. As a consequence, we are justified in replacing the term "axial perspective" with "two-point perspective."

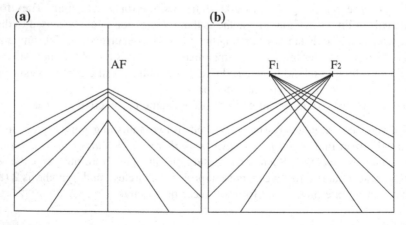

Fig. 9.4 Resolution of the vanishing axis (**a**) into two vanishing points (**b**). Author's drawing

These diagrams are equivalent so long as one does not invoke the hypothesis of projection circles, because the vanishing lines of a curvilinear perspective cannot form pencils of lines. Here the error does not consist so much in the recognition of the vanishing axis, as in the assumption that it is a consequence of the curvilinear perspective. The visual equivalence of the two diagrams—apart from the property that the vanishing lines of a curvilinear perspective never concur in a single point— no doubt lies at the origin of Panofsky's speculation about "projection circles."

Perspectives in which two pencils of lines meet in two points, which Kern and Panofsky termed "vanishing axis perspectives," do not derive from the principles of curvilinear perspective. This conclusion reinforces the doubts expressed by Elkins and Tobin[16] regarding whether the curvilinear system was ever used before the nineteenth century.

Moreover, none of these perspectives with two principal vanishing points can be linked—whatever intermediate transformations might be envisaged—to the usual form of linear perspective. Do these constructions therefore circumvent entirely the principles of linear perspective? Their only particularity in this regard is their use of the postulate of binocular vision. It would be more precise to refer to this construction as "binocular linear perspective," analogous to the *perspectiva artificialis* adopted by painters and architects during the Renaissance. It is this elementary relationship that unites and at the same time distinguishes the two systems of representation.

---

[16]Elkins, "Did Leonardo develop a theory of curvilinear perspective?" and Tobin, "Ancient Perspective and Euclid's Optics," op. cit.

Chapter 10
The White–Carter Conjecture on Synthetic Perspective

Abstract In the wake of Panofsky's work on perspective, John White and Bernard Carter identified the "axial construction" as the product of a system (denominated synthetic perspective) in which the measurements of an object on a projection circle are transferred to the picture plane by means of transfer lines. While this construction has been criticized for its complexity and therefore the limited possibility of its being put to use, it has never been studied in detail although it is regularly mentioned in discussions of perspective systems. We will show that none of the works of art examined here satisfies the mathematical property exhibited by synthetic perspective, namely, that three vanishing lines on the same side of the axis cannot be parallel nor can they meet in a point of concurrence. This refutation of the White–Carter conjecture should bring to a close the long succession of contradictory evaluations that have appeared over the course of the years; it also reinforces the explication presented in Chap. 8

In his celebrated essay "Die Perspektive als symbolische Form," Erwin Panofsky[1] retraced the history of studies on perspective and proposed that constructions based on a "vanishing axis" could be classified as examples of curvilinear perspective. This hypothesis was taken up by numerous art historians, notably Alan Little, Miriam Bunim, and John White.[2]

---

[1]Erwin Panofsky, "Die Perspektive als symbolische Form," *Vorträge der Bibliothek Warburg* 4 (1924/5): 258–331, *Perspective as Symbolic Form*, pp. 31–45.
[2]Alan M.G. Little, "Scaenographia," *The Art Bulletin* 18 (1936): 407–418; *idem*, "Perspective and scene painting," *The Art Bulletin* 20 (1937): 487–495; Miriam S. Bunim, *Space in Medieval Painting and the Forerunners of Perspective* (New York, 1940); John White, *Perspective in Ancient Drawing and Painting* (London, 1956); idem, *The Birth and Rebirth of Pictorial Space* (London, 1967). See also Gezienus ten Doesschate, *Perspective. Fundamentals, Controversials, History* (Nieuwkoop, 1964), p. 85–99 and 105–118.

© Springer International Publishing Switzerland 2016
D. Raynaud, *Studies on Binocular Vision*,
Archimedes 47, DOI 10.1007/978-3-319-42721-8_10

Little[3] supported Panofsky's ideas, maintaining that many Romanesque paintings were constructed around a vanishing axis with a clear geometric structure and this geometric property served to demonstrate that synthetic perspective was being used by the artists of the period. Others disagree, recognizing in these works only the signs of empirically drawn perspective lines.

Miriam Bunim, while acknowledging certain objections to Panofsky's hypothesis,[4] accorded some credit to the idea of a generalized usage of the axial construction:

> Panofsky's arguments against the assumption of the existence in ancient painting of a focused system of perspective... are derived from the ancient theory of optics itself and are extremely convincing. The fact that the visual field was noted to be spherical, that the size of objects was said to depend on the visual angle rather than on the proportional distance from the eye [Euclid, *Optica*, 8], and that curvilinear distortions in straight lines were observed, all tend to disprove the possibility of a focused system of perspective, for these are the very factors disregarded in such a method... The need for an adequate method of perspective which would give a convincing optical illusion of tridimensional space and at the same time promote the compositional unity of the scene was met by the systematized use of the vanishing axis.[5]

The British art historian John White places constructions based on a vanishing axis in the category of attempts by artists to combine "artificial perspective" (linear perspective) and "natural perspective" (which he identifies with curvilinear perspective).[6] He calls such views "synthetic perspectives,"[7] because transferring the measurements of angles to a picture plane produces a representation that is a combination of these two systems. In effect, if the construction of a synthetic perspective implies the manipulation of a curvilinear picture plane, the final result cannot be considered *stricto* sensu a curvilinear perspective, which is based on a geometric exercise in which the vanishing lines are derived from the arcs of a circle.[8] Therefore, White's construction hypothesis—like Panofsky's—involves a hybrid system in which angular measurements are made on projection circles before being carried over to the picture plane (Fig. 10.1).

---

[3]Little, "Scaenographia" and "Perspective and scene painting," op. cit.

[4]Miriam Bunim admits that, "In Giotto's paintings, the vanishing-axis procedure for vertical planes in depth was not used according to the clearly defined systematic form of converging pairs of parallel receding lines," Bunim, *Space in Medieval Painting*, p. 141. All the same, numerous cases in favor of the theory of the vanishing axis were then given apropos of Simone Martini, Barna da Siena, Bernardi Daddi and Italian painting in the Trecento generally; see pp. 148 and 154, 157, 166, 174, respectively.

[5]Bunim, *Space in Medieval Painting*, pp. 24–25.

[6]John White, *Perspective in Ancient Drawing and Painting*, London, 1956; idem, *The Birth and Rebirth of Pictorial Space*, London, Faber and Faber, 1967, French transl., *Naissance et renaissance de l'espace pictural*, Paris, Adam Biro, 1992.

[7]White, *Naissance et renaissance de l'espace pictural*, p. 213.

[8]André Flocon and André Barre, *La Perspective curviligne*, Paris, 1968. These principles were supposedly applied by Jean Fouquet in *Entrée de l'Empereur Charles IV à Saint-Denis*, ca. 1460. With regard to the theoretical foundations of this construction, we can only formulate conjectures.

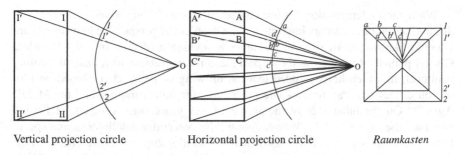

Vertical projection circle   Horizontal projection circle   *Raumkasten*

Fig. 10.1 The construction of the *Raumkasten* after Panofsky, *Perspective as Symbolic Form*, p. 74. Author's drawing

As indicated previously, there are two ways of transferring projection circles onto a picture plane. One can apply the process of *developing* projection circles, in which case the length of segment *ab* in the figure will turn out to be equal to the length of the homonymous arc *ab* (Chap. 9). Alternatively, one could transfer the measurements onto the perspective view by means of *transfer lines*. In this case the length of the segments and arcs will not be equivalent. Historically the second alternative has been used due to difficulties associated with the first. Indeed, what does the development of a projection circle signify in concrete terms? If the projection circle had a tangible reality, we could make our measurements by spreading a strip of paper over its curved surface, marking it with the various lengths, and then extending it over the panel or wall segment on which the image is to be depicted. But it is difficult to imagine what such a circle would be made of. And if the projection has no tangible reality, it is impossible to make measurements along its surface. The construction based on transfer lines therefore appears to be better suited to the practical requirements of artists,[9] and it is to this that we will be referring as we discuss the "White–Carter conjecture."[10]

---

[9]Bernard A.R. Carter presents the construction based on transfer lines in this light in the entry on "Perspective" in *The Oxford Companion to Western Art*, ed. H. Osborne, Oxford, 1987, pp. 840–861.

[10]The distinction between the conjectures of Hauck–Panofsky and White–Carter was made for the purposes of convenience. Panofsky recognized his debt to Hauck, who "suggested a procedure based on the projection of the object on a round cylinder with vertical generatrices emanating from a point on the axis, and the *development* of the image thus obtained on a picture plane." Marisa Dalai Emiliani, "La question de la perspective," introduction to Erwin Panofsky, *La Perspective comme forme symbolique*, Paris, 1975, p. 22. White and Carter did not link their construction hypothesis based on transfer lines to Hauck's work on perspective. The difference between the two conjectured constructions was based on no more than a detail, John White following Panofsky's lead regarding the signification of this alternative to linear perspective.

While their interpretation has been much criticized,[11] it is revived periodically in discussions among scholars interested in the problem of perspective in medieval art. I would like to cite two recent examples as evidence of this. In 1989 De Mesa Gisbert published a paper on the *ex post facto* reconstruction of fourteenth-century paintings, in which he concluded that the drawing of perspectives based on projection circles was "an unthinkable procedure, for both antiquity and the Middle Ages."[12] On the other side of the argument, six years later Antonella Ballardini endorsed the views of John White, noting "the perceptive intelligence and speculative precision with which he highlighted the relationship that existed between the theory of perspective and its different applications during the Middle Ages."[13] In the face of such diametrically opposed opinions, it is clear that new and—if possible, definitive—arguments must be brought to the debate.

Anyone who has taken an interest in the problem of axial constructions knows how difficult it is to evaluate whether the lines in a perspective drawing correspond to the proposed White–Carter conjecture. Even Panofsky, finding himself unable to identify lines that concurred with his own hypothesis, admitted the existence of a "more schematic, but more practicable, form of a more or less pure parallelism of oblique orthogonals."[14] If one follows this indication, there are in fact two types of constructions that correspond to the principles of synthetic perspective: (1) representations in which the vanishing lines are parallel; and (2) representations in which the vanishing lines are convergent.

The geometric concept of parallelism is clear and the vanishing lines in certain works, such as the *Life of Saint Cecilia*, do meet the condition of parallelism (Fig. 10.4c).

The concept of convergence, which is not as clearly defined, is ambiguous when applied to more than three vanishing lines. Two cases can be distinguished. Convergence *stricto sensu* signifies that the vanishing lines aa', bb', cc' ... end in a single point of concurrence M. Convergence *lato sensu* signifies a series of paired

[11]Hendrick G. Beyen, *Die pompejanische Wanddekoration vom zweiten bis zum vierten Stil*, 2 vols., Den Haag, 1938; idem "Die antike Zentralperspecktive," *Jahrbuch des deutschen archäologischen Instituts* 54 (1939): 47–72; Maurice Pirenne, "The scientific basis of Leonardo da Vinci's theory of perspective," *British Journal for the Philosophy of Science* 3 (1952): 165–185; Gezenius ten Doesschate, *Perspective. Fundamentals, Controversials, History*, Nieuwkoop, 1964; Luigi Vagnetti, "De naturali et artificiali perspectiva," *Studi e Documenti di Architettura* 9/10 (1979): 3–520; James Elkins, *The Poetics of Perspective*, Ithaca, Cornell University Press, 1994; Gérard Simon, "Optique et perspective: Ptolémée, Alhazen, Alberti," *Revue d'Histoire des Sciences* 54 (2001): 325–350.

[12]Andrés de Mesa Gisbert, "El 'fantasma' del punto de fuga en los estudios sobre la sistematización geométrica de la pintura del siglo XIV," *D'Art* 15 (1989), p. 49.

[13]Antonella Ballardini, "Lo spazio pittorico medievale: Studi e prospettive di ricerca," in Rocco Sinisgalli, ed., *La Prospettiva. Fondamenti teorici ed esperienze figurative dall'Antichità al mondo moderno*, Fiesole, 1998, p. 281. Like the views of Panofsky, those of White circulated widely; *The Birth and Rebirth of Pictorial Space* was reprinted three times (1967, 1970, 1972, 1987) and was translated into many languages.

[14]Panofsky, *Perspective as Symbolic Form*, p. 40.

points of concurrence: aa' and bb' cross at point M, bb' and cc' cross at point N, and so on. It may be noted that it is easier to construct a set of vanishing lines that meet the conditions of strict convergence, because once the point of concurrence is defined, all the vanishing lines will pass through that point. The lines of the perspective drawing in numerous works with an axial perspective clearly indicate the type of convergence used by the painters. In all of the works examined by us,[15] the vanishing lines issuing from one side of the axis meet on the opposite side at a point beyond the vanishing axis. This elementary property has been generally overlooked due to the influence of Kern[16] and Panofsky, according to whom vanishing lines converge from the left and right in pairs along the "vanishing axis." All the same, it suffices to extend the vanishing lines beyond the axis to find the points of concurrence. This is illustrated by the fresco *Christ Among the Doctors* by Giusto de' Menabuoi (1376–1378), in which the vanishing lines concur in points F and F' (see Fig. 10.5 at the end of this chapter). The White–Carter conjecture should take into account all constructions in which the vanishing lines issuing from the left meet in a point of concurrence on the right and the vanishing lines issuing from the right meet in a point of concurrence on the left.

In the present chapter we will attempt to bring to a close this long-running debate with its concatenation of critical analyses and contradictory conclusions, by establishing the mathematical impossibility of the axial construction. We will demonstrate that the White–Carter conjecture does not allow parallelism in the vanishing lines nor the convergence of vanishing lines in a point of concurrence. Hence works viewed as examples of synthetic perspective must be linked to a hitherto unrecognized principle of construction.

10.1   The Mathematical Properties of Synthetic Perspective

Let us return to our construction and compare the perspective with its underlying plan and elevation (see Fig. 10.3). Assuming that the 'space box' (*Raumkasten*) is symmetrical, it suffices to examine the vanishing lines situated to the right of the line of sight OK, O being the eye of the viewer.

Given that two non-parallel lines must eventually meet in a point of concurrence, a minimum of three vanishing lines should be studied. Let us therefore take the right side of the space box $ABEF$, of which three sides—AB, CD and EF—are parallel to the line of vision OK. A line from a given point on this box (designated C) intercepts the curvilinear surface IJ (called the "projection circle") at a point c_1

---

[15]The results depend on the errors in the perspective drawing. In order to reach clear conclusions, we conducted a verification of possible errors following a procedure that is outlined in Appendix A.

[16]Kern, "Die Anfänge der zentralperspektivischen Konstruktion."

which is then projected onto the picture plane GH at point c_2. The rabatment of the picture plane GH provides the actual perspective view, in which the line RS represents the upper, anterior edge BF of the box and the line PQ represents the image of the upper posterior edge AE of the box. Since point C belongs to AE, the transfer line issuing from c_1 intercepts a point c associated with PQ. We have adopted a uniform system of notation throughout: the quadrilateral $abef$ is the image of the upper face of the box $ABEF$.

Let us examine the conditions associated with the convergence or parallelism of the vanishing lines. If the lines ba, dc, fe … are strictly convergent (and end in a single point M) or alternatively are parallel to each other, Thales' theorem states:

$$\frac{ac}{ae} = \frac{bd}{bf} \tag{1}$$

The cases of parallelism and convergence respect the same property. The pairs of segments ac and bd, and ae and bf are proportional, where either $df = ce$ (parallelism) or $df \neq ce$ (convergence).

10.1.1 The Case of Linear Perspective

In a linear perspective, the lines CO, DO, EO and FO are directly intercepted by the picture plane GH (Fig. 10.2). It follows that the ratio (1) is equivalent to:

$$\frac{Kc_1}{Ke_1} = \frac{Kd_1}{Kf_1} \tag{2}$$

Since OAC and OKc_1 are similar triangles,

$$\frac{Kc_1}{OK} = \frac{AC}{OA} \tag{3}$$

Similarly:

$$\frac{Ke_1}{OK} = \frac{AE}{OA}; \quad \frac{Kf_1}{OK} = \frac{BF}{OB}; \quad \frac{Kd_1}{OK} = \frac{BD}{OB}$$

The relationship (2) could therefore be expressed as:

$$\frac{AC/OA}{AE/OA} = \frac{BD/OB}{BF/OB} \tag{4}$$

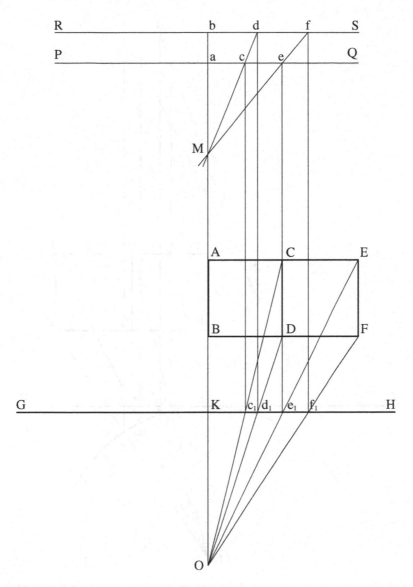

Fig. 10.2 Linear perspective. Author's drawing

After simplification one has

$$\frac{AC}{AE} = \frac{BD}{BF} \tag{5}$$

This will always hold true, since by definition $AC = BD$ and $AE = BF$. This signifies that in a central linear perspective all the vanishing lines corresponding to the orthogonals of the object converge toward a single vanishing point M.

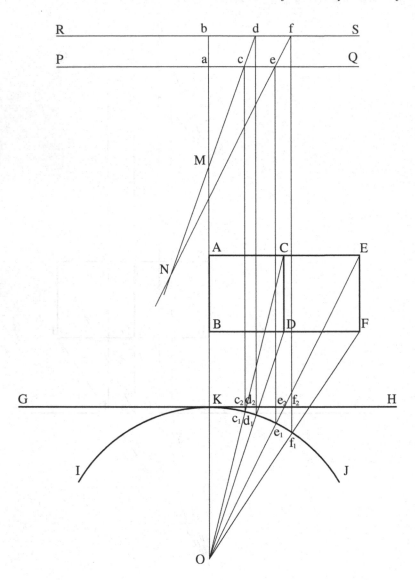

Fig. 10.3 Synthetic perspective in which the measurements of an object are transferred to the picture plane by means of transfer lines. Author's drawing

10.1.2   The Case of Synthetic Perspective

For this property to hold true for synthetic perspective, it is necessary that the vanishing lines *ba*, *dc*, *fe* … satisfy equivalent relationships. Here the lines *CO*, *DO*, *EO*, *FO* are intercepted by the curvilinear surface *IJ*, on which there are a number of ways of measuring the intercepted arcs (Fig. 10.3).

Panofsky made explicit that "the arcs of the circle are replaced by the corresponding chords" (Panofsky 1991: 38). Panofsky's choice of chords Kc_1, Ke_1, Kd_1, Kf_1 ... repeated by White and Carter must be rejected because it does not retain segment additivity. If in the space box $AE = AC + CE$, on the curvilinear surface it occurs that $Ke_1 < Kc_1 + c_1e_1$. The lenght of chord Ke_1 is indeterminate. Strictly speaking, White-Carter construction is impossible. For further analysis, the previous construction should be interpreted sensu lato. The simplest solution is then to replace the chords by the sines Kc_2, Ke_2, Kd_2, Kf_2 ... As the sines are measured on the picture plane GH, segment additivity is kept. However this interpretation is not better than the previous one, since it can be proved impossible (Raynaud 2004). Since then, Michel Baillet, a mathematician from the University of Orléans, found a demonstration shorter than mine, which I reproduce below with his permission.

Can condition $(1 \equiv 6)$ be satisfied in a synthetic perspective? (Fig. 10.3).

$$\frac{ac}{ae} = \frac{bd}{bf} \tag{6}$$

Each term of condition (6) can be expressed as

$$ac = \frac{AC}{\sqrt{AC^2 + OA^2}}$$
$$ae = \frac{AE}{\sqrt{AE^2 + OA^2}}$$
$$bd = \frac{AC}{\sqrt{AC^2 + OB^2}} \tag{7}$$
$$bf = \frac{AE}{\sqrt{AE^2 + OB^2}}$$

Values (7) are reintroduced into condition (6) which can, after simplification by AC and AE and squaring, be written:

$$\frac{AC^2 + OA^2}{AE^2 + OA^2} = \frac{AC^2 + OB^2}{AE^2 + OB^2} \tag{8}$$

We make the cross product:

$$\left(AC^2 + OA^2\right)\left(AE^2 + OB^2\right) = \left(AE^2 + OA^2\right)\left(AC^2 + OB^2\right) \tag{9}$$

This expression can be developed:

$$AC^2AE^2 + AC^2OB^2 + OA^2AE^2 + OA^2OB^2$$
$$= AE^2AC^2 + AE^2OB^2 + OA^2AC^2 + OA^2OB^2 \tag{10}$$

We simplify by AC^2AE^2 and OA^2OB^2:

$$AC^2OB^2 + OA^2AE^2 = AE^2OB^2 + OA^2AC^2 \tag{11}$$

Factoring, we get:

$$\left(AE^2 - AC^2\right) \cdot OA^2 = \left(AE^2 - AC^2\right) \cdot OB^2 \tag{12}$$

which includes the solution because, on the one hand, $AC < AE$, thus $(AE^2 - AC^2) \neq 0$, and on the other hand, $OB < OA$.

Therefore equality (12) never holds true.

10.2 Conclusion

As the synthetic perspective never fulfills condition $(6 \equiv 12)$, there can be no strict convergence of vanishing lines in this construction. This refutes the White–Carter conjecture.[17] The synthetic perspective only allows for an approximate convergence of the vanishing lines (comparable in optics to the section of a caustic curve produced by spherical aberration). The following theorems can therefore be enunciated:

Theorem 1. In a synthetic perspective, the vanishing lines on either side of the axis never meet in a point of concurrence and are never parallel to each other.

Reciprocally:

Theorem 2. Any perspective representation that has at least three vanishing lines on the same side of the axis which are parallel or meet in a point of concurrence cannot be considered a synthetic perspective.

Let us examine the consequences of these properties for practical perspective:

Corollary 1. There is no parallelism or focal point to facilitate the drawing of a synthetic perspective; it must be constructed on the basis of geometric plans. (In contrast, a perspective that has a vanishing point or parallel vanishing lines can be constructed without a geometric projection).

If one considers the preparatory drawings for certain works, this corollary indicates that the perspective lines could have been traced by the artist directly onto the picture plane before he began painting, without resorting to a plan or elevation. This only calls into question the *application to perspective* of geometric plans—early examples of which can be found in works such as the *Elevation for the Campanile of the Florence Cathedral* drawn by Giotto di Bondone in 1334 (Siena,

---

[17]The principles of synthetic perspective were therefore unknown in the Middle Ages and it is highly unlikely that they were applied before the nineteenth century: Arthur Parsey, *Perspective Rectified*, London, 1836; William G. Herdman, *A Treatise on the Curvilinear Perspective of Nature*, London, 1853; Guido Hauck, *Die subjektive Perspektive und die horizontalen Curvaturen des dorischen Styls*, Stuttgart, 1879.

Cathedral Museum), or even in the distinction made by Vitruvius between *ichno-graphia* (plan), *orthographia* (elevation) and *scaenographia* (perspective) in *De Architectura*.[18] But these examples do not in any way prove that geometric plans were actually *utilized* to construct perspective views.

Corollary 2. Since the refutation of the White–Carter conjecture holds for the two cases of convergence and parallelism, these constructions can no longer be considered to be related. Parallel perspectives and two-point perspectives must be interpreted separately.

The case of parallel vanishing lines. One could in the case of parallel vanishing lines identify a construction that is much simpler than Panofsky–White's. Let us begin with the fact that in the Middle Ages some artists utilized the cavalier per-spective—for example, Giotto in his fresco *Extasis* (Fig. 10.4a). By turning the uppermost face of the object onto its posterior horizontal edge, one obtains a splayed cavalier perspective comparable to the *Death of the Knight of Celano* (Fig. 10.4b). In the same way, a two-fold splayed cavalier perspective is obtained if one pivots the left face of the object onto its posterior vertical edge, as in the *Wedding of St. Cecilia and Valerian* painted by an anonymous master in the opening years of the Trecento (Fig. 10.4c). The result is a space box, all of whose interior faces are splayed outward "so as to maximize visibility."[19] In none of these cases did the artist rely on a mathematical construction; instead the perspective was drawn directly on the picture plane.

The case of converging vanishing lines. All of the works in our corpus—such as Giusto de' Menabuoi's *Christ Disputing with the Doctors of the Temple* (Fig. 10.5) —exhibit vanishing lines that converge in two vanishing points. They cannot be classified as "bifocal perspectives," a term reserved for perspectives that utilize distance points (as in the case of a box, none of whose faces are perpendicular to the axis of vision). The two points that can be observed in the works in our corpus correspond to two "principal vanishing points" with no link to the bifocal per-spective. It was noted above that these constructions—which were quite heterodox in view of the conventions adopted during the Renaissance—could have resulted from a *qualitative* application of the principles of binocular vision. The distance separating the architectural frame (the background) from the principal subject which serves as the fixation point (the foreground), triggers in the viewer a "homonymous diplopia" of the objects lying beyond the fixation point; that is to say, a diplopia

---

[18]"Ichnography (plan) demands the competent use of compass and rule; by these plans are laid out upon the sites provided. Orthography (elevation), however, is the vertical image of the front, and a figure slightly tinted to show the lines of the future work. Scenography (perspective) also is the shading of the front and the retreating sides, and the correspondence of all lines to the center of the circle [literally, the center of the compass]/Ichnographia est circini regulaeque modice continens usus, e qua capiuntur formarum in solis arearum descriptiones. Orthographia autem est erecta frontis imago modiceque picta rationibus operis futuri figura. Item scaenographia est frontis et laterum abscendentium adumbratio ad circinique centrum omnium linearum responsus," *De Architectura*, I, II, ed. Granger, vol. 1, pp. 24–26.

[19]On this, Ian Verstegen, "A classification of perceptual corrections of perspective distortions in Renaissance painting," *Perception* 39 (2010), p. 689.

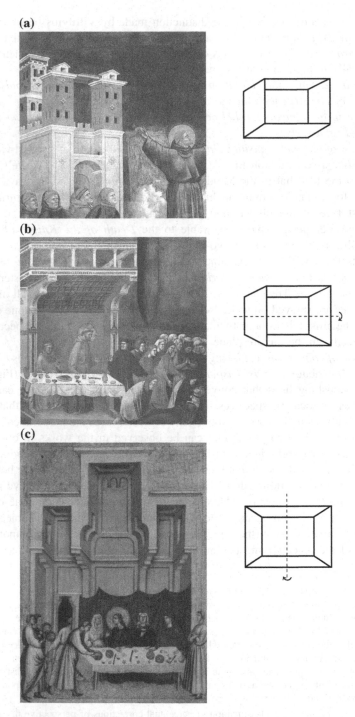

Fig. 10.4 Alternative constructions: **a** Cavalier perspective by Giotto, *Extasis*, ca. 1296; **b** Splayed cavalier perspective by Giotto, *The Death of the Knight of Celano*, ca. 1296; **c** Two-fold splayed cavalier perspective by the Maestro della S Cecilia, *Wedding of St. Cecilia and Valerian*, tempera on wood, ca. 1304

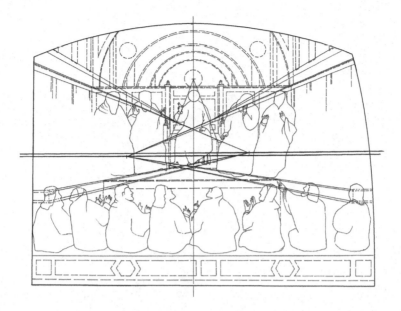

Fig. 10.5 Giusto de' Menabuoi, *Christ Disputing with the Doctors of the Temple*, fresco, ca. 1376–1378 (Padua, Baptistery of the Cathedral). Author's reconstruction

with regard to the rectilinear edges that provide the vanishing lines of the perspective.[20] The two principal vanishing points turn out to be the result of a doubling of the two pencils of vanishing lines.

As we have seen in Chap. 6, the plausibility of this hypothesis depends on the fact that the treatises on optics of Alhacen and his Latin successors allowed for such an application: (1) all of these authors discussed the phenomenon of the fusion of quasi-images, a point that is central to binocular vision; (2) none of them posited the thesis of suppression, which was formulated for the first time by Giovanni Battista della Porta in 1593; and (3) these treatises were widely read in Italy, circulating in the same milieux as those in which the axial construction was utilized by artists.[21]

---

[20]Hermann von Helmholtz, *Optique physiologique*, Paris, 1867; Yves Le Grand, *Optique physiologique*, 3 vols., Paris, 1948–1956.

[21]Luca Baggio, "Sperimentazioni prospettiche e ricerche scientifiche a Padova nel secondo Trecento," *Il Santo* 34 (1994): 173–232; Francesca Cecchini, "Artisti, committenti e perspectiva in Italia alla fine del Duecento," Rocco Sinisgalli, ed., *La prospettiva. Fondamenti teorici ed esperienze figurative dall'Antichità al mondo moderno*, Fiesole, 1998, pp. 56–74; Dominique Raynaud, *L'Hypothèse d'Oxford*, Paris, 1998.

Chapter 11
De Mesa's Hypothesis Regarding the Arithmetic Construction of Perspective

Abstract Andrés De Mesa Gisbert proposes that the perspectives in paintings from the Duecento and Trecento were drawn arithmetically, i.e. without resorting to vanishing points. The most convincing argument for this hypothesis is that the division of two parallel lines by straight lines intersecting each other at a vanishing point (the geometric method) is equivalent to the division of these parallel lines into proportional parts (the arithmetic method). If the arithmetic method was indeed used by medieval artists, then the vanishing points exhibited ex post should be purely fortuitous. There are sound objections to this assertion, however: the lack of simple multiples and submultiples of the measurement units, the absence of proportionality ratios, the lengths of the operating series, and the correspondence of the vanishing points to visible loci in the pictures. The application of optics and the geometric method is a more probative thesis, although it does not imply that painters were using concepts of linear perspective, which would have been an anachronism.

Perspective is a legitimate topic of investigation for historians of science when the layouts of paintings are mathematically constructed. With the putative contributions by Brunelleschi and Alberti as a preamble, the history of perspective *as a science* actually starts with Piero della Francesca's *De prospectiva pingendi* (1470).[1] This chapter considers the paintings of the Duecento and Trecento from the angle of the history of science. We deliberately chose to focus on an example with little coeval documentation: the frescoes in the nave of the Upper Church of the Basilica of Saint Francis in Assisi. Here are the reasons justifying this choice:

1. Practically no contemporary documents exist concerning these mural paintings. The names of the painters who were working on the site are only assumed. As in the case of many early large-scale projects, numerous hands have been suggested: Jacopo Torriti, Cimabue, the Master of the Capture, the Master of Isaac, Giotto di Bondone, etc. For the scenes of *The Legend of St. Francis* the

---

[1] Judith V. Field, "Alberti, the abacus and Piero della Francesca's proof of perspective," *Renaissance Studies* 11/2 (1997): 61–88.

© Springer International Publishing Switzerland 2016

D. Raynaud, *Studies on Binocular Vision*,
Archimedes 47, DOI 10.1007/978-3-319-42721-8_11

spectrum of attributions ranges from Giotto[2] and the Master of Isaac (sometimes identified as Arnolfo di Cambio),[3] to an unknown Roman painter (perhaps Pietro Cavallini or Filippo Rusuti).[4]

2. Equally little is known about the dating of this decorative cycle. The church rose between 17 July 1228, when the foundation stone was laid by Pope Gregory IX (17 July 1228) and 11 June 1253, the date of its consecration by Pope Innocent IV. The period of its decoration cannot be pinpointed as accurately, but ranges from about 1254 to 1338. According to Vasari,[5] Giotto painted the thirty-two scenes of the *Legend of St. Francis* in the lower register of the Upper Church around 1296–1305. The *terminus post quem* (1296) coincides with the election of Giovanni da Morrovalle as the general minister of the Franciscan order. The *terminus ante quem* (1305) can be deduced from the fact that the tower of the Palazzo del Capitano, which appears to be under construction in the frescoes, was completed in that year. Some art historians have narrowed the interval to 1296–1299 or even 1296–1297 on stylistic grounds.[6]

The present tendency is to antedate the paintings. The rationales for this are: (1) Vasari's well-known bias in favor of the contributions of Florence and the Medici to the development of Renaissance art; (2) growing evidence of the influence of Roman art on the frescoes in Assisi; (3) the vacancy of the papal throne between 1292 and 1294, which would not corroborate the handing down of any key decisions regarding the papal church of Assisi in this period; and (4) the recent discovery of a fourteenth-century Franciscan manuscript that documents the decision of Pope Nicholas IV to commission sumptuous, large-scale paintings for the basilica of Assisi ("*nec videmus in ecclesiis fratrum sumptuositatem magnam picturarum nisi in ecclesia Assisii, quas picturas dominus Nicolaus IV fieri precepit…*"). There are sound reasons therefore to date the initiation of the work to ca. 1290–1292.[7]

---

[2]Luciano Bellosi, Giovanna Ragionieri, "Giotto e le storie di San Francesco nella basilica superiore di Assisi," in *Assisi anno 1300*, edited by S. Brufani and E. Menestò, Assisi, Edizioni Porziuncola, 2002, pp. 455–473.

[3]Angiola Maria Romanini, "Arnolfo pittore: pittura e spazio virtuale nel cantiere gotico," *Arte medievale* 11 (1997): 3–33.

[4]Hayden B.J. Maginnis, Andrew Ladis, "Assisi today: the upper church," *Source* 18 (1998): 1–6.

[5]"Having finished these works [in Arezzo, Giotto] betook himself to Assisi, a city of Umbria, being called thither by Fra Giovanni di Muro della Marca, then General of the Friars of St. Francis, where in the upper church he painted *a fresco*, under the gallery that crosses the windows, on both sides of the church, thirty-two scenes from the life and acts of St. Francis/ Finite queste cose [in Arezzo, Giotto] si condusse in Ascesi città dell'Umbria, essendovi chiamato da fra' Giovanni di Muro della Marca, allora Generale de' frati di san Francesco, dove nella chiesa di sopra dipinse a fresco, sotto il corridore che attraversa le finestre, dai due lati della chiesa, trentadue storie della vita e fatti di San Francesco," Giorgio Vasari, *Vite de' più eccellenti Pittori Scultori e Architettori*, edited by R. Bertanini and P. Barocchi, Florence, 1967, II, p. 100.

[6]Elvio Lunghi, *La Basilica di San Francesco di Assisi*, Antella, 1996, pp. 66–67.

[7]Bellosi and Ragionieri, "Giotto e le storie di San Francesco," op. cit.; Donal Cooper, Janet Robson, "Pope Nicholas IV and the upper church at Assisi," *Apollo* 157 (2003): 31–35.

3. Knowledge is scant regarding the fresco techniques of the time. The only written accounts we have are those by Vasari and Cennini,[8] which date to several centuries later. The material examination[9] of the paintings themselves partially compensates for this lacuna, but the many steps involved in the complex process of fresco painting must be inferred from scattered clues. Nevertheless, three principal stages can be identified: (1) sketching the drawing (*disegno*); (2) squaring the pattern (*gratta*) onto the rough plaster underlayer (*arriccio*); and (3) transferring the pattern onto the layer of fresh plaster (*intonaco*), drawing the straight lines with a ruler (and sometimes retracing them with a *puntaruolo* or awl) and drawing the circles with a compass. Curves were sketched freehand on the fresh plaster, while the figures' faces were transferred by means of *patroni* (drawings that served as a kind of template). In general, the preliminary drawings were not conserved by the atelier.[10]

The geometric lines of the frescoes of the Upper Church at Assisi are scarcely visible, but beneath the crumbling plaster the reddish brown underdrawing (*sinopia*) on the left side of the scene depicting *St. Francis Preaching Before Pope Honorius III* has been discovered.[11] Examination under low-angled lighting has disclosed signs of the line drawn in the fresh coating, the perforations left by nails to which a cord was attached to trace the vanishing lines, and at times even the painter's fingerprints, as in the *Extasis*.[12] The fact that the *disegno* had to be transferred twice—first in the form of the *sinopia* underdrawing and then onto the freshly laid *intonaco*—creates a difficult problem for historians of perspective, because the layout sketched in the first stage is partially effaced during the second transfer.

---

[8]Vasari, *Vite*, op. cit., p. 199; *Cennino Cennini, Il Libro dell'arte*, edited by Gaetano and Carlo Milanesi, Florence, 1859, p. 60.

[9]Giuseppe Basile, ed., *Pittura a fresco. Tecniche esecutive, cause di degrado, restauro*, Florence, 1989; Bruno Zanardi, Chiara Frugoni and Federico Zeri, *Il Cantiere di Giotto*, Milan, 1996.

[10]Bernhard Degenhart, Annegrit Schmitt, *Einleitung a Corpus der Italienischen Zeichnungen 1300–1450*, I-1. Sud- und Mittelitalien, Berlin, 1968, p. xix.

[11]"Immediately above the aureola of Francis, one can see a very small, completely indecipherable trace of a drawing in red sinopia. But, above and beyond this infinitesimal fragment of proof is the vast complexity of the iconographic cycle that precludes the hypothesis of an execution *a fresco* of the scenes without an extremely detailed preparatory drawing on paper (or parchment) that could be transferred directly to the *arriccio* [the rough underlayer of plaster] in the form of a sinopia... These material data as well confirm the fact that there must have been quite detailed preliminary planning for the Franciscan fresco cycle, first with drawings on paper (or parchment) and then on the *arriccio*/ Subito sopra l'aureola di Francesco è visibile una piccolissima traccia, del tutto indecifrabile, d'un disegno in rosso sinopia. Ma, al di là di questa minima prova è l'enorme complessità iconografica del ciclo a rendere impossibile l'ipotesi di una esecuzione a fresco delle scene in assenza d'un detagliatissimo progetto su carta (o pergamena) da riportare al vere sull'arriccio in forma di sinopia... Questi dati materiali impongono di nuovo di dar per certo un assai dettagliato lavoro di progettazione del ciclo francescano, prima con disegni su carta (o pergamena) e poi sull'arriccio," Zanardi, *Il Cantiere di Giotto*, pp. 24, 32.

[12]Zanardi, *Il Cantiere di Giotto*, pp. 24–32.

Under low-angled lighting what can be seen in the *sinopia* is generally limited to the edges of the architectural elements that anchor the vanishing lines.[13]

4. Gaps in the documentation preclude a direct apprehension of the extent of the artists' understanding of perspective during this early period, which is why these mural paintings have given rise to multifarious interpretations encompassing a spectrum of approaches from intellectualism to empiricism. There is no single best approach to adopt in the analysis of their perspectives. Intellectualism runs the risk of anachronism by using concepts such as 'horizon' and 'vanishing point', which were undefined at that time, whereas empiricism may underestimate the overt or tacit knowledge that is necessary to draw in perspective.

Notwithstanding these considerations, the frescoes in the Upper Church at Assisi can unquestionably be considered pioneering trials in the rationalization of the visual space in painting. They await an interpretation that will reconcile the empirical evidence with a convincing conceptual foundation.[14]

11.1   The Arithmetic Construction Hypothesis

To explain the advances made in the representation of depth in Duecento and Trecento painting, historians at first assumed that artists were in possession even at this early date of a basic knowledge of convergence, infinity, etc. Well-known objections have been raised against this interpretation:[15]

(1)   Little is known and therefore we can only hypothesize about the mathematical knowledge to which the artists and artisans would have had access;
(2)   There are scant examples of convergence on a defined vanishing point in medieval painting. More often we find cases of convergence on a "vanishing axis" or an even less precisely defined "vanishing region";
(3)   Experiments in the representation of pictorial space led to many competing systems in the Duecento and Trecento, thus reducing the importance of linear perspective as such. This undermines the credibility of Panofsky's hypothesis that painters such as Lorenzetti had a mathematical understanding of the

---

[13]Zanardi, *Il Cantiere di Giotto*, p. 29 (Figs. 18, 19, 20), p. 31 (Figs. 22, 23, 24).

[14]John White, *The Birth and Rebirth of Pictorial Space*, London, 1967, p. 32.

[15]See not only Panofsky's groundbreaking "Die Perspective als symbolische Form," *Vorträge der Bibliothek Warburg*, 4 (1924/5): 258–331, *Perspective as Symbolic Form*, Zone Books, New York, 1991, but also Guido J. Kern, "Die Anfänge der zentralperspektivischen Konstruktion in der italianischen Malerei des 14. Jahrhunderts," *Mitteilungen des Kunsthistorischen Instituts in Florenz* 2 (1913): 39–65, and Decio Gioseffi, *Perspectiva artificialis. Per la storia della prospettiva, spigolature e appunti*, Trieste, 1957. This thesis was amended by Rocco Sinisgalli, *Per la storia della prospettiva, 1405–1605*, Rome, 1978.

vanishing point, which the art historian saw as "the concrete symbol of the discovery of infinity itself."[16]

Empiricism is the very opposite of intellectualism. Among the empiricist positions, that of Andrés de Mesa Gisbert demands consideration because it combines rigor with minimalism. The author proposes a hypothesis regarding the convergence of vanishing lines in Trecento paintings[17] which, in fact, constitutes a method for building a perspective arithmetically.

> Let two parallel lines be drawn at any distance from each other; if one of them is divided into any number of parts and the second parallel is treated similarly, by keeping exactly the same proportions that we used initially, the extension of the straight lines passing through the corresponding points *will generate a convergence of all lines on one single point with no need of having operated with it.*[18]

The author's fundamental insight was that the convergence of segments AD, BE, CF... may be obtained either by drawing the lines OA, OB, OC... whose visible segments are AD, BE, CF... (the *geometric method*), or by drawing segments DE, EF... in proportion to segments AB, BC..., the proportionality ratio between AB and DE, between BC and EF..., being sufficient to ensure the existence of the virtual vanishing point O (the *arithmetic method*). Among the geometric relationships in similar triangles, craftsmen would have extracted relationship (1), setting aside relationships (2), (3), etc. which would have required the inclusion of the concurrence point O (Fig. 11.1).[19]

$$(1)\frac{DE}{EF} = \frac{AB}{BC} \quad (2)\frac{OE}{ED} = \frac{OB}{BA} \quad (3)\frac{OE}{EF} = \frac{OB}{BC}$$

The arithmetic hypothesis provides a simple explanation as to why ex post layouts exhibit a point at infinity; this point would be the by-product of an analysis that presupposes the use of geometric instruments. But in fact painters could instead have used a proportionality rule to divide the frontal-horizontal lines and distribute them in depth.

---

[16]Panofsky, *Perspective as Symbolic Form*, p. 57.

[17]Andrés de Mesa Gisbert, "El 'fantasma' del punto de fuga en los estudios sobre la sistematización geométrica de la pintura del siglo XIV," *D'Art* 15 (1989): 29–50. The author has since specialized in architectural surveys.

[18]"Si disponemos dos rectas paralelas con cualquier distancia entre sí, y luego de dividir una de ellas en un número cualquiera de partes lo hacemos en forma similar sobre la segunda paralela, guardando exactamente las mismas proporciones con las que se lo ha hecho inicialmente, al unir los puntos correspondientes con líneas rectas, en su prolongación *obtendremos la convergencia de todas ellas sobre un solo y único punto sin necesidad de haber operado con él*," De Mesa Gisbert, "El 'fantasma' del punto de fuga," p. 33 (italics mine).

[19]De Mesa Gisbert, "El 'fantasma' del punto de fuga," pp. 33–34. This idea is discussed by Ian Verstegen, "Viewer, Viewpoint, and Space in the Legend of St. Francis: A Viennese-Structural Reading," preprint 2011.

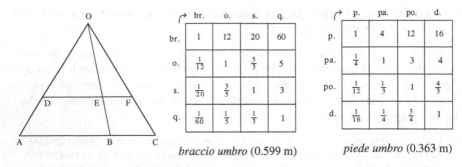

br.	br.	o.	s.	q.
br.	1	12	20	60
o.	$\frac{1}{12}$	1	$\frac{5}{3}$	5
s.	$\frac{1}{20}$	$\frac{3}{5}$	1	3
q.	$\frac{1}{60}$	$\frac{1}{5}$	$\frac{1}{3}$	1

p.	p.	pa.	po.	d.
p.	1	4	12	16
pa.	$\frac{1}{4}$	1	3	4
po.	$\frac{1}{12}$	$\frac{1}{3}$	1	$\frac{4}{3}$
d.	$\frac{1}{16}$	$\frac{1}{4}$	$\frac{3}{4}$	1

braccio umbro (0.599 m)   *piede umbro* (0.363 m)

Fig. 11.1 The arithmetic hypothesis. Author's drawing after De Mesa Gisbert, "El 'fantasma' del punto de fuga," p. 33

In addition, de Mesa's hypothesis resolves three problems associated with pre-Renaissance paintings: (1) it accounts for the arbitrary behavior of certain lines by stating that proportional ratios were not defined for all the lines (observe, for example, the parallel edges of the abaci of the capitals in *St. Francis Preaching Before Pope Honorius III*); (2) in the same way, it explains the additional presence of a vanishing axis in works such as *The Pentecost*; and (3) the convergence on a vanishing region is seen as a secondary effect of errors made during the transfer of proportional segments (any inaccuracies in the positioning of points *A*, *B*, *C* will induce a deviation in the vanishing lines *AD*, *BE*, *CF*). This arithmetic scheme is often cited to emphasize the point that any attempt to trace the beginnings of linear perspective to the Middle Ages would be an anachronism.[20]

Despite its obvious ingenuity and apparent applicability, the arithmetic method gives rise to difficulties that have never been systematically explored. This is understandable because when a conclusion appears to be correct, rarely do we thoroughly examine its premises. Moreover, the arithmetic hypothesis is readily acceptable, because nothing proves that medieval craftsmen made use of vanishing points. However, the arithmetic method has up to now been a hypothesis rather than a firmly established concept, and it needs to be subjected to careful scrutiny.

11.2 The Absence of Multiples or Submultiples of the Measurement Units

In a perspective that has been composed using the arithmetic method, the frontal lines should be divisible into multiples or submultiples of standard measurement units. The choice of a unit of measurement is never arbitrary, whether it has a

---

[20]The arithmetical method supports the idea that perspective was a Renaissance invention. De Mesa speaks of Brunelleschi's contribution in "El 'fantasma' del punto de fuga," p. 35. For a critique, see D. Raynaud, *L'Hypothèse d'Oxford*, Paris, 1998, pp. 4–9 and 132–150.

symbolic or a practical significance. Many examples can be found in the history of architecture. Take, for example, the hypothesized geometric scheme of the Cappella Pazzi designed by the architect Brunelleschi.[21] Konrad Hecht[22] inaugurated the critical approach by drawing attention to the discrepancies between the seventeen different regulating layouts published between 1867 and 1957. In the same vein Jean Guillaume[23] has shown that the regulating layouts proposed to explain the architectural design do not even match the actual measurements of the Cappella dei Pazzi, to which Brunelleschi always assigned integer or simple values—for example, he chose pilasters measuring 1½ *braccia* in width.

Transposed to the case of wall paintings, the discovery of unit multiples would offer corroboration that the arithmetic method had been used because: (1) it is easier to calculate proportional ratios on simple measurements; (2) it is also easier to remember and apply the measurements required to construct the perspective. Do early perspective paintings exhibit multiples or submultiples of specific measurement units?

In Umbria two measurement systems were used in the late Middle Ages: the *braccio* and the *piede*. In Perugia, Foligno, Orvieto, Spoleto and elsewhere, there were two values for the *braccio*, i.e. the *braccio lungo* (0.668 m) and the *braccio corto* (0.599 m). We decided to use the *braccio corto*, which was referred to as *da legname* (for wood) and *da muratori* (for masons) rather than the *braccio lungo*, which was the unit *da lana* (for wool), *da panno* (for cloth) and *da seta* (for silk). The *braccio corto* system consisted of the *braccio* (599 mm), the *oncia* (49.92 mm), and the *soldo* (29.95 mm). The Umbrian value for the *piede da legname e da fabbrica* (*piede* for wood and for building) was equivalent to 0.363 m, from which a second system of measurements can be deduced: the *piede* (363 mm), *palmo* (90.75 mm), *pollice* (30.25 mm), and *dito* (22.69 mm).[24]

Let us now examine *The Approval of the Franciscan Rule*, which de Mesa (*op. cit.*, Figures 11–14) cites as an exemplary application of the arithmetic method. The procedure that he followed is outlined here.

---

[21] A similar example may be found in Lescot's façade for the Louvre, whose measurements are multiples of the *pied du Roi* (326.6 mm) used in Paris ca. 1546, Jean-Paul Saint Aubin, "Photogrammétrie et étude des ordres: le Louvre de Lescot," in *L'Emploi des ordres à la Renaissance*, ed. Jean Guillaume, Actes du colloque de Tours (9–14 juin 1986), Paris, 1992, pp. 219–226.

[22] Konrad Hecht, "Maßverhältnisse und Maße der Cappella Pazzi," *Architectura* 6 (1976): 148–174.

[23] Jean Guillaume, "Désaccord parfait: ordres et mesures dans la chapelle des Pazzi," *Annali di Architettura* 2 (1991): 9–23.

[24] "Brachium continet 12 vntias," "Pes palmorum quattuor, pollicum seu vnciarum duodecim, digitorum vero sexdecim." We leave aside the *quattrino*, whose narrow step (9.98 mm) is not sufficiently differential. Ronald E. Zupko, *Italian Weights and Measures from the Middle Ages to the Nineteenth Century*, Philadelphia, The American Philosophical Society, 1981, pp. 47–48 (*braccio*), 197 (*piede*).

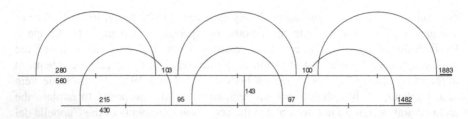

Fig. 11.2 *The Approval of the Franciscan Rule*, fresco by Giotto in the Upper Church of the Basilica of Assisi, ca. 1296–1299. Author's reconstruction after Zanardi, *Il Cantiere di Giotto*, p. 128

First, draw the layout of the fresco,[25] and record its principal measurements. Then mark on the layout: above the line on the left side, the radius of the arch; below this line, the diameter of the arch; and above the line on the right side, the total length of the vaulted ceiling (underlined) (Fig. 11.2).

Braccio-soldi system

	Horizontal dimensions (mm)		*Vertical dimensions* (mm)
549.08 < 560 < 569.05	419.30 < 430 < 449.25	1856.90 < 1883 < 1886.85	119.80 < 143 < 149.75
269.55 < 280 < 299.50	209.65 < 215 < 239.60	1467.55 < 1482 < 1497.50	
99.83 < 103 < 119.80	89.85 < 95 < 99.83		
99.83 < 100 < 119.80	89.85 < 97 < 99.83		

Number of concordances plus or minus the error: 2
Concordances: 99.83 mm = 2 o

Piede-palmi system

	Horizontal dimensions (mm)		*Vertical dimensions* (mm)
544.50 < 560 < 567.19	423.50 < <u>430</u> < 431.06	1875.50 < 1883 < 1883.06	136.13 < 143 < 151.25
272.25 < 280 < 294.94	211.75 < 215 < 226.88	1474.69 < <u>1482</u> < 1482.25	
90.75 < 103 < 113.44	90.75 < 95 < 113.44		
90.75 < 100 < 113.44	90.75 < 97 < 113.44		

Number of concordances plus or minus the error: 3
Concordances: 431.06 mm = 1 br. 3 d., 1482.25 mm = 4 br. 1 po., 1883.06 mm = 5 br. 3 d

We discover that the values expressed in either of these units are quite cumbersome, making their use by artists unlikely. Only 2 of the 11 measurements expressed using the *braccio-soldi* system are integer values ±3 mm and only 3 out of 11 in the *piede-palmi* system. This fact raises doubts as to the existence of an underlying system of measurements.

---

[25]We have relied on the photographic survey by Zanardi, Frugoni and Zeri in *Il Cantiere di Giotto*, p. 128. The survey scale can be deduced from the dimensions of the fresco (363 × 357 cm), the height of the standing figure of St Francis (122 cm), the height of the other Figs. (122, 111, 123 cm), and the diameter of the saint's halo (35.5 cm). We systematically checked the parallelism and absence of distorsion in the fresco, following Raynaud, "La théorie des erreurs et son application à la reconstruction des tracés perspectifs," see Appendix A.

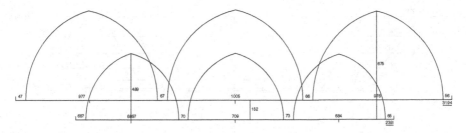

Fig. 11.3 *St. Francis Preaching Before Pope Honorius III*, fresco by Giotto in the Upper Church of the Basilica of Assisi, ca. 1296–1299. Author's reconstruction after Zanardi, p. 242

Let us now apply the same analysis to the fresco *St. Francis Preaching Before Pope Honorius III*, in which a space with a groined vault roof is represented.[26] (Fig. 11.3).

Braccio-soldi system			
	Horizontal dimensions (mm)		Vertical dimensions (mm)
998.33 < 1005 < 1018.30	698.83 < 709 < 718.80	3174.70 < 3194 < 3194.67	658.90 < 675 < 688.85
958.40 < 977 < 988.35	658.90 < 685 < 688.85	2346.08 < 2351 < 2366.05	479.20 < 489 < 499.17
958.40 < 976 < 988.35	658.90 < 684 < 688.85		149.75 < 152 < 179.70
59.90 < 67 < 89.85	59.90 < 73 < 89.85		
59.90 < 66 < 89.85	59.90 < 70 < 89.85		
49.92 < 56 < 59.90	59.90 < 65 < 89.85		
29.95 < 47 < 49.92	59.90 < 65 < 89.85		

Number of concordances plus or minus the error: 3
Concordances: 49.92 mm = 1 o., 149.75 mm = 5 s., 3194.67 mm = 5 br. 4 o

Piede-palmi system			
	Horizontal dim. (mm)		Vertical dim. (mm)
998.25 < 1005 < 1020.94	703.31 < 709 < 726.00	3176.25 < 3194 < 3198.97	665.50 < 675 < 680.63
975.56 < 977 < 998.25	680.63 < 685 < 695.75	2336.83 < 2351 < 2359.52	484.00 < 489 < 499.13
975.56 < 976 < 998.25	680.63 < 684 < 695.75		151.25 < 152 < 158.81
60.50 < 67 < 68.06	68.06 < 73 < 90.75		
60.50 < 66 < 68.06	68.06 < 70 < 90.75		
45.38 < 56 < 60.50	60.50 < 65 < 68.06		
45.38 < 47 < 60.50	60.50 < 65 < 68.06		

Number of concordances plus or minus the error: 7
Concordances: 45.38 mm = 2 d., 68.06 mm = 3 d., 151.25 mm = 1 pa. 2 po., 975.56 mm = 2 br. 2 pa. 3 d

Once again the values for the measurements of this space are improbably complicated, with 3 out of 19 approaching an integer value ±3 mm using the *braccio-soldi* system and 7 out of 19 using the *piede-palmi* system.

---

[26]Zanardi, *Il Cantiere di Giotto*, p. 242.

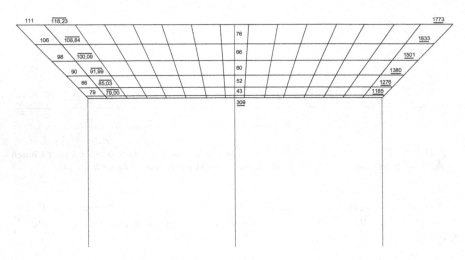

Fig. 11.4 *The Recovery of the Wounded Man of Lerida*, fresco by Giotto in the Upper Church of the Basilica of Assisi, ca. 1296–1299. Author's reconstruction after Zanardi, p. 332

Consider now *The Recovery of the Wounded Man of Lerida*, whose flat-coffered ceiling is similar to a checkerboard pattern drawn in perspective.[27] In the reconstructed perspective below (Fig. 11.4), each number in the first column on the left records the width of the narrowest coffer in that horizontal row, while the second column of numbers records the mean width of the coffers in the row; the column closest to the central axis gives the height of the central coffers; the numbers in the column on the right record the total width of each row of coffers; and finally the number below the axis gives the total height of the ceiling.

Braccio-soldi system			
	Horizontal dimensions (mm)		*Vertical dimensions* (mm)
99.83 < 111 < 119.80	99.83 < 118.23 < 119.80	1767.05 < 1773 < 1797.00	59.90 < 76 < 89.85
99.83 < 106 < 119.80	99.83 < 108.84 < 119.80	1617.30 < 1633 < 1647.25	59.90 < 66 < 89.85
89.85 < 98 < 99.83	99.83 < 100.09 < 119.80	1497.50 < 1501 < 1527.45	59.90 < 60 < 89.85

(continued)

---

[27]Zanardi, *Il Cantiere di Giotto*, p. 332. *The Recovery of the Wounded Man* is one of the earliest works to present a correct foreshortening of the intervals, but it does not represent a case of linear perspective because the correct perspective is limited to the coffered ceiling. (1) The side ceilings are depicted in oblique perspective, while the main ceiling is depicted in a central perspective. (2) The horizon is situated 722 mm above the level of the eye, with which it should instead coincide. (3) There is a lack of consistency in the foreshortening. The most remote horizontal line of the ceiling produces an interval that is as high as the one immediately preceding it, probably because of some confusion between this line and the one that marks the boundary of the coffered space. In perspective, however, two equal intervals ought to be of different heights. (4) The fresco displays some minor errors of drawing. For example, the axis of the ceiling is shifted 12 mm to the right compared to the axis of the composition.

(continued)

Braccio-soldi system			
89.85 < 90 < 99.83	89.85 < 91.99 < 99.83	1377.70 < 1380 < 1397.67	49.92 < 52 < 59.90
59.90 < 86 < 89.85	59.90 < 85.03 < 89.85	1257.90 < 1275 < 1287.85	29.95 < 43 < 49.92
59.90 < 79 < 89.85	59.90 < 79.00 < 89.85	1168.05 < 1185 < 1198.00	299.50 < 309 < 329.45

Number of concordances plus or minus the error: 8
Concordances: 49.92 mm = 1 o., 59.90 mm = 2 s., 89.85 mm = 3 s., 99.83 mm = 2 o., 119.80 mm = 4 s.,
1377.70 mm = 2 br. 5 s

Piede-palmi system			
	Horizontal dimensions (mm)		Vertical dimensions (mm)
90.75 < 111 < 113.44	113.44 < 118.23 < 121.00	1769.63 < 1773 < 1784.75	68.06 < 76 < 90.75
90.75 < 106 < 113.44	90.75 < 108.84 < 113.44	1610.81 < 1633 < 1633.50	60.50 < 66 < 68.06
90.75 < 98 < 113.44	90.75 < 100.09 < 113.44	1497.38 < 1501 < 1512.50	45.38 < 60 < 60.50
68.06 < 90 < 90.75	90.75 < 91.99 < 113.44	1361.25 < 1380 < 1383.94	45.38 < 52 < 60.50
68.06 < 86 < 90.75	68.06 < 85.03 < 90.75	1270.50 < 1275 < 1293.19	30.25 < 43 < 45.38
68.06 < 79 < 90.75	68.06 < 79.00 < 90.75	1179.75 < 1185 < 1202.44	302.50 < 309 < 317.63

Number of concordances plus or minus the error: 8
Concordances: 45.38 mm = 2 d., 60.50 mm = 2 po., 68.06 mm = 3 d., 90.75 mm = 1 pa., 113.44 mm = 1 pa. 1 d.,
121.00 mm = 1 pa. 1 po., 1633.50 mm = 4 br. 2 pa

In the three frescoes just examined, none of the dimensions measured turn out to be a simple combination of multiples or submultiples of the measurement units of that time.

1. In *The Recovery of the Wounded Man of Lerida*, the mean dimensions of the coffers (column 2), which are presumably a more reliable measure than the individual values, are no closer to integer numbers than the most erroneous values directly measured on the fresco (column 1).
2. For each series (width, mean width and height of the individual coffers, total width of a row, total height of the ceiling), many of the measurements are far from integer values (59.90 < 79 < 89.85; 299.50 < 309 < 329.45; 1257.90 < 1275 < 1287.85, etc.) and hence the dimensions of the perspective elements cannot be expressed in either *braccio-soldi* or *piede-palmi* units.
3. Very few of the individual measurements expressed in medieval units approach integer values. In *The Approval of the Franciscan Rule*, only 2 out of 11 in the *braccio-soldi* system and 3 out of 11 in the *piede-palmi* system do so. In *St. Francis Preaching Before Pope Honorius III*, 3 out of 19 in the *braccio-soldi* system and 7 out of 19 in the *piede-palmi* system approach integer values. In *The Recovery of the Wounded Man of Lerida*, there are 8 out of 24 in the *braccio-soldi* system, and 8 out of 24 in the *piede-palmi* system. Taken together, 13 out of 54 (24 %) measurements are close to integer values in the first system, and 18 out of 54 (33 %) in the second. The total number of values being greater than 50, the rate is comparable to a random set of occurrences.

In the *braccio-soldi* system, integer values fall on every 29.95 mm and 49.92 mm ± 3 mm interval. One *braccio* (LCM of *oncia* and *soldo*) contains twelve *oncie* error intervals and twenty *soldi* error intervals, four of which overlap. The probability of obtaining an integer value ± 3 mm in the LCM interval is therefore equal to $\frac{6(20 + 12 - 4)}{599.00} = \frac{168}{599} = 28\,\%$.

In the *piede-palmi* system, integer values fall every 22.69 mm and 30.25 mm $\pm$ 3 mm. One *palmo* (LCM of *pollice* and *dito*) contains three *pollici* error intervals and four *dita minuti* error intervals, among which one is in common. The probability of measuring an integer value in the LCM interval is thus $\frac{6(3 + 4 - 1)}{90.75} = \frac{36}{90.75} = 39\,\%$.

The occurrences of integer values for the theoretical and empirical measurements are of the same order of magnitude: 24 % ≈ 28 % and 33 % ≈ 39 %. Therefore the empirical concordances do not exceed the level of randomness. The slightly higher result in the *piede-palmi* system is due to the fact that there is a narrower gap between the *pollici* and the *dita minuti*.

4. Finally, let us hypothesize a variation in the historically documented standard units used by craftsmen. It is necessary to introduce such a variation, because the medieval units were not as precise as present measurement units. This variation is the result of several factors:

   (1) Historical fluctuations. For example, architectural surveys in the Quattrocento set the *braccio fiorentino* at 0.5875, 0.5860 or 0.5836 m.[28]
   (2) Differences between professions. *Braccio* and *piede* values were specific to different crafts, as demonstrated by their names: *agrimensorio, da legname, da muratori, da panno, da mercatori*, etc.
   (3) Fluctuations within and between regions. In Tuscany alone, the *braccio* took on the common value of 0.584 m in Arezzo, Florence, San Miniato, Pistoia, Siena, Montepulciano, Lucca, Pisa, Volterra, etc., but shorter or longer values were used in Fivizanno (0.486 m), Massa (0.4.95 m), Montecarlo (0.593 m), and Pontremoli (0.692 m).[29] The actual or supposed presence of master artists from other parts of Italy (Cavallini, Rusuti, Giotto, etc.) working at the site of the Upper Church prevents us from excluding one or another of the units that we know were being used in Assisi (*br.* 0.599 m; *p.* 0.363 m), Rome (*br.* 0.636 m; *p.* 0.298 m) or Florence (*br.* 0.584 m).

In order to take into account these historical, professional and regional fluctuations, let us produce a continuous variation of the *braccio* from 525.6 mm (*br. fl.* −10 %) to 699.6 mm (*br. rom.* +10 %) and then study the fluctuation in the number of integer values y as a function of this extensible *braccio* x. If the function

---

[28]These *braccio* values are given, for instance, by Konrad Hecht, "Maßverhältnisse und Maße der Cappella Pazzi"; Leonardo Benevolo, Stefano Chieffi e Giulio Mezzetti, "Indagine sul S. Spirito di Brunelleschi," *Istituto di Storia dell'Architettura. Quaderni* 85/90 (1968): 1–52; and Christoph L. Frommel, *Der Römische Palastbau der Hochrenaissance*, 3 Bde, Tübingen, 1973.

[29]Zupko, *Italian Weights and Measures*, p. 46.

$y = f(x)$ admits a maximum that is almost equal to the total number of fresco measurements, this maximum will represent the value of the unit being searched for. An optimization algorithm enables us to detect the maxima. The function admits 10 values (550.01 $\le$ *br.* $\le$ 550.24) as a *minimum minimorum*, and 27 values (652.40 $\le$ *br.* $\le$ 652.44) as a *maximum maximorum*.

The integer values plus or minus the margin of error range from $\frac{10}{54} = 18$ to $\frac{27}{54} = 50\%$, i.e. less than one-half of the total. Consider now a continuous variation in the value expressed in *piede* from 268.2 mm (*p. rom.* -10%) to 399.3 mm (*p. umbro* $+10\%$). The number of integer values fluctuates from 12 (375.34 $\le$ *p.* $\le$ 375.54) to 31 (273.30 $\le$ *p.* $\le$ 273.35). Thus, the range of integer values plus or minus the margin of error extends from $\frac{12}{54} = 22$ to $\frac{31}{54} = 57\%$. It follows that neither of the medieval unit systems can convert the dimensions measured on the frescoes into integer or simple values. The fresco dimensions reveal no numerical consistency and in most cases they come to a standstill after their division by the smallest subunit. This conclusion exhibits a significant discrepancy with respect to the arithmetic hypothesis, which requires a unit to be repeated at regular intervals.

11.3   The Absence of Simple Proportional Ratios

There is another possibility that could resolve the inherent problems and allow us to salvage the arithmetic hypothesis. Let us imagine that painters relied on arithmetic formulas but applied them using non-numerical instruments such as lengths of cord that could be folded into equal parts to determine any given ratio. In this case the arithmetic method would work even in the absence of any standard units. Suppose that, like their predecessors in classical antiquity, Duecento and Trecento painters used dimensionless modules. With the arithmetic method, homologous parts should nevertheless stand in the simple proportional ratio a_n/a_{n+1}. This is what Andrés de Mesa assumes when he takes $a_1/a_2 = 2$.[30] But in fact the ratios are entirely random in the frescoes examined by us. In *The Approval of the Franciscan Rule*, the ratios

$$\frac{430}{560} = 0.7678\ldots, \frac{1482}{1883} = 0.7870\ldots$$

do not coincide with the elementary fractions $\frac{3}{4} = 0.75$ or $\frac{4}{5} = 0.8$.

Likewise, in *St. Francis Preaching Before Pope Honorius III*,

---

[30]De Mesa, "El 'fantasma' del punto de fuga," pp. 34 (Fig. 8). The concept of proportional ratios was repeated, without success, by Pietro Roccasecca, "La prospettiva lineare nel Quattrocento: dalle proporzioni continuata e ordinata alla proporzione degradatta," S. Rommevaux et al., eds., *Proportions. Science—Musique—Peinture & Architecture*, Turnhout, 2011, pp. 277–297.

$$\frac{685}{977} = 0.7011\ldots, \frac{709}{1005} = 0.7054\ldots\frac{684}{976} = 0.7008\ldots\frac{2351}{3194} = 0.7360\ldots$$

differs from $\frac{3}{2} = 0.6666\ldots$ or $\frac{3}{4} = 0.75\ldots$

The result is even clearer in the third fresco. *The Recovery of the Wounded Man of Lerida* allows us to calculate similar ratios from more accurate mean values. However, the ratios

$$\frac{79.00}{85.03} = 0.9290\ldots, \frac{85.03}{91.99} = 0.9243\ldots, \frac{91.99}{100.09} = 0.9190\ldots$$

$$\frac{100.09}{108.84} = 0.9196\ldots \text{ and } \frac{108.84}{118.23} = 0.9205\ldots$$

fluctuate around $\frac{12}{13} = 0.9230\ldots$, which is an unlikely fraction owing to its denominator. Therefore, craftsmen seem to have set aside elementary fractions as well.

It is also necessary to examine de Mesa's argument (*op. cit.*, pp. 43–45) that painters used the *superbipartiens* rule to plan their foreshortening. This proportionality rule consists in tracing an interval two-thirds as high as the previous one. Alberti had already criticized this method:

> Here some would draw a transverse line parallel to the base line of the quadrangle. The distance which is now between the two lines they would divide into three parts and, moving away a distance equal to two of them, add on another line. They would add to this one another and yet another, always measuring in the same way so that the space divided in thirds which was between the first and second always advances the space a determined amount. Thus continuing, the spaces would always be, as the mathematicians say, *superbipartiens* to the following spaces…[31]

This passage can be interpreted in several ways. The narrowest interpretation consists in attaching more importance to the two-thirds rule than to the name—*superbipartiens* proportion—that Alberti assigned to it. The broader interpretation, in the contrary sense, considers the two-thirds rule as merely an instance of the general case of *superbipartiens* proportions.

The narrow interpretation does not apply to the works studied in this chapter. Let us return to *The Recovery of the Wounded Man of Lerida* with its 15×5 checkerboard ceiling. We can compare the intervals to two *superbipartiens* series resulting either from the foreshortening of the largest interval:

---

[31]"Hic essent nonnulli qui unam ab divisa aequedistantem lineam intra quadrangulum ducerent, spatiumque, quod inter utrasque lineas adsit, in tres partes dividerent. Tum huic secundae aequedistanti lineae aliam item aequedistantem hac lege adderent, ut spatium quod inter primam divisam et secundam aequedistantem lineam est, in tres partes divisum una parte sui excedat spatium id quod sit inter secundam et tertiam lineam, ac deinceps reliquas lineas adderent ut semper sequens inter lineas esset spatium ad antecedens, ut verbo mathematicorum loquar, superbipartiens…" Leon Battista Alberti, *De la peinture/De pictura* (1435), eds. Schefer and Deswarte-Rosa, Paris, 1992, pp. 116–117, commentary p. 242.

$76\frac{2}{3} = 50.7$ $76\frac{4}{9} = 33.8$ $76\frac{8}{27} = 22.5$ $76\frac{16}{81} = 15.0$ (series 1)

or the enlargement of the smallest interval:

$43\frac{3}{2} = 64.5$ $43\frac{9}{4} = 96.7$ $43\frac{27}{8} = 145.1$ $43\frac{81}{16} = 217.7$ (series 2)

Neither of the two series corresponds to the intervals in *The Recovery* (series 0):

Series 1	76	50.7	33.8	22.5	15.0
Series 0	76	66	60	52	43
Series 2	217.7	145.1	96.7	64.5	43

The broad interpretation rests on the fact that the *superbipartiens* proportion concept concerns a class of ratios. The concept itself stems from the medieval theory of proportions.[32]

The *superpartiens* genus characterizes every proportion $\left(n + \frac{p}{q}:1\right)$; n, p, and q being natural integers, with $n = 1, p < q, p \geq 2$. Among the species of this genus, the *superbipartiens*[33] proportion meets the additional condition that $p = 2$. The two-thirds rule $\left(1 + \frac{2}{3}:1\right)$ is then given for $q = 3$, which is a subspecies of the *superbipartiens* species. It is thus possible to read in *De Pictura* a foreshortening rule that extends to any proportional ratio $\left(\left(1 + \frac{2}{q} : 1\right), q \in \mathbb{N}, q > 2\right)$. Unfortunately, no such ratios correspond to the intervals of *The Recovery of the Wounded Man*. The series diverges as q increases, so the best match occurs for $q = 3$, which is an unsatisfactory case.[34] Consequently there is no proof that the *superbipartiens* rule was used.

---

[32]The most influential texts were those by Boethius, *De Institutione arithmetica*, ed. by J.-Y. Guillaumin, Paris, 1995; Jordanus de Nemore, *De Elementis arithmetice artis. A Medieval Treatise on Number Theory*, ed. by Hubert L.L. Busard, Stuttgart, 1991; and Hubert L.L. Busard, "Die Traktate 'De proportionibus' von Jordanus Nemorarius und Campanus," *Centaurus* 15 (1971): 193–227. Correct definitions can also be found in less well-known treatises, such as an anonymous *Tractatus proportionum*: "The first species of the superpartiens genus is the superbipartiens proportion, which is produced when the greatest number contains the entire smallest plus two parts, such as 5 to 3, 7 to 5. The second above-mentioned species is the supertripartiens proportion, which is produced when the greatest number contains the entire smallest plus three parts, such as 7 to 4, 11 to 8, etc./ Prima species superpartienti generis est proportio superbipartiens, que fit quando maior numerus continet totum minorem et insuper eius duas partes, ut 5 ad 3, 7 ad 5. Secunda species supradicti generis est proportion supertripartiens, que fit quando maior numerus continet in se totum minorem in se et insuper eius tres partes, ut sunt 7 ad 4, 11 ad 8, etc.," Saint-Dié, Bibliothèque municipale, MS. 42, fol. 119r.

[33]Supertripartiens, superquadripartiens, and superquinquepartiens proportions were formed on the same pattern.

[34]To match the observed series to the superbipartiens series, a non-integer q would be required (the optimal matching would be for $q = 2.3$ that provides the terms 43.45, 49.97, 57.47, 66.09, 76.00). This, however, is impossible by definition.

Fig. 11.5 Foreshortening in
constant ratio. Author's
drawing

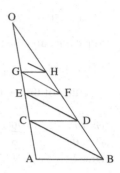

We can nevertheless imagine a *latissimo sensu* interpretation of the rule by extending it to all cases in which a given term of the series is a constant ratio of the previous one. The rule is thus widened to *multiplex, superparticularis, superpartiens, multiplex superparticularis* and *multiplex superpartiens* proportions. This interpretation fails to provide a better match to *The Recovery*'s intervals, because they do not follow a constant ratio. The series 0.827 (43/52); 0.867 (52/60); 0.909 (60/66); 0.868 (66/76) can be compared to the corresponding ratios of a checkered pattern in linear perspective. Let us establish first that constant ratios are inconsistent with linear perspective.

(1) Lines AB, CD, EF, GH... are horizontal, and points A, C, E, G... are collinear (Fig. 11.5). Therefore, $\angle BAC = \angle DCE = \angle FEG$...
(2) By hypothesis, the intervals are in a constant ratio to each other; hence $AC/AB = CE/CD = EG/EF$...
(3) As stated in Euclid's *Elements*, Book VI, Prop. 7, it follows that $\angle ABC = \angle CDE = \angle EFG$...
(4) Since AB, CD, EF, GH... are parallels, the diagonals BC, DE, FG... are parallel to each other and cannot intersect, whereas a linear perspective requires a concurrent point (distance point). Consequently, successive equal intervals in a linear perspective cannot be in a constant ratio. Reciprocally, the perspective in which successive intervals stand in a constant ratio cannot be a linear perspective.

Knowing that the intervals in *The Recovery* are not in a constant ratio, its ceiling foreshortening should be analyzed and compared to a linear perspective. The empirical values are already known and the theoretical values can be found by means of analytic geometry. Let us begin with a simplified schema of the fresco, on which the coordinates (x, y) of the points that can be used to obtain the theoretical values are marked (Fig. 11.6).

Mark in system H_1xy, the coordinates of points C_1, D_1, E_1, F_1 and G_1; the coordinates of the central point O; and the distance point T that should be used to create a linear perspective. Find the ordinates of points C_6, D_5, E_4, F_3, G_2 that set the height of the intervals. Each point belongs to two lines: $C_6 = OC_1 \cap H_1T$; $D_5 = OD_1 \cap H_1T$... Therefore each point solves a system of equations that

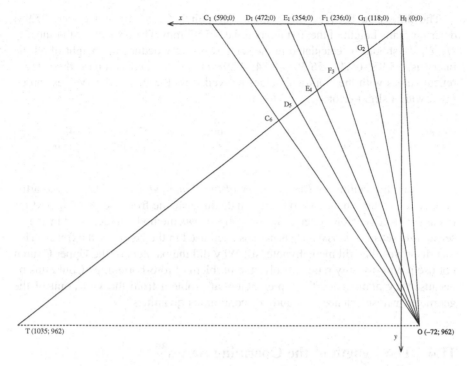

Fig. 11.6 Schema for *The Recovery of the Wounded Man of Lerida*. Author's drawing

describe the lines to which the point belongs. For example, point G_2 solves the system

$$\begin{cases} y = \frac{962}{1305}x & (1 : eq.H_1T) \\ y = -\frac{962}{118+72}x + 597.45 & (2 : eq.G_1O) \end{cases}$$

By introducing the value x of (1) in (2), we obtain

$$-\frac{962}{118+72}\left(\frac{1305}{962}y\right) + 597.45 - y = 0$$

and after simplification and factorization

$$-y\left(\frac{1305}{118+72} + 1\right) + 597.45 = 0$$

whence

$$y = \frac{597.345}{1 + \frac{1305}{118+72}} = \frac{597.45}{7.869} = 75.93$$

The difference between the ordinates of points G_1 $(x; 0)$ and G_2 $(x; 75.93)$ determines the height of the first interval, i.e. 75.93 mm. The ordinates of points C_6, D_5, E_4, F_3 should be calculated in the same way. Now deduce the height of all the intervals: 75.93; 64.44; 56.36; 48.84; 42.98 (series 3), and compare these theoretical values with the empirical values derived from the fresco. The values can be fitted with a slight error $\epsilon_{max} = 3.64$ mm:

Series 0	76	66	60	52	43
Series 3	75.93	64.44	56.36	48.84	42.98

To draw the ceiling in *The Recovery of the Wounded Man of Lerida* the artist utilized a foreshortening method that is indistinguishable from the one required for linear perspective. Consequently no *superbipartiens* method—understood in *stricto sensu, lato* sensu or *latissimo* sensu—has been used in the layout of the fresco. This fact disproves the arithmetic hypothesis. Why did the painters of the Upper Church not use proportionality ratios to solve the problem of foreshortening? Could this be because they approached the representational problem from the viewpoint of the geometer, whose science is a study of continuous quantities?

11.4   The Length of the Operating Series[35]

The apparent simplicity of the arithmetic method is due in part to the fact that certain operations are left deliberately vague. For instance, the proportionality of the intervals is established on parallels lines, but the existence of these parallels is taken for granted, whereas they need to be constructed. Many pre-perspective paintings, such as *The Approval of the Franciscan Rule, St. Francis Preaching Before Pope Honorius III,* or *Christ among the Doctors*, present an axis of symmetry. But to draw an axis of symmetry based on proportional ratios is no easy matter. It is therefore necessary to describe *all the operations* of the construction process in order to fairly compare the geometric and arithmetic methods for creating a perspective.

Consider once again *The Recovery of the Wounded Man of Lerida* and, in order to lay to rest all previous objections, suppose that the intervals obey the rule of simple dimensions and proportions. Assume the layout to be a ruler-and-compass construction, in accordance with the usual devices of geometry. The operating series can be described on different scales. One should distinguish between *macro-operations* (*m.o.*) (draw a perpendicular, divide a line into n equal parts...) and *elementary operations* (*e.o.*) (take a given aperture of the compass, join two points with a ruler...).

---

[35]For details, see D. Raynaud, "Las primeras perspectivas de los siglos XIII y XIV según el enfoque del modus operandi," in Magno Mello (ed.), *Perspectiva: fundamentação teórica e cultural*, Belo Horizonte, 2009, pp. 41–62.

Fig. 11.7 Steps of
construction of a
perpendicular to a given line.
Author's drawing

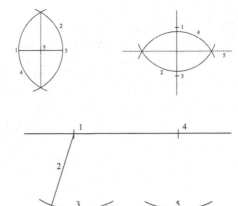

Fig. 11.8 Steps of
construction of a parallel to a
given line. Author's drawing

For example, the *m.o.* "draw a perpendicular to a given line" contains five *e.o.*:
"fix the needle of the compass on a point of the given line," "draw a circle of any
aperture," "fix the needle on another point of the given line," "draw a circle of same
aperture," and "join the intersecting points of the circles with a ruler" *(Elements,* I,
11, Fig. 11.7).

Similarly, the *m.o.* "draw a parallel to a given line" contains 6 *e.o.* (Fig. 11.8).

We can compare the lengths of the different operating series by comparing the
minimum number of operations required (the *m.m.o.* and *m.e.o.,* respectively).

Operating series using the arithmetic method (Fig. 11.9)

1. Draw the axis OS (take two marks, draw a vertical), 7 e.o.
2. Draw the first horizontal $A_1P_1 \perp OS$ (draw a perpendicular), 5 e.o.
3. Draw the horizontal A_2P_2 at a distance A_1P_1 from H_1H_2 (draw a parallel), 6 e.o.
4. Calculate $H_2H_3 = $ k. H_1H_2 (apply a proportional ratio), 2 e.o.
5. Draw the horizontal A_3P_3 at a distance A_2P_2 from H_2H_3, 6 e.o.
   Repeat operations 4 and 5 three times to obtain lines $A_4P_4 \dots A_6P_6$, 24 e.o.
12. Calculate $\frac{1}{2}H_1I_1$ (apply a ratio), 3 e.o.

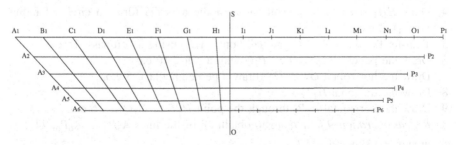

Fig. 11.9 *The Recovery of the Wounded Man of Lerida.* Author's perspective reconstruction
based on the arithmetic method

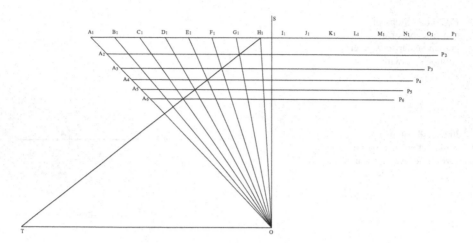

Fig. 11.10 *The Recovery of the Wounded Man of Lerida.* Author's perspective reconstruction according to the geometric method

13. Draw H_1I_1 on A_1P_1 on both sides of axis OS (draw a circle of radius $\frac{1}{2}H_1I_1$), 2 e.o.
14. Transfer the interval H_1I_1 gradually onto A_1P_1 using a compass, 15 e.o.
15. Calculate $K = k^5$, 5 e.o.
16. Calculate $\frac{1}{2}H_6I_6 = K.\frac{1}{2}H_1I_1$ (apply a ratio), 3 e.o.
17. Draw H_6I_6 on A_6P_6 on both sides of the axis OS, 2 e.o.
18. Transfer the interval H_6I_6 segment by segment onto A_6P_6 using a compass, 15 e.o.
19. Join A_1A_6 with a ruler, 1 e.o.
    Repeat operation 19 fifteen times to obtain all the segments $B_1B_6 \ldots P_1P_6$, 15 e.o.
    m.m.o. = 34, m.e.o. = 111

Operating series using the geometric method (Fig. 11.10)

1. Draw the axis OS (draw a vertical), 3 e.o.
2. Draw the first horizontal $A_1P_1 \perp OS$ (draw a perpendicular), 5 e.o.
3. Calculate $\frac{1}{2}H_1I_1$ (apply a ratio), 3 e.o.
4. Draw H_1I_1 over A_1P_1 on both sides of the axis OS (draw a circle of radius $\frac{1}{2}H_1I_1$), 2 e.o.
5. Transfer the interval H_1I_1 gradually onto A_1P_1 using a compass, 15 e.o.
6. Draw the pencil lines $A_1O \ldots P_1O$ with a ruler, 16 e.o.
7. Draw the horizontal $OT \perp OS$ (draw a perpendicular), 5 e.o.
8. Draw the diagonal H_1T, 3 e.o.
9. Draw the horizontal A_2P_2 through the point $H_1T \cap G_1O \ldots$, 6 e.o.
    Repeat operation 9 four times to obtain all of the lines $A_3P_3 \ldots A_6P_6$, 24 e.o.
    m.m.o. = 13, m.e.o. = 83

A comparison of the two operating series shows that the arithmetic method is less advantageous because it requires a larger number of operations than the geometric method (34 vs. 13 *m.m.o.* and 111 vs. 83 *m.e.o.* respectively). Furthermore, one should note that the present comparison has been limited to the devices of classical geometry,[36] which assumes that the perspective is constructed with the ruler and compass alone. But practical geometry[37] freed the operator from such constraints by giving him recourse to many geometric instruments, a fact known since earliest antiquity. Let us add to the ruler and compass three other instruments whose use in the thirteenth century is well documented—the square, the level, and the *libella*. Thus, the arithmetic method appears to involve twice as many steps as the geometrical, instrument-based method (34 vs. 13 *m.m.o.* and 111 vs. 63 *m.e.o.* respectively). This difference is not due to the complexity of the arithmetic method, but rather to the fact that it includes many background operations. If one includes these hidden operations, the method is much more burdensome.

11.5   The Coincidence of Points at Infinity with Visible Loci

Since points at infinity (the central and lateral vanishing points) are useless according to the arithmetic method, it follows from de Mesa's thesis that painters would not have used them. Consequently, there is no reason why our reconstructed points at infinity should coincide with the visible loci of the pictorial composition. In contrast, if a systematic and accurate correspondence is found between them, it would support the thesis that Duecento and Trecento painters actually used geometric devices instead of arithmetic ones.

---

[36]The most popular text was Adelard of Bath's version of Euclid's *Elements*; for the purposes of our discussion see H.L.L. Busard (ed.), *The First Latin Translation of Euclid's Elements commonly ascribed to Adelard of Bath*, Books I–VIII and Books X.36–XV.2, Toronto, The Pontifical Institute of Medieval Studies, 1983. For a mathematical commentary, see Euclid, *Les Éléments*, ed. by Bernard Vitrac, 4 vols., Paris, PUF, 1990–2001; also of interest is the commentary on Euclid's *Elements* by al-Nayrīzī, *Anaritii in decem libros priores Elementorum Euclidis commentarii ex interpretatione Gherardi Cremonensis*, edidit M. Curze, Leipzig, 1899.

[37]See the treatises by Boethius (Menso Folkerts, « *Boetius* » *Geometrie II: Ein mathematisches Lehrbuch des Mittelalters*, Wiesbaden, 1970); Abraham bar Ḥiyya (Maximilian Curze, *Der Liber Embadorum des Abraham bar Chijja Savasorda in der Übersetzung des Plato von Tivoli*, Leipzig, 1902); the anonymous *Artis cuiuslibet consummation* (Stephen K. Victor, *Practical Geometry in the High Middle Ages*, Philadelphia, 1979); Leonardo Fibonacci (Baldassare Boncompagni, *Leonardi Pisani. Practica geometriae*, Roma, 1862); Abū al-Wafā' al-Būzjānī (*Kitāb fī mā yahtāju ilayhi al-sāni' min a 'māl al-handasa*, ed. by Svetlana Krasnova, "Abu-l-Vafa al-Buzdjani, Kniga o tom, chto neobhodimo remeslenniku iz geometricheskih postroenij" in *Fiziko-Matematicheskie Nauki v Stranah Vostoka* 1 (1966): 56–130 [text], 131–140 [commentary]); and Nicolas Chuquet (*La Géométrie. Première géométrie algébrique en langue française* (1484), ed. Hervé L'Huillier, Paris, 1979).

The systematic correspondence of the vanishing points with visible loci was already a feature of certain compositions from the period of Giotto. For instance, in *The Recovery of the Wounded Man* the central point O and the lateral points $T\,T'$ are aligned, thus forming a line: the so-called 'horizon line' of classical perspective. But with the side ceiling being divided into four rows of six coffers, it appears that this line coincides with a perfectly visible line that divides the lateral ceilings into two equal parts. In addition, the checkerboard ceiling is marked by the diagonals $A_1B_2C_3...$ $B_1C_2D_3...$ $C_1D_2E_3...$ on the left side, and by the diagonals $P_1O_2N_3...$ $O_1N_2M_3...$ $N_1M_2L_3...$ on the right side. The diagonals converging on the lateral points $T\,T'$ are precisely the ones that should be used to obtain the perspective foreshortening. If medieval painters did ever follow the arithmetic method, why are the horizon and diagonals lines so neatly visible? This correspondence is not accidental. A squared coffer divided by a diagonal is quite usual in the Assisi frescoes. It works as a spatial roofing pattern in several scenes: *St Francis Honored by a Simple Man*, *St Francis before the Sultan (Trial by Fire)*, *The Death of the Knight of Celano*, and *Christ among the Doctors*. Thus we cannot exclude the possibility that such diagonals could have served as an empirical rule to place the receding lines in depth.

11.6   Reinterpreting Perspective Anomalies

A comparison of the arithmetic and geometric methods in early perspective painting shows the widespread use of geometric devices in the Duecento and Trecento. This is a somewhat counterintuitive finding. We must attempt to clarify the discrepancy between the present results and generally accepted views on perspective in medieval painting by interpreting afresh the main errors affecting medieval paintings.

1. The somewhat arbitrary representation of small architectural elements, such as the abaci of capitals, does not particularly support the arithmetic method. Such anomalies are understandable considering the material constraints imposed by the process of fresco painting. The quick-drying plaster required the rapid transfer of the drawing onto the *intonaco*. Consequently only the principal lines were transferred, and not the smaller elements.
2. The convergence of edges on a vanishing region hardly offers better proof in favor of an arithmetic formula. Methodological analysis shows that all perspectives—whether derived from an arithmetic or geometric formula—are subject to error. An error of parallelism in the perspective is sufficient to cause the vanishing lines to deviate.
3. The presence of a vanishing axis is not so easily attributable to errors of construction. This pictorial scheme does indeed present a systematization that departs from the purely random. As previously mentioned, the axial

composition could have arisen from the theory of binocular vision expounded by Ibn al-Haytham *(De aspectibus*, III, 2) and his Latin commentators; many works whose edges converge on two central vanishing points correspond to the case of homonymous diplopia identified by Ibn al-Haytham.[38]

---

[38]See Appendix A.

Conclusion

The intention of this book was to test the interdependence between medieval theories of optics and the practice of perspective. More particularly we chose to examine the question of physiological optics, because the proof of a connection between optics and perspective would be less probative if it took geometric optics as its starting point (for example, it is not possible to separate the respective influences of optics and geometry in the use of similar triangles, a method shared by the two sciences). In contrast, the theory of binocular vision is critically dependent on the particular way in which optics was addressed in the Middle Ages, combining as it did geometric and physical optics with anatomy and ocular physiology.

It is within this framework that the differing interpretations of "axial perspective" and "two-point perspective" were examined here. As has been seen, these forms of representations do not correspond to any type of parallel projection that places the spectator at infinity, therefore excluding the possibility that they might constitute axonometric (isometric, dimetric, trimetric) or oblique (cavalier and military) projections. One could go even further in the critical analysis of the two-point system of representation:

1. Two-point perspective is not a linear perspective, because the two vanishing points are separated by distances that are many times greater than the maximal metric error.
2. Two-point perspective bears no relationship to bifocal perspective (Parronchi's hypothesis).
3. Two-point perspective cannot be considered a form of trifocal perspective, whose object verticals are always represented by parallel lines.
4. The two vanishing points are not the result of an absence of parallelism in the side walls nor of an alteration in the parallelism intended to highlight the subject of the scene.
5. Two-point perspectives do not fall into the category of curvilinear perspective, the object lines being represented by lines rather than by the arcs of a circle (Chap. 8).
6. Two-point perspectives do not represent the application of a modified curvilinear perspective, since the development of projection circles prevents the

© Springer International Publishing Switzerland 2016
D. Raynaud, *Studies on Binocular Vision*,
Archimedes 47, DOI 10.1007/978-3-319-42721-8

vanishing lines from concurring in a single point (the Hauck–Panofsky conjecture, Chap. 9).

7. Two-point perspective cannot be categorized as a form of synthetic perspective, because the latter precludes the strict convergence of vanishing lines in one point (the White–Carter conjecture, Chap. 10).

8. Two-point perspectives are not based on the arithmetic method, because the distances measured in the compositions studied here are not the multiples or sub-multiples of any known units, nor can they be reduced to multiples of each other (the conjecture proposed by De Mesa Gisbert, Chap. 10).

This critical review of the current interpretations of axial perspective supports the thesis that the principles of binocular vision were applied in the early stages of the development of artificial perspective. Two-point perspective was in general use during the period 1295–1450, and the exact same construction can be found much later, in the works of Passerotti, Caracci, Tibaldi, Le Brun and de Sève. In them we find two pencils of lines that cross the central axis before concurring in two vanishing points, F and F', on the horizon line.

Three concluding questions remain to be examined:

1. Who were the actors responsible for introducing the theories of optics to the first practitioners of perspective?
2. To which disciplines does this cross between optics and perspective belong?
3. Why was the use of binocular perspective abandoned during the classical period?

The Birth of the Optics–Perspective Nexus in the Late Duecento

It is useful to begin by asking who introduced optics to the earliest practitioners of perspective, and notably including the artists commissioned to paint the fresco cycles for the Basilica of Assisi beginning in 1296. The theory of optics could have been transmitted in two ways:

1. A *direct path* through *De aspectibus*, the Latin translation of Ibn al-Haytham's treatise on optics that began to circulate at the beginning of the Duecento. An Italian translation, *De li aspecti*, became available later in the Trecento.[1]
2. An *indirect path*, via the numerous Latin treatises and commentaries on Ibn al-Haytham's optics produced during the course of the Middle Ages. The

---

[1] Enrico Narducci, "Nota intorno a una traduzione italiana fatta nel secolo decimoquarto del trattato d'ottica d'Alhazen," *Bollettino di bibliografia e di storia delle scienze matematiche e fisiche* 4 (1871): 1–40; Graziella Federici Vescovini, "Contributo per la storia della fortuna di Alhazen in Italia. Il volgarizzamento del ms. vat. 4595 e il 'Commentario Terzo' del Ghiberti", *Rinascimento* 5 (1965): 17–49; Eadem, "Alhazen vulgarisé: Le 'De li aspecti' d'un manuscrit du Vatican (moitié du XIVe siècle) et le troisième 'Commentaire sur l'optique' de Lorenzo Ghiberti," *Arabic Sciences and Philosophy* 8 (1998): 67–96.

principal treatises—Roger Bacon's *Perspectiva*,[2] Witelo's *Perspectiva*,[3] and two works by John Pecham, *Tractatus de perspectiva*[4] and *Perspectiva communis*[5]—dealt primarily with physiological optics. As we have shown in Chap. 6, all of these texts refer explicitly to the experiments carried out by Ibn al-Haytham using his binocular tablet and all differentiate between the cases of homonymous and crossed diplopia.

One necessary condition for the transmission of this knowledge was that the texts had to be accessible to the earliest perspectivists. And in fact works on optics appear in the inventories of many libraries in Rome, Florence and Padua,[6] intellectual centers where optics was studied and taught in the late Middle Ages, and where artists were beginning to use two-point perspective. The profusion of texts in circulation would certainly have multiplied the opportunities to derive a system of perspective from the laws of optics. This system—which appears unorthodox in the light of the modern conventions of linear perspective—becomes perfectly intelligible if examined in the context of the optical theories of Ibn al-Haytham and his Latin commentators.

Access to authoritative texts is a necessary, but not sufficient condition for the transfer of knowledge. Key works might have been *available* to craftsmen but ignored for various reasons. In fact, one of the greatest obstacles to their transmission was the fact that these texts were—with the exception of *De li aspecti*—written in Latin, a language that was beyond the reach of artisans, who received their training in the atelier rather than at university. Various hypotheses regarding the actual processes of knowledge transfer have been advanced, ranging from personal contacts between savants and artists (for example, between Toscanelli and Brunelleschi), to the instruction lavishly dispensed in the *scuole d'abaco* by masters such as Grazia de' Castellani in Florence. Alongside these hypotheses must be placed the rarely mentioned but central role of the *art commission*.[7]

Let us limit ourselves here to the earliest experiments in perspective—the frescoes in the Basilica of Assisi realized between the end of the Duecento and the

---

[2]*The Opus maius of Roger Bacon*, ed. J.H. Bridges, Frankfurt am Main, 1964, pp. 92–99, new edition: David C. Lindberg, *Roger Bacon and the Origins of Perspectiva in the Middle Ages: A Critical Edition and English Translation of Bacon's* Perspectiva *with Introduction and Notes*, Oxford, 1996.

[3]Witelo, *Opticae Thesaurus libri X*, edited and published by Friedrich Risner, Basel, 1572; modern edition published by David Lindberg, New York, 1972, pp. 98–108. It should be recalled that many more copies of Bacon's *Perspectiva* and Pecham's *Perspectiva* were in circulation than Witelo's *Perspectiva*.

[4]John Pecham, *Tractatus de perspectiva*, ed. D.C. Lindberg, New York, 1972, pp. 56–57.

[5]David C. Lindberg, *John Pecham and the Science of Optics*, Madison, 1970, pp. 116–118.

[6]David. C. Lindberg, *A Catalogue of Medieval and Renaissance Optical Manuscripts*, Toronto, 1975.

[7]Dominique Raynaud, *L'Hypothèse d'Oxford: Essai sur les origines de la perspective*, Paris, p. 319 and passim; this hypothesis was taken up in some detail by Jean-Philippe Genet, "Revisiter Assise: la lisibilité de l'image médiévale," in *Itinéraires du savoir de l'Italie à la Scandinavie (Xe–XVIe siècle). Études offertes à Élisabeth Mornet*, ed. Corinne Péneau, Paris, 2009, pp. 415–417.

beginning of the Trecento in central Italy. During this period artists constructed perspectives based on a vanishing axis (*The Approval of the Franciscan Rule* or *St. Francis Preaching before Honorius III*) and worked out the correct reduction of receding intervals (for example, in *Christ among the Doctors* and *The Recovery of the Wounded Man of Lerida*).[8] How can these significant advances, which took place during the course of a single decade from 1296 to 1305, be explained?

We have shown elsewhere[9] the direct contribution made by members of the Franciscan order to the spread of the science of optics. To mention just a few names, Bartholomew of England, Roger Bacon, and John Pecham, all of whom cited Alhacen's *De aspectibus* in their works, were Friars Minor. Within the order many of them exercised the functions of Master Regent before becoming provincial ministers or Ministers-General. In their shift from academic to administrative duties, the friars were in no way required to repudiate the knowledge acquired during their formation as scholars. Furthermore, one of the responsibilities of the Ministers-General would have been to commission and supervise the work of architects and artists in building and embellishing their churches.

Renato Bonelli[10] has identified various inflections in the construction of the Upper Church of the Basilica of Assisi introduced by architects working under successive Ministers-General. In 1240–1244 a transept in the Angevin Gothic style was commissioned by Aymon de Faversham. In 1247–1257 Giovanni da Parma oversaw the construction of the primary walls and supporting structures of the church. During his tenure as Minister-General between 1258 and 1273, Bonaventure commissioned the building of the church's Gothic-style groined vault. Each of these ministers was a former academic: Aymon de Faversham first taught at the universities in Tours, Padua and Bologna; Giovanni da Parma completed his studies in Paris and served as Master Regent at the universities of Bologna and Naples; and Bonaventure had been a celebrated professor at the *studium* in Paris.

Similarly, the Ministers-General were involved in supervising the decoration of the order's churches with fresco cycles by master artists. Matteo d'Acquasparta (who had studied under John Pecham in Paris before becoming Master Regent in Bologna and Paris and then *lector Sacri Palatii*) was appointed General of the Franciscan Order at the Narbonne chapter and, while serving in this position between 1287 and 1289, chose the iconographic program for the lower register of the nave of the Upper Church in Assisi.[11] Giovanni Minio da Morrovalle (formerly

---

[8]The correctness of the perspective in this fresco (single vanishing point, correct reduction in the intervals) was first noted by Decio Gioseffi, *Perspectiva artificialis*, Trieste, 1957, pp. 60–73.

[9]Dominique Raynaud, *Optics and the Rise of Perspective. A Study in Network Knowledge Diffusion*, Oxford, 2014.

[10]Renato Bonelli, "Basilica di Assisi: i committenti," *Antichità Viva / Mélanges Luisa Becherucci* 24 (1985): 174–179.

[11]Elvio Lunghi, *La Basilica di San Francesco di Assisi*, Antella, 1996, p. 63; Bruno Zanardi, *Giotto and Pietro Cavallini: La questione di Assisi e il cantiere medievale della pittura a fresco*, Milan, 2002; Donal Cooper and Janet Robson, "Pope Nicholas IV and the upper church at Assisi," *Apollo* 157 (2003): 31–35.

Master Regent in Paris and then *lector Sacri Palatii*) served as Minister-General during the period 1296–1304 and it was he who summoned Giotto to work on the decoration of the Basilica of Assisi.[12]

Given these facts, the possibility cannot be excluded that a series of Ministers-General, all academics with a thorough grounding and abiding interest in optics—e.g., Bonaventure, John Pecham, and Pecham's former students Matteo d'Acquasparta, Bartolomeo da Bologna and Roger Marston[13]—may have made a direct contribution to the development of *perspectiva artificialis* by encouraging artists to experiment with the representation of depth based on the laws of optics. These artists might then have propagated the new techniques in their own ateliers.

One central figure in this movement could well have been Matteo d'Acquasparta (ca. 1237–1302), the *socius* of John Pecham and Master Regent of the *studium* in Paris at the time that Roger Bacon happened to be sojourning there (1256–1280). In *Questiones disputate de anima XIII* Matteo d'Acquasparta sustained, based on his study of the work of Pseudo-Aristotle (*Problemata XXXI*, 4, 7), the thesis of the fusion of visual sensations (Chap. 6), in which perception is a function of the number of agents, rather than the number of instruments, of vision. If there is only a single agent, the following integration can be derived:

> For if you have one principal agent and two instruments, there is one agent but two actions... And even if there is more than one eye, the sense of sight will nevertheless be one, and there will be many visions but one seeing/Si enim fuerit unum principale agens et duo instrumenta, est unum agens sed duae actiones... Si igitur fuerint aliquorum plures oculi, tamen unus sensus visus erit, erunt quidem plures visiones, sed unus videns.[14]

---

[12]The commissioning of the frescoes for the Upper Church of Assisi is mentioned by Vasari: "Having finished these works [in Arezzo, Giotto] betook himself to Assisi, a city of Umbria, being called thither by Fra' Giovanni di Muro della Marca, then General of the Friars of St. Francis/Finite queste cose [in Arezzo, Giotto] si condusse in Ascesi città dell'Umbria, essendovi chiamato da fra' Giovanni di Muro della Marca, allora Generale de' frati di san Francesco," Vasari, *Vite*, edited by R. Bertanini, p. 100.

[13]Bonaventura deals with the properties of light in his "Commentarium" in II Sententiarum, dist. XIII, art. 2, qq. 1–2, art. 3, qq. 1–2, *Opera Omnia S. Bonaventurae*, vol. 2, Grottaferrata,1885, pp. 323–326; Roger Marston was also the author of a scholastic *Questio disputata de lux naturalis*; Bartolomeo da Bologna, who served as Minister Provincial of the Province of Bologna, wrote a treatise entitled *Tractatus de luce*, of which a modern edition was published by Ireneo Squadrani in *Antonianum* 7 (1932): 201–238, 465–494.

[14]Matteo d'Acquasparta, *Quaestiones Disputatae de Anima XIII*, ed. A.J. Gondras, Paris, 1961, p. 132.

A complete edition of the works of Matteo d'Acquasparta, of which only an infinitesimal part has been published so far,[15] could reveal much about his scientific and artistic interests.

Optics and Perspective in the Institutional Context

Examining the earliest notions of perspective in the context of how techniques for the representation of space were disseminated and taught allows us to test the hypothesis that craftsmen incorporated mathematics into their workshop practices. The departure point for perspective as a science can be traced to the middle of the Quattrocento and Piero della Francesca's first geometrical proofs. Before this date there was—strictly speaking—nothing more than an 'affinity' between practical perspective and the sciences.

1. It is generally assumed that, since painters and architects were not admitted to the universities, the only training available to them was provided by the institution of the abacus schools. This might appear to support the arithmetic hypothesis, since *abaco* meant 'arithmetic' in medieval Latin, but a differentiation must be made between *abaco* and *scuola d'abaco*. Arithmetic was not the only subject taught in these schools. *Arismetricha, geometria,*[16] *edifichare* and *prospettiva*[17] also formed part of the curriculum, as is known from the *scuola d'abaco* established by Paolo

---

[15]Still awaiting analysis are the manuscripts of the *Commentarius in I–IV Sententiarum* written in 1271–1272, Todi, Biblioteca comunale 122 (Book I), Assisi, S. Francesco, fondo antico 132 (Book II and parts of Book IV), together with the manuscript of the *VI Quodlibeta* composed in 1276–1279, Grottaferrata, Collegio San Bonaventura (Frati Quaracchi). For a commentary, see Martin Grabmann, *Die philosophische und theologische Erkenntnislehre des Kardinal Matthaeus von Acquasparta*, Vienna, 1906; Giulio Bonafede, "Il problema del 'lumen' nel pensiero di M. d'Acquasparta," *Rivista rosminiana* 31 (1937): 186–200; Helen M. Beha, "Matthew of Acquasparta's Theory of Cognition," *Franciscan Studies* 20 (1960): 161–204 and 21 (1961): 1–79, 383–465.

[16]On practical geometry, see Annalisa Simi and Laura Toti-Rigatelli, "Some 14th- and 15th-century texts on practical geometry," in *Vestigia Mathematica. Studies in Medieval and Early Modern Mathematics in Honour of H.L.L. Busard*, eds. M. Folkerts and J.P. Hogendijk, Amsterdam/Atlanta, 1993, pp. 453–470; Annalisa Simi, "Problemi caratteristici della geometria pratica nei secoli XIV–XVI," in *Scienze mathematiche e insegnamento in epoca medioevale*, eds. P. Freguglia, L. Pellegrini and R. Paciocco, conference proceedings, Chieti, 2–3 May 1996, Naples, 2000, pp. 153–200; Dominique Raynaud, ed., *Géométrie pratique. Géomètres, ingénieurs, architectes, XVIe–XVIIIe siècle*, Besançon, 2015.

[17]On practical perspective, see Gino Arrighi, "Un estratto dal "De visu" di M° Grazia de' Castellani (dal Codice Ottoboniano latino 3307 della Biblioteca Apostolica Vaticana)," *Atti della Fondazione Giorgio Ronchi* 22 (1967): 44–58; Filippo Camerota, "Misurare 'per perspectiva': geometria practica e prospectiva pingendi," in *La prospettiva*, ed. R. Sinisgalli, Fiesole, 1998, pp. 340–378; Francesca Cecchini, "Ambiti di diffusione del sapere ottico nel Duecento," M. Dalai Emiliani et al., *L'Artiste et l'Œuvre à l'épreuve de la perspective*, Rome, 2006, pp. 19–42.

Dagomari dell'Abaco close to the Church of the Santa Trinità in Florence. Hence, there is some justification for thinking that medieval painters may have acquired the basic notions of geometry and optics at such schools, which they could afterwards have put into practice.

Opportunities for architecture (*edifichare*) and perspective to cross-fertilize each other also arose from the dissemination of Vitruvius' *De architectura* in the Middle Ages.[18] According to Vitruvius, the architect "must be well-read, expert in drawing, learned in geometry [and not ignorant in optics]/*peritus graphidos, eruditus geometria [et optices non ignarus].*"[19] He explicitly treated the subject of perspective (*scaenographia*), referring to the work of Agatharcus, Democritus, and Anaxagorus.[20] With the exception of Agatharcus, who was a painter, his sources were therefore savants in the fields of optics and geometry. Vitruvius attributed only a minor role to arithmetic: "*By arithmetic, the cost of building is summed up; the methods of mensuration are indicated; while the difficult problems of symmetry are solved by geometrical rules and methods/Per arithmeticen vero sumptus aedificiorum consummantur, mensurarum rationes explicantur difficilesque symmetriarum quaestiones geometricis rationibus et methodis inveniuntur.*"[21] Medieval artisans seem to have understood the affinities between perspective, optics and geometry in much the same way. For instance, an epigraph carved on the pulpit of the church of Sant' Andrea in Pistoia (Tuscany) states that the sculptor and architect Giovanni Pisano was learned in optics: "*Giovanni carved it... knowledgeable over all visible things/Sculpsit Johannes... doctum super omnia visa.*" Similarly, Villard de Honnecourt used geometry to introduce the art of drawing: "*Here begins the method of representation as taught by the art of geometry, to facilitate work/Ci comence li force des trais de portraiture si con li ars de iometrie les ensaigne, por legierement ovrer.*"[22] On folios 203–21r Villard presents various devices to measure inaccessible heights or distances, a standard problem in perspective.

---

[18]"Le texte de Vitruve n'a cessé d'être connu (et donc recopié) de l'Antiquité à la Renaissance. Aussi ne faut-il pas s'étonner du nombre relativement important (près d'une centaine) de manuscrits aujourd'hui recensés qui contiennent des extraits, des parties ou l'ensemble du *De architectura*," Vitruvius, *De l'architecture*, Book I, edited by Ph. Fleury, Paris, 1990, p. liii. The *editio princeps* dates back to the fifteenth century, *L. Vitruuii Polionis ad Cesarem Augustum de architectura libri decem*, Rome, Johannes Sulpicius, 1487.

[19]Vitruvius, *De l'architecture*, p. 5 (I, 3). The portion enclosed in brackets may be found in a few manuscripts.

[20]Vitruvius, *De l'architecture*, Book VII, op. cit., p. 5 (praef. 12).

[21]Vitruvius, *De l'architecture*, Book I, op. cit., p. 5–6 (I, 4). This passage renders arithmetic thoroughly useless for the purposes of perspective if we translate *symmetria* as "modularity" or "common scale of measures."

[22]*Villard de Honnecourt. Kritische Gesamtausgabe des Bauhüttenbuches*, ed. Hans R. Hahnloser, Vienna, 1935, Taf. 36 (fol. 18v). I have adopted the translation of Theodore Bowie.

The End of Binocular Perspective in Classical Europe

Linear perspective—the only system that gained universal acceptance and remains in use to this day—is firmly associated with the postulate of monocular vision. Linear perspective makes use of the section of the visual pyramid whose rays concur in a single point—the eye of the spectator. In this way it set itself apart from medieval optics, which reserved—as we have seen—a prominent place for normal binocular vision (i.e., physiological diplopia). How did linear perspective finally manage to gain the ascendancy over a multiplicity of concurrent and competing systems?

First reason. Linear perspective has often claimed for itself the status of an ideal mathematical system, thus opposing any consideration pertaining to the physiological functioning of the eye. Yet this status of mathematical perfection was acquired only gradually, as is testified to by the numerous points regarding physiological optics that were raised in treatises on perspective. Scholars themselves exhibited an equal interest in optics and perspective until a very late date.[23] If one were to hazard a chronology, it might be suggested that perspective retained an undefined and fluid status, oscillating between pure science and workshop practices up until the groundbreaking treatises published by Commandino, Danti and Guidobaldo del Monte at the end of the sixteenth century.[24] It was only then that perspective achieved the explicit status of a geometric science. The ineluctable repercussion of this association was the embracing of a purified and essentially mathematical conception of perspective to the exclusion of all questions relating to the physiological foundations of vision.

Second reason. If one compares the construction of two paintings depicting the same scene, the first in linear perspective and the second in binocular perspective, one can immediately see that the latter requires a much larger number of geometric operations than the first. Since in absolute terms a linear perspective is much simpler to construct, pragmatic reasons may also have played a role in the ascendancy of linear perspective. These reasons often surfaced as a result of the flawed arguments against binocular vision made by Cardano, Della Porta, Danti, Bassi, Huret and Le Clerc. How was it possible that authors such as Della Porta[25] and Danti,[26] who were interested in optics and studied the works of the greatest authorities on the subject, rejected out of hand the theory of binocular vision in

---

[23]See, for example, Edme Mariotte, "La scénographie ou perspective," *Procès verbaux de l'Académie royale des Sciences, Registre de Mathématiques,* vol. 4 (14 April to 24 December 1668), regarding the session held on 20th June 1668, fol. 62r–73r; Philippe De la Hire, "Traité sur les differens accidens de la veüe," (1694), *Mémoires de l'Académie royale des Sciences* 9 (1730): 530–634.

[24]Francesco Commandino, *Claudii Ptolomaei liber de analemnate,* Rome, 1562; Egnatio Danti, *Le Due Regole della prospettiva pratica di M. Iacomo Barozzi da Vignola,* Roma, 1583; Guidobaldo del Monte, *Perspectivae libri sex,* Pesaro, 1600.

[25]Giambattista Della Porta, *De refractione optices parte libri novem,* Naples, 1593, pp. 142–143.

[26]Egnatio Danti, *Le Due Regole della prospettiva pratica,* Rome, 1583, pp. 53–55.

favor of more simplistic conceptions? The existence of pragmatic motives allows us to conjecture that this discrepancy was not due to a misapprehension of the texts, but to the preference for a simpler and more functional solution to the problem of perspective.

Third reason. Finally, we must remind ourselves that the themes of mathematical perfection and simplicity of construction would not have resonated to such a degree and exerted such an influence on the history of perspective if every practitioner had felt free to apply the procedure he thought most fit. The emerging homogeneity was due to a codification of perspective practices stemming from the manner in which they were taught. It is no accident that lessons on perspective were introduced at the Accademia del Disegno in Florence and the Accademia di San Luca in Rome (which were founded in 1563 and 1577, respectively) just as linear perspective was consolidating its position among the geometric sciences. The affirmation of perspective practices in the academies was the institutional factor that made it possible for linear perspective—with its characteristics of operational simplicity and mathematical clarity—to assume a position of supremacy and eclipse all other systems. In the final analysis this may be the most plausible explanation for the disappearance of two-point perspective and its anchor—the theory of binocular vision which had dominated medieval optics.

By revealing the foundations of a system of perspective that spanned the period from the end of the Duecento to the Cinquecento (Chaps. 5–7), and refuting past interpretations that have linked this system to a classical form of perspective (Chaps. 8–11), this book not only establishes the fundamental contribution made by the medieval theory of binocular vision to the science of optics and the psychology of visual perception, but also shows its impact on the conception of representational systems. The theory of binocular vision being specific to optics, it may be concluded that the main source of perspective theory lies in medieval optics rather than other, more marginal fields such as practical geometry (with its exercises in *ars mesurandi*, astronomy and the use of the astrolabe) or cartography (with its systems for the projection of geographic data). The systematic use of two-point perspective in works dating as fatback as the end of the Duecento suggests the existence of a much earlier link between *perspectiva naturalis* and *perspectiva artificialis* than had ever been imagined, signifying that the first experiments in artificial perspective coincided exactly with the *floruit* of Latin optics.

Appendix A
Error Analysis and Perspective Reconstruction

Abstract Perspective reconstructions are not always reliable. Proof lies in the variety of interpretations that have been proposed for the same work. A rational method for reducing the number of interpretations by decreasing the number of errors in the reconstruction will be presented here. It consists on the one hand of following a rigorous protocol and on the other hand of proceeding with an error calculation that allows one to choose the perspective scheme that minimizes the error. The method will be applied to a fresco by the fourteenth-century artist Giusto de' Menabuoi, which presents two principal vanishing points that can be defined with a very small error (4–7 mm in situ). The two vanishing points coincide with elements in the architectural framework and since they are separated by a distance that is thirty times greater than the error, this fresco can be used to formulate a robust hypothesis on how to implement the principles of binocular vision.

The errors involved in the reconstruction of perspective lines is a question that has been little addressed in the literature. It nonetheless deserves our attention because it is not possible to draw valid conclusions regarding the use of perspective methods without taking into account the errors that may be introduced by the operator.

This Appendix is organized as follows. First, the types of disparities that are observable in the reconstruction of a perspective painting will be analyzed. The fundamental principles of error theory that can be applied to such reconstructions will then be reviewed. Finally, the perspective of a fresco by Giusto de' Menabuoi will be analyzed by way of illustration.

The exposition that follows will be limited to determining the vanishing point in a painting, from which the foreshortening methods used should be extrapolated. Other well-known principles of perspective will not be discussed. Since every operation is a potential source of error, there is an advantage to be gained in reducing the number of steps between the original work and the reconstructed drawing. We therefore judged it useful to illustrate our methodology under the less favorable condition in which the reconstruction is carried out manually based on a

© Springer International Publishing Switzerland 2016
D. Raynaud, *Studies on Binocular Vision*,
Archimedes 47, DOI 10.1007/978-3-319-42721-8

reproduction of the original work. The same methodology can easily be adapted to different cases and working conditions:

(a) If the perspective is reconstructed in situ, certain sources of error and intermediate verification steps can be eliminated, but other errors will be introduced because the perspective lines are traced on a vertical plane and the instruments handled by the operator have a discrete weight that may affect the measurement;

(b) Photogrammetric techniques are no less subject to error. As was shown in Chap. 4, even the most sophisticated analyses of Masaccio's *Trinity* have entirely ignored the accidental and systematic errors that may affect any perspective lines. The same problems are encountered in the computer-based reconstructions that are presently in vogue. These techniques efface certain errors, but generate new ones. Many reconstructions are carried out using digitized photographs. To the errors inherent to the photographs must be added the fact that it is impossible to check an alignment precisely when the entire work does not fit on the screen. In short, these new tools and approaches do nothing to abolish the intrinsic errors, which mainly depend on judgments made by the individual operator.

The Problem of Errors in Reconstruction

One often finds significant divergences in the reconstruction of the same perspective by different experts. *This finding signifies that every reconstruction is operator-dependent.* Let us consider the example of *The Miracle of the Profaned Host* by Paolo Uccello, a panel painting dated ca. 1468 (Urbino, Galleria Nazionale delle Marche), and focus on three reconstructions of the scene of the *Holocaust* (Fig. A.1a–c).[27]

1. Ennio Sindona assumed that there is a single vanishing point, whereas the two other operators (Martin Kemp and myself) have identified an approximate zone of convergence.
2. Kemp distinguished a pencil of lines converging on a distance point Z situated on the horizon line, indicating that Uccello's method of reduction was correct, whereas Sindona deduced the existence of a diagonal, used by the artist to verify

---

[27]Ennio Sindona, "Una conferma uccellesca," *L'Arte* 9 (1970): 67–107; Martin Kemp, *The Science of Art*, New Haven, 1990, p. 50; Dominique Raynaud, *L'Hypothèse d'Oxford*, Paris, 1998, p. 87, Fig. 24. Some of the best studies of Renaissance fresco underdrawings include: Roberto Bellucci and Cecilia Frosinini, "Cum suis debitis proportionibus," M. Israëls, ed., *Sassetta: The Borgo San Sepolcro Altarpiece*, vol. 1. Florence, 2009, pp. 359–370; and more recently "Underdrawing in Paintings," A. Sgamellotti et al., eds., *Science and Art: The Painted Surface*, London, 2014, pp. 269–286.

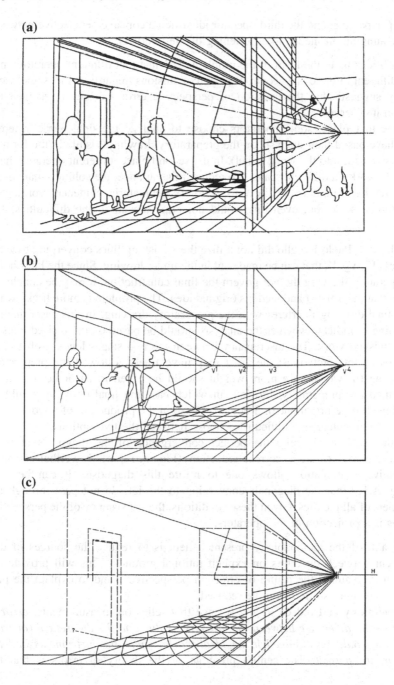

Fig. A.1 Paolo Uccello, *Miracle of the Profaned Host*, painting on wood, ca. 1468 (Urbino, Galleria Nazionale delle Marche). Author's reconstructions after **a** Sindona, "Una conferma uccellesca," p. 83; **b** Kemp, *The Science of Art*, p. 50; **c** Raynaud, *L'Hypothèse d'Oxford*, p. 87

the perspective, and the third operator identified a concave perspective network, indicating an erroneous foreshortening of the intervals.

This brings us to the heart of the problem. The fact that different operators may draw different, not to say contradictory, conclusions regarding the same work strongly suggests that the question of perspective error has never received the attention it deserves.

In the first place, some operators choose to analyze the final work, whereas others have based their analysis on the preparatory drawing (visible in the form of the incisions cut into the wood panel). In the second place, different operators have chosen to work on different scales. To reduce this variance, the solution that comes most naturally to mind is the direct inspection of the material evidence, working on a one-to-one scale, but even this would not eradicate all of the difficulties. For example:

1. In the end, Paolo Uccello did not utilize the vanishing lines converging towards points V_1, V_2, V_3 that can be made out in the underdrawing. Since the lines in the preparatory drawing did not govern the final construction, it may be concluded here that the artist abandoned his original idea. The problem is more tricky when the final drawing interferes with the preparatory drawing. In this case, there is nothing to indicate whether the analysis should be based on the incised lines or the finished work. The discrepancy between the two stages lends itself to two opposing interpretations: the artist may have freely followed the preparatory drawing (in which case more weight should be given to it) or he may have rectified the preparatory drawing (in which case the final drawing should be assigned more importance). Since this phase in the production of a work is not well documented, it is difficult to choose between the two options.

2. In the work by Uccello the perfect convergence of the diagonals at point Z seems to indicate a correct reduction (Fig. A.1a, b). Nevertheless, more attentive examination allows one to refute this diagnosis. It can be seen (Fig. A.1c) that the diagonals converging to the left do not pass through the corners of all the tiles. Given these conditions, the consistency of the perspective lines is a projection of the operators.

The aim of the methodology presented here is to reduce the sources of disagreement between operators on explicit rational grounds. This will provide the occasion to examine the nature of errors in perspective[28] and to explain the procedure by which its effects can be reduced.

In metrology and indeed in any science that relies on measurements, *error is defined as the difference between the measured value and the theoretical (or true) value that would be obtained in an ideal world where the instruments and the operator are perfect*. The central principle of the theory of errors is that the

---

[28]This idea was first presented in *L'Hypothèse d'Oxford*, pp. 52–53.

operator, despite all his efforts, can never arrive at an ideal result. Error cannot be eliminated from the quantification of any physical magnitude, because no measurement—not even the most accurate—can achieve an infinite degree of precision. Therefore the operator must integrate this error into his reasoning.

Let us designate the true value x_0 and the approximate value x. An ideal measurement would be $x = x_0$. Since the measuring device is never perfect and the operator will introduce an element of perturbation, it follows that $x \neq x_0$. If one wishes to obtain a more precise result, one would have to make multiple measurements that produce a series of values $x_1, x_2, ..., x_n$, with the final result being expressed as a single value $x \pm \Delta x$, where Δx denominates the *uncertainty* within which the true value is asserted to lie with a given level of confidence. There are two principal types of error.

Random or accidental error (unpredictable in a series of measurements) is an error that can be attributed to faulty execution rather than to the measurement protocol. Random errors can generally be traced to the operator and are identifiable because they occur independently in either direction, producing a similar dispersion of values above and below an average value. This type of error can be corrected for by reiterating the measurement. Assuming the absence of any systematic error, when a series of measurements is made the arithmetic average of the values obtained is called *the most probable value* μ or $\bar{x}$. An estimate of the dispersion around the true value is expressed by the *mean deviation e* or, most often, by the *standard deviation* σ. Since the frequencies f_i satisfy the condition $\sum_i f_i = 1$, one has:

$$\bar{x} = \sum_i f_i x_i \quad e = \sum_i f_i |x_i - \bar{x}| \quad \sigma = \sum_i f_i (x_i - \bar{x})^2$$

In error theory, one well known result is that if the data is plotted with the values measured along the abscissa and the number of results comprised within a fixed interval of values along the ordinates, a normal Laplace–Gauss distribution is obtained. The most probable value μ or $\bar{x}$ is the abscissa or x-coordinate at the apex of the bell curve. The error is generally described as $\Delta x = \pm 2\sigma$, because 95.4 % of the values are comprised within the interval $(\bar{x} - 2\sigma, \bar{x} + 2\sigma)$ (Fig. A.2).

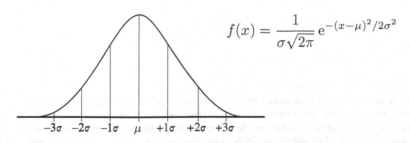

$$f(x) = \frac{1}{\sigma\sqrt{2\pi}} e^{-(x-\mu)^2/2\sigma^2}$$

$-3\sigma \quad -2\sigma \quad -1\sigma \quad \mu \quad +1\sigma \quad +2\sigma \quad +3\sigma$

Fig. A.2 The normal distribution of errors. Author's drawing

Systematic error (which is constant within a series of measurements) is the error that stems from a permanent cause, i.e., one inherent to the methodology chosen or the incorrect setting of the instruments used. Systematic errors can be recognized because they always occur in the same direction. Unlike random errors, which can be eliminated by making repeated measurements, systematic errors will persist even in the most scrupulously determined values. The only known strategy to detect and eliminate this type of error consists in reversing or inverting every possible element in the experimental setup before repeating the measurement.

Precision of a measurement is defined in relation to the random error; the smaller the deviation between the series of measurement results and the mean value, the higher the precision.

Trueness is defined concurrently in relation to the systematic error, and expresses the closeness of the mean of a set of results to the true value.

Accuracy of a measurement is the combination of its precision and trueness.

The Work Protocol

Many *extrinsic errors* can be avoided if the operator adheres to a strict set of conditions, which must be adapted to the reconstruction techniques utilized. Here we will examine the relatively unfavorable case of a perspective reconstructed using a ruler. However, instead of listing all of the errors associated with this operating context, it is assumed that the operator is working under ideal conditions that will be laid out in the form of a working protocol.

Correction of Random Errors

1. If the operator is using a photographic reproduction, he must verify that it contains no *geometric (monochromatic)* or *chromatic*[29] *aberrations*.
   There are five types of geometric aberrations, which may be defined by the appropriate coefficients contained in the formula of the deviation from

---

[29]*Chromatic aberration* is produced by white (non-monochromatic) light when the blue wavelengths, which converge more rapidly than the red wavelengths, result in a spread of the focal point along the optic axis. As a consequence one observes a dispersion of light as fringes of color around the edges of objects. This aberration can be corrected by an achromatic system of crown and flint glass lenses.

Gaussian optics.[30] The coefficient ρ_0^3 characterizes spherical aberration.[31] The two $r_0\rho_0^2$ coefficients express the coma aberration.[32] The two $r_0^2\rho_0$ coefficients characterize astigmatism[33] and the curvature of the field.[34] Finally, the r_0^3 coefficient represents the distortion of the image.[35]

2. The operator can then proceed to the reconstruction of the perspective, for which he will require a well-lit worktable (receiving a nominal illuminance of $E \geq$ 1500 lux).[36]

---

[30] An optical system is *rigorously stigmatic* if it transforms one homocentric beam into another (their centers being conjugate points). From a practical standpoint, the condition of rigorous stigmatism is of little interest. It suffices that the optical system transforms an object point into a luminous spot with a diameter that is smaller than the resolution limit of the receptor. This is why one can use spherical diopters that meet the condition of approximate stigmatism. The Gaussian approximation is the set of conditions that allows one to obtain a high-quality image; a system is approximately stigmatic if it receives paraxial incident light rays that are only a short distance away from and form only a small angle with the optical axis. The deviation from Gaussian optics is expressed as a sum of complex numbers involving r_0 (the distance between the object and the optical axis) and ρ_0 (the angle of the incident light wave relative to the optical axis).

[31] *Spherical aberration* varies to the third power of the angle between the incident light ray and the optical axis; it occurs when the angle is very large. The edge rays converge more rapidly than the central rays, producing a caustic envelope around the focus. Before the point of concurrence of the paraxial rays, the image of an object point is represented by a centered circle (the projection of the tangential and sagittal layers of the caustic envelope on the image plane). Spherical aberration can be reduced by the use of doublets in which a convergent lens is paired with a divergent lens.

[32] *Coma aberration* appears when the object is located at a considerable distance from the optic axis and the incident light rays form a wide angle with the axis. It is the sum of two factors, which are perceived as one because they operate on the same image plane. The object point takes the form of the tail of a comet, hence the name.

[33] *Astigmatism* manifests itself when the incident rays are inclined relative to the optical axis and the object is located far from the axis. The rays no longer meet at a single point, but instead converge on two perpendicular segments (the sagittal focus and the tangential focus), which are separated by the distance of the astigmatism. Based on the position of the image plane, an object point will assume a more or less markedly elliptical form.

[34] The *curvature of the field of vision* occurs under the same conditions as astigmatism. It signifies that the geometric image of a planar object is a curved surface that does not coincide with the focal plane except at the point where it intercepts the optical axis. In this case the central part of the image is focused while the marginal zone is blurred.

[35] The *distortion* in the Gaussian image is proportional to the distance, raised to the third power, between the object and the optical axis, and will depend on the apparent size of the object. This aberration becomes visible if one places a diaphragm along the optical axis that allows one to observe the curvature of the tangential lines of the object, specifically either a negative 'barrel' distortion when the diaphragm is placed in front of the lens or a positive 'pincushion' distortion when it is placed behind the lens. This aberration can be corrected by a combination of lenses arranged on either side of the diaphragm.

[36] Insufficient lighting can be a significant source of error. It is known that only the cone cells of the eye, which function in photopic vision, allow one to visualize details. Under photopic conditions, strong lighting produces myosis and as a consequence the restricted use of the foveal retinal cells; visual acuity is therefore enhanced. In parallel, the myosis causes a reduction in the visual field that eliminates the spherical aberration.

3. The lighting in the work area must be homogeneous, with shadows reduced to a minimum and all sources of contrast, reflection and glare eliminated.
4. The operator should use a sheet of transparent plastic film to retrace the lines of the perspective, because the opaque quality and sensitivity to humidity of cellulose-based tracing paper makes it difficult to copy the original line accurately and avoid its displacement from the true value.
5. The sheet should be securely fixed to the painting, and the operator's hand should not come into contact with the sheet.[37]

Correcting for Systematic Errors

1. Before beginning a reconstruction, the operator must make certain that his work surface is flat and that his ruler is true.
2. He should work on a horizontal surface to avoid effects due to the weight of the instruments.[38]
3. He must avoid using a printer or photocopier in his reconstruction.[39]
4. While tracing the perspective lines, the operator should maintain his line of sight perpendicular to the work surface and to his ruler.[40]
5. He should hold his pen perpendicular to the work sheet.

---

[37]The hand is a source of heat that can influence the physical geometry of the transparent sheet on which one is tracing the line.

[38]When working on an inclined surface, there is a tendency for the vanishing point to fall due to the weight of the ruler. This error factor is proportional to the dimensions of the ruler and the degree of inclination of the working surface and will limit, for example, the accuracy of the reconstruction of a fresco *in situ*. The results of experimental studies support the notion that reconstructions should be carried out on a horizontal surface.

Suppose (Figure App.6) that the operator must extend a segment AB (the visible edge of the object) as far as F (the vanishing point) with the help of a ruler weighing 600 g (assuming a uniformly distributed tare) and a pen with a normalized nib width $e = 0.20$ mm. Let $AB = 40$ mm and $AF = 1000$ mm. The position of F is determined by aligning the ruler using the equal light gap method (cf. *infra*. 3.1.1), marking the point F freehand, and removing the ruler after each measurement. Carrying out two series of measurements—on the horizontal plane (A) and then on a plane inclined at a 45° angle (B)—one notes that the measurements in the first series follow a centered normal distribution (A), whereas those in the second series exhibit a bimodal distribution (B): a significant group of values are subject to gravity, while by all appearances the values in the other group reflect the tonic compensation of gravity (Figure A.10).

[39]The use of photocopiers and printers should be avoided, because the drive system can cause a distension, in the direction of the roller, of the lines on the sheet of paper. The operator must at least verify the geometry of the machine by copying a page with a diagonal line and a large square and seeing whether the diagonal is reproduced as a straight line and the sides of the square are equal.

[40]Otherwise a parallax error will occur.

(a) equal dark gap method   **(b)** equal light gap method

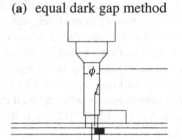
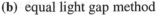
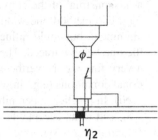

Fig. A.3 The two main methods of reconstruction: **a** the equal dark gap method and **b** the equal light gap method. Author's drawing

6. The vanishing lines should always be traced from the visible edge of the object toward the vanishing point, and not the reverse.[41]

The Reduction of Simple Errors

Even when the operator follows an ideal working protocol, his drawing will be affected by certain errors. These are *intrinsic errors* that cannot be avoided—no matter which method is used—but they can be minimized.

(1) *Systematic errors*

    1. *A line drawn with a ruler* will introduce a specific error: the width of the pen point will cause a gap between the line drawn and the edge of the ruler which is aligned with the perspective line in the work. Therefore, the only property retained by the operator's drawing is the parallelism between the line in the original work and the line which is being retraced. Furthermore, this error is dependent on the method followed, of which there are two. The first consists in covering the line of the work with the edge of the ruler and then displacing the edge by the smallest and most constant distance possible within the line's thickness (*the equal dark gap method*; Fig. A.3a). The second is to position the ruler alongside the line of the work, and then to move the ruler close to this line, leaving the narrowest possible distance between the edge and the line (*the equal light gap method*) (Fig. A.3b). In both cases the width of the gap l depends on the separating power of the eye—which is approximately 0.075 mm [cf. *infra*, Random errors].

---

[41]If the operator works backward, drawing a line from the vanishing point, he runs the risk of constructing a biased, although coherent perspective.

2. *The width of the pen nib* results in a thickening of the line that could lead to a misconstrual of the convergence of the vanishing lines. Using the equal light gap method, the width of the line will not reduce the accuracy of the alignment, but could influence the gap between the optical alignment and the line being retraced. The precision can be increased by using a pen with a very fine nib. Nevertheless, this problem will only appear under exceptional conditions (e.g., in works of large dimensions or when a broad pen is used). In practice, a pen nib with a normalized width ($0.10 \le e \le 0.20$ mm) will lead to an acceptable reconstruction.

3. *Calculating the systematic error*. If η is designated as the systematic error, e the width of the normalized line, ϕ the thickness of the nib ($\phi > e$), and e' the width of the line in the work, the following relationships emerge:

$$ \text{(A)} \ \ \eta_1 = \frac{e' + \phi}{2} - l \quad \text{(B)} \ \ \eta_2 = \frac{e' - \phi}{2} + l $$

Since under typical operating conditions $e' = 0.15$ mm, $\phi = 0.25$ mm and $l = 0.075$ mm, the systematic error will be: $\eta_1 = 0.125$ mm, $\eta_2 = 0.025$ mm. *Therefore, in order to minimize the systematic error one should employ the equal light gap method.*

(2) *Random errors*

1. *The scale of the work being examined* is another source of error. It is obvious that details will be difficult to distinguish in small-scale photographs. But it is also true that large works requiring the manipulation of cumbersome materials can give rise to specific errors linked to the size and weight of the instruments used (cf. *supra, Correcting for systematic errors*) or to the fact that the operator cannot simultaneously verify the optical alignment of two points that lie a significant distance apart. Contrary to a widely held notion, working on one-to-one scale does not eliminate all possibility of error. If, on the other hand, the operator is obliged to use a reproduction, he can calculate the scale of the reduction and take it into account; the width of the line traced by his pen and the width of the line *in situ* must be commensurate.

Suppose that the work being analyzed is a fresco whose lines have an average width of $e' = 2.5$ mm and that the width of the pen nib is $e = 0.20$ mm. One would accept a scale of reduction of $K < 12.5$ for which the width of the retraced line is narrower than that of the line in the work itself (i.e., $K < e'/e$). Reciprocally, if one is working with a photograph reduced by $K = 15$, one would select a pen nib whose thickness is $e < e'/K$, that is $e = 0.15$ mm, to meet the same conditions.

Fig. A.4 The Vernier optotype. Author's drawing

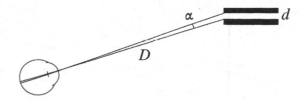

2. Another significant source of error is a *deficit in visual acuity*. Every operator must be aware of the limits of his own visual acuity before embarking on a reconstruction. For a person with normal emmetropia,[42] the acuity will be on average 1 arc min. This can be measured using an optotype chart, which must be carefully chosen because different factors can cause variations in the results.[43] Since parallels play a central role in any reconstruction, the most suitable chart to test the visual acuity in an observer would be a horizontal Vernier optotype.[44] The distance between the horizontal lines must be estimated in a photopic context and without any stenopeic hole (Fig. A.4).

In my own case, for an interval $d = 1.5$ mm, ten successive measurements produced an average separation distance D of 594 cm. Therefore $\alpha \approx \tan \alpha = d/D \approx 52$ arc sec. For a reconstruction in which the eyes are located 30 cm from the tracing sheet, this is equivalent to a separation between the lines of 0.075 mm, which would fix the value of the gap l.[45]

3. *Calculating the random error.* Since estimating random error is difficult, one will preferably proceed with a comparison of the theoretical and experimental values for the error. One might adopt the following device. Assume that the operator extends the segment $AB = 40$ mm to F, such that

---

[42]We apply our reasoning to the case of an operator who is not affected by astigmatism, a condition that is a significant source of error because perspective requires one to work with lines that are inclined at various angles. In the case of regular astigmatism, the cornea does not present meridians with an identical curvature and there is sharp vision only for the frontal lines in a given direction. This direction is measured by the astigmatic dial, on which the astigmatic subject distinctly perceives only the meridian corresponding to the focus that is closest to the retinal surface.

[43]The factors of variation in acuity are both extrinsic and intrinsic. Visual acuity increases under the following conditions: with increasing luminance, reaching a plateau towards +1.4 millilamberts; with increasing contrast between the test and the background; when the presentation distance of the test diminishes, because the power of the eye increases during focusing; for yellow and red radiation; and for vertical rather than horizontal lines. Visual acuity increases in the fovea with myosis (minimizing the spherical aberration) and with binocular vision, but diminishes with age.

[44]The Foucault resolution target overestimates the scores, because it allows one to discriminate between several lines simultaneously.

[45]In this case the value is an overestimate; when viewing things close up, visual acuity increases slightly because the eye changes focus by increasing the curvature of the crystalline lens (in a mechanism involving the ciliary muscle, the zonule and the capsular bag) and the refraction index (through the centripetal movement of the crystalline micelles). The retinal image is enlarged by about one-fifth at a presentation distance of 25 cm.

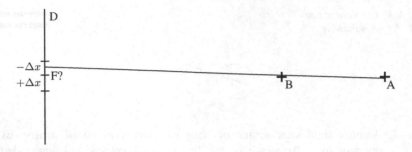

Fig. A.5 The production or random errors. Author's drawing

> AF = 1000 mm, the point F being traced blindly each time. Multiple measurements could lead to an estimate of the error produced at the level of point F on line D (Fig. A.5).

Let us first of all determine the theoretical value of the error according to the equal light gap method. The gap l, which is based on the resolving power of the eye, can be considered a *minimum separabile*, as the eye is not capable of seeing a narrower interval. But because during the process of verifying the parallelism the eye passes imperceptibly from one point to another along the segment AB, it may be assumed that the maximal error will appear at the two extremes, A and B (Fig. A.6).

In this case the theoretical value of the reconstruction error will depend exclusively on the width of the gap l and the ratio between the visible portion of the vanishing line AB and its total length AF. The angular and metric errors α and ϵ can be calculated from the equations:

$$\alpha = \pm \arctan(l/AB) \qquad \epsilon = \pm AF \tan \alpha = l\frac{AF}{AB}$$

For AB = 40 mm and AF = 1000 mm, α = 6′ 27″ and ϵ = ±1.87 mm.

Let us now determine the experimental value for the error. This can be calculated beginning with about fifty measurements according to the protocol described above. If the measurements for point F are plotted against the frequencies, the values will follow a normal distribution (Fig. A.7). The histogram on the left can then be transformed into the normal distribution on the right. A simple change in scale (0.95 mm = 1 unit) will provide the value of the random error: $\Delta x = \pm 2\sigma = \pm 1.9$ mm.

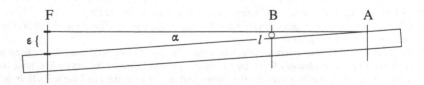

Fig. A.6 The metric and angular errors. Author's drawing

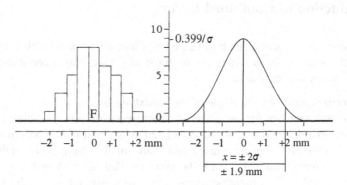

Fig. A.7 An histogram of errors. Author's drawing

This value turns out to be an excellent approximation of the theoretical value calculated above, and therefore demonstrates the concordance between the estimates for the theoretical and experimental errors when the equal light gap method is applied.

The most notable difference between the study of measurement errors in the physical sciences and in perspective reconstructions is that in the latter *the operator proceeds one line at a time.* In the most extreme of cases, the vanishing point may be affected by a maximal deviation on the left or on the right. This is why one must double the error of the reconstruction calculated from a series of measurements,[46] resulting in:

$$\Delta x = \pm 2l \frac{AF}{AB}$$

Consequently, it may be concluded that for an operator following a rigorous protocol based on the equal light gap method and working on a small scale with a fine-nibbed pen, if $AB = 40$ mm and $AF = 1000$ mm the systematic error will be $\eta = +0.025$ mm and the random error will be $\epsilon = \pm 1.87$ mm. The systematic error is thus negligible. *It follows that the main source of error is the angular deviation, that is, the failure to assess the parallelism in the composition. Not being dependent on the size of the work, this error will affect any reconstruction, whether realized in situ or calculated from small-scale reproductions.*

---

[46]This is a refinement of the estimate presented in my paper "Perspective curviligne et vision binoculaire." In the determination of F one can, using a pen with a nib thickness of e, redraw the entire vanishing line covering segment AB as it recedes towards F. The error will be $\epsilon = \pm e\, AF\,/ AB$. This value is greater than the experimental value ($\epsilon = \pm 5.0$ mm $> \Delta x = \pm 1.9$ mm), but since the latter must be doubled ($2\Delta x = \pm 3.8$ mm), either value can be retained.

The Reduction of Combined Errors

When an operator retraces a perspective he finds himself working with not just one, but several vanishing lines. Therefore he must also take into account the combinations of errors that may arise.

(1) *The error caused by the angle of the vanishing lines*
When one searches for the point of concurrence of two vanishing lines, the random errors ϵ and ϵ' that are associated with them define an error quadrilateral in the picture plane. Since ϵ and ϵ' are small in comparison to the length of the vanishing lines AF, it can be assumed that $ad \approx cb$ and $ca \approx bd$. The error quadrilateral is therefore comparable to a parallelogram, the length of whose sides and diagonals can be calculated (Fig. A.8a).

$$\text{sides:} \quad ca \approx bd \approx \frac{\epsilon}{\cos\left(a - \frac{\pi}{2}\right)} \quad ad \approx cb \approx \frac{\epsilon'}{\cos\left(a - \frac{\pi}{2}\right)}$$

$$\text{diagonals:} \quad ab \approx \frac{\sqrt{\epsilon^2 + \epsilon'^2 - 2\epsilon\epsilon'\cos(\pi - \alpha)}}{\cos\left(a - \frac{\pi}{2}\right)}$$

$$cd \approx \frac{\sqrt{\epsilon^2 + \epsilon'^2 - 2\epsilon\epsilon'\cos(\alpha)}}{\cos\left(a - \frac{\pi}{2}\right)}$$

Since the length of the diagonals of the error quadrilateral will vary as a function of the angle of the vanishing lines, the operator must seek to minimize this combined error. It is easy to see that if one varies the angle α, the minimum is obtained when the vanishing lines are perpendicular (Fig. A.8b).

The problem amounts in fact to looking for the minima of ab and cd. When $\alpha = \pi/2$, the denominator is equal to 1 and the cosine term in the numerator is cancelled out. The diagonals become equal:

$$ab \approx cd \approx \sqrt{\epsilon^2 + \epsilon'^2}$$

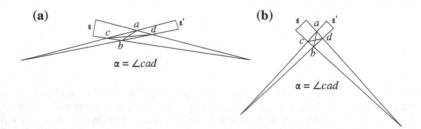

(a) $\alpha = \angle cad$ **(b)** $\alpha = \angle cad$

Fig. A.8 The composition of errors as a function of the angle of the lines. Author's drawing.

It follows that the determination of the vanishing point F—situated at the center of the error polygon—is all the more precise when the two vanishing lines converge to form a right angle. As the operator can do nothing to modify this angle, he will reduce the error by choosing, two by two, the pairs of vanishing lines that form an angle approaching 90°.

(2) *The error caused by the length of the vanishing lines*
    There is another source of combined error, which depends on the length of the visible edges from which the vanishing lines are traced. Consider the theoretical error:

$$\Delta x = \pm 2l \frac{AF}{AB}$$

The error will vary as a direct function of the ratio AF/AB (or as an inverse function of AB/AF). When reconstructing the perspective, the operator must regularly ascertain whether a new vanishing line passes through the error polygon of a known vanishing point. Yet if one began the construction using vanishing lines with a small AB/AF ratio, this initial choice would distort the completed drawing. It follows that the larger the ratio AF/AB between the first vanishing lines chosen, the more precise the determination of the vanishing point F will be. From this stems the necessity to arrange all of the vanishing lines in descending order based on the ratio AB/AF and to reconstruct the perspective in that order.

The Tracing Procedure

From this examination of the simple and combined errors that may arise when reconstructing perspective lines, one can draw up a procedure designed to minimize these errors. The tracing of the lines is carried out in two stages; first, all the measurements are made and the error is calculated, and then the drawing is constructed.

The preparatory drawing (measurements)

1. Trace the visible portions AB ... of the vanishing lines;
2. Make an approximate determination of the eventual vanishing point(s);
3. Calculate the ratio AB/AF ... for each vanishing line;
4. Arrange the vanishing lines in descending order based on this relationship: A_1B_1/A_1F, A_2B_2/A_2F ...;
5. Calculate the random error at the level of the vanishing point F for each vanishing line.

The final drawing (the traced lines)

6. Choose vanishing line No.1 (the first in the series of AB/AF ratios);

7. Take the optimal vanishing line, i.e. the line with the highest ranking and whose angle in relation to line No.1 approaches 90°;
8. Determine the point of concurrence F_1 of these two lines;
9. Choose vanishing line No.2. If it passes inside the error polygon of point F_1, join line No.2 to F_1; otherwise, apply rules 7 and 8 to obtain a new vanishing point F_2;
10. Choose vanishing line No.3; repeat step 9, and continue this process until the drawing is completed.

A Fresco by Giusto de' Menabuoi

I will now apply this protocol to the study of a fresco attributed to Giusto de' Menabuoi, *The Saint Enthroned* (110 × 126 cm), which was painted between 1370 and 1380 on the north wall of the Palazzo della Ragione in Padua, and of which there is a photograph[47] with a reduction coefficient of $K = 7$.

Using a pen with normalized nib width of $e = 0.20$ mm that produces a 1.4 mm line in situ, the lines in the preparatory drawing can be constructed based on the visible edges of the objects *D1*, *D2*, *D3*... (Fig. A.9a) and their extension to a hypothetical vanishing point (Fig. A.9b). The potential error ϵ at the level of the vanishing point being a function of the ratio *AB/AF*, the vanishing lines should be arranged in descending order based on this ratio (Table A.1). The optimal lines for the reconstruction can be deduced from this ranking (Table A.2). The vanishing lines *G5* (No.1) and *D2* (No.10), which form an angle of 115°, allow one to determine the vanishing point *F'* with a combined error of 1.01 mm (which is the diagonal of the error quadrilateral *cd*). In the same way the vanishing lines *D5* (No.2) and *G2* (No.6), which form an angle of 117°, result in a vanishing point *F* with a combined error of 0.64 mm. These values fix the maximal errors in the retracing of the perspective. Only the vanishing lines *D3*, *D7* and *G7*, being affected by errors of convergence, are to be rejected.

The operator will then specify and describe as far as possible the error affecting the perspective reconstruction. The preparatory drawing for *The Saint Enthroned* was traced with an error of 1.01 mm for the vanishing lines from the left and an error of 0.64 mm for the vanishing lines from the right, which represent in situ errors of about 7 and 4 mm at the level of the vanishing points (Fig. A.9b). The distance between the two vanishing points is *thirty times greater* than the maximal error of the reconstruction. The study therefore yields a result that is as robust as it is surprising: the two vanishing points are not the result of inaccurate retracing. They are not accidental.

---

[47]Giampiero Bozzolato, *Il Palazzo della ragione a Padova. Gli affreschi*, Roma, 1992, Plate LXII.

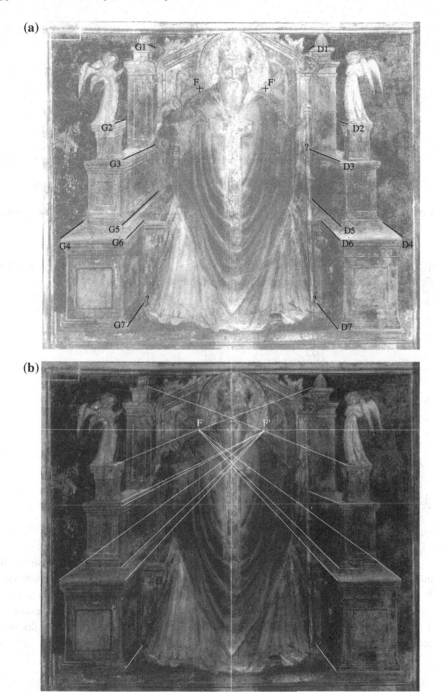

Fig. A.9 **a** Preparatory drawing of the *Saint Enthroned* by Giusto de' Menabuoi and **b** Final drawing of the *Saint Enthroned* by Giusto de' Menabuoi, fresco, ca. 1370–80 (Padua, Palazzo della Ragione, north wall). Author's reconstructions after Bozzolato, *Il Palazzo della Ragione a Padova*, plate LXII

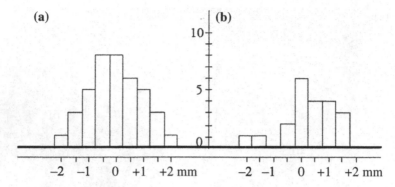

Fig. A.10 An histogram of errors with no gravity (*left*) and with gravity (*right*). Author's drawing.

Table A.1 The arrangement of the vanishing lines in Giusto de' Menabuoi's fresco *The Saint Enthroned*, shown in descending order based on the error ratio *AB/AF*

Rank	line	AB/AF	ϵ (mm)
1	G5	0.250	0.30
2	D5	0.223	0.34
3	G3	0.200	0.37
4	D3	0.184	0.41
5	D6	0.144	0.52
6	G2	0.139	0.54
7	G6	0.126	0.59
8	D4	0.124	0.60
9	G4	0.111	0.67
10	D2	0.100	0.75
11	G1	0.067	1.12
12	D1	0.065	1.15

The methodology that has been described in detail here could transform the analytical study of perspective and our understanding of the human errors involved. It also invites one to question the utility of representing irregularities in the perspective reconstruction, a solution that has sometimes been adopted by operators as a proof of their good faith and objectivity. Such a practice inevitably casts doubt on reconstructions that have a more regular appearance. And yet, for all scholars

Table A.2 The ordering of the pairs of vanishing lines in *The Saint Enthroned* (from most to least optimal for the reconstruction of the perspective) based on their angle α

Step	Combin.	Point	Angle α	cd (mm)
1	(G5,D2)	F'	115°	1.01
2	(D5,G2)	F	117°	0.64
3	(G3,1)	F'	132°	0.37 < 1.01
4	(D3,2)	F	133°	0.41 < 0.64
5	(D6,2)	F	114°	0.52 < 0.64
6	(G6,1)	F'	112°	0.59 < 1.01
7	(D4,2)	F	125°	0.60 < 0.64
8	(G4,1)	F'	123°	0.67 < 1.01
9	(G1,1)	F'	180°	1.12 > 1.01*
10	(D1,2)	F	180°	1.15 >0.64°

* allowed based on the alignment *G1F2D2*
° allowed based on the alignment *G2FD1*

acquainted with the theory of errors, it is preferable to produce a regular drawing accompanied by an error calculation rather than a drawing that simulates the artist's presumed errors, which in some cases are neither detected nor quantified. In essence, the methodology proposed in this Appendix is aimed at eliminating any element of rhetoric in the treatment of error.

Appendix B
Catalogue of the Works

No.	Author	Title of the work	Date	Height (cm)	Width (cm)
1.	Isaac Master	*The Pentecost*	ca. 1295	390	286
2.	Giotto di Bondone	*The Approval of the Franciscan Rule*	1296–99	270	230
3.	Giotto di Bondone	*St. Francis Preaching before Honorius III*	1296–99	270	230
4.	Giotto di Bondone	*Christ before Annas and Caiaphas*	1304–06	185	200
5.	Giotto di Bondone	*Virgin in Majesty*	ca. 1306	325	204
6.	Giotto di Bondone	*Justicia*	ca. 1308	120	60
7.	Duccio di Buoninsegna	*The Wedding at Cana*	1308–11	43.5	46.5
8.	Duccio di Buoninsegna	*The Appearance of Christ behind Closed Doors*	1308–11	39.5	51.5
9.	Duccio di Buoninsegna	*The Last Supper*	1308–11	100	53.5
10.	Simone Martini	*The Death of St. Francis*	1315–17	284	230
11.	Simone Martini	*The Funeral of St. Francis*	1315–17	284	230
12.	Maestro senesegiante	*Christ among the Doctors*	1315–20	230	270
13.	Giotto di Bondone	*Madonna and Saints*	1328	54	29
14.	Pietro Lorenzetti	*The Nativity of the Virgin*	1342	188	183
15.	Giusto de' Menabuoi	*Madonna Enthroned*	1349	—	—

(continued)

© Springer International Publishing Switzerland 2016
D. Raynaud, *Studies on Binocular Vision*,
Archimedes 47, DOI 10.1007/978-3-319-42721-8

(continued)

No.	Author	Title of the work	Date	Height (cm)	Width (cm)
16.	Barna da Siena	The Kiss of Judas	ca. 1350	259	236
17.	Tommaso da Modena	St. Romuald	1352	—	—
18.	Lorenzo Veneziano	Madonna with Child	1372	124	50
19.	Giusto de' Menabuoi	Saint Enthroned	ca. 1370	110	126
20.	Altichiero	The Council of King Ramiro	1374–79	—	—
21.	Giusto de' Menabuoi	Christ among the Doctors	1376–78	190	310
22.	Stefano di Sant'Agnese	Madonna with Child	ca. 1390	128	58
23.	Taddeo di Bartolo	The Last Supper	1394–01	45	31.5
24.	Lorenzo Monaco	The Adoration of the Magi	ca. 1421	30	50
25.	Lorenzo Ghiberti	Christ among the Doctors	ca. 1415	31	31
26.	Niccolò di Pietro	St. Benedict Exorcising a Monk	ca. 1420	110	66
27.	Gentile da Fabriano	The Crippled Cured at the Tomb of St. Nicholas	1425	36	35
28.	Giovanni di Ugolino	Madonna with Child	1436	40	26
29.	Donatello	Miracle of the Newborn Child	1447–48	57	123
30.	Fra Angelico	The Mocking of Christ	1450	38.5	37

Appendix C
Errors of Reconstruction

No.	Title of the work	K	FF' (mm)	AB/AF_{left} (mm)	AB/AF_{right} (mm)	ϵ_{max} (mm)	FF'/ϵ_{max}
1.	The Pentecost	15.0	15	9/54	8/53	1.3	11.5
2.	The Approval of the Franciscan Rule	12.8	179	26/222	26/222	1.7	105.3
3.	St. Francis Preaching before Honorius III	12.0	16	29/105	30/107	0.7	22.8
4.	Christ before Annas and Caiaphas	9.3	140	25/148	34/200	1.2	116.7
5.	Virgin in Majesty	15.9	72	8/84	8/84	2.1	34.3
6.	Justicia	5.2	117	18/136	18/136	1.5	78.0
7.	The Wedding at Cana	2.2	72	12/108	12/120	2.0	36.0
8.	The Appearance of Christ behind Closed Doors	2.5	71	8/81	8/80	2.0	35.5
9.	The Last Supper	3.9	76	13/103	13/101	1.6	47.5
10.	The Death of St. Francis	15.5	6	10/52	7/57	1.6	3.7
11.	The Funeral of St. Francis	14.6	19	9/75	9/75	1.7	11.2
12.	Christ among the Doctors	14.3	4	9/47	10/51	1.0	4.0
13.	Madonna and Saints	2.4	73	11/92	12/89	1.7	43.7
14.	The Nativity of the Virgin	7.9	7	39/141	40/146	0.7	10.0

(continued)

© Springer International Publishing Switzerland 2016
D. Raynaud, *Studies on Binocular Vision*,
Archimedes 47, DOI 10.1007/978-3-319-42721-8

(continued)

No.	Title of the work	K	FF' (mm)	AB/AF$_{left}$ (mm)	AB/AF$_{right}$ (mm)	ϵ_{max} (mm)	FF'/ϵ_{max}
15.	Madonna Enthroned	c.4	67	7/103	13/107	2.9	23.1
16.	The Kiss of Judas	17.2	28	13/61	14/64	0.9	31.1
17.	St. Romuald	c.5	35	14/75	13/70	1.1	31.8
18.	Madonna with Child	5.7	20	21/77	20/77	0.8	25.0
19.	Saint Enthroned	7.0	27	17/76	15/71	0.9	30.0
20.	The Council of King Ramiro	c.10	52	16/187	16/187	2.3	22.6
21.	Christ among the Doctors	14.1	77	30/125	41/144	1.0	77.0
22.	Madonna with Child	5.0	31	16/87	13/85	1.3	23.8
23.	The Last Supper	1.8	58	25/138	23/137	1.2	48.3
24.	Adoration of the Magi	1.9	111	17/152	27/146	1.8	61.7
25.	Christ among the Doctors	1.7	31	10/46	9/45	1.0	31.0
26.	St. Benedict Exorcising a Monk	4.6	116	37/119	30/104	0.7	165.7
27.	The Crippled Cured at the Tomb of St. Nicholas	1.7	143	50/174	57/190	0.7	204.3
28.	Madonna with Child	1.7	104	24/128	25/127	1.1	94.5
29.	The Miracle of the Newborn Child	5.7	19	11/123	12/127	2.1	9.0
30.	The Mocking of Christ	1.6	6	28/187	16/97	1.3	4.6

Appendix D
Distance Between the Vanishing Points

No.	Title of the work	XH (cm)	XP (cm)	XP/XH	FF'$_{obs}$ (cm)
1.	The Pentecost	278	134	0.482	36
2.	The Approval of the Franciscan Rule	219	165	0.753	376
3.	St. Francis Preaching before Honorius III	381	82	0.215	39
4.	Christ before Annas et Caiaphas	312	245	0.785	331
5.	Virgin in Majesty	158	41	0.259	108
6.	Justicia	147	36	0.245	172
7.	The Wedding at Cana	696	194	0.279	120
8.	The Appearance of Christ behind Closed Doors	581	127	0.219	122
9.	The Last Supper	463	99	0.214	195
10.	The Death of St. Francis	246	117	0.476	15
11.	The Funeral of St. Francis	748	76	0.102	36
12.	Christ among the Doctors	680	137	0.201	13
13.	Madonna and Saints	231	25	0.108	65
14.	The Nativity of the Virgin	317	69	0.218	16
15.	Madonna Enthroned	193	37	0.192	100
16.	The Kiss of Judas	540	92	0.170	56
17.	St. Romuald	147	37	0.252	40
18.	Madonna with Child	240	36	0.150	29
19.	Saint Enthroned	213	45	0.211	42
20.	The Council of King Ramiro	492	107	0.217	125
21.	Christ among the Doctors	656	859	1.309	440
22.	Madonna with Child	128	38	0.297	38
23.	The Last Supper	380	210	0.553	106

(continued)

© Springer International Publishing Switzerland 2016
D. Raynaud, *Studies on Binocular Vision*,
Archimedes 47, DOI 10.1007/978-3-319-42721-8

(continued)

No.	Title of the work	XH (cm)	XP (cm)	XP/XH	FF'_{obs} (cm)
24.	Adoration of the Magi	412	116	0.282	198
25.	Christ among the Doctors	355	53	0.149	75
26.	St. Benedict Exorcising a Monk	494	384	0.777	278
27.	The Crippled Cured at the Tomb of St. Nicholas	738	597	0.809	429
28.	Madonna with Child	458	202	0.441	166
29.	The Miracle of the Newborn Child	783	96	0.123	72
30.	The Mocking of Christ	558	64	0.115	12

Appendix E
Plates

See Plates E1, E2, E3, E4, E5, E6, E7, E8, E9, E10, E11, E12, E13, and E14.

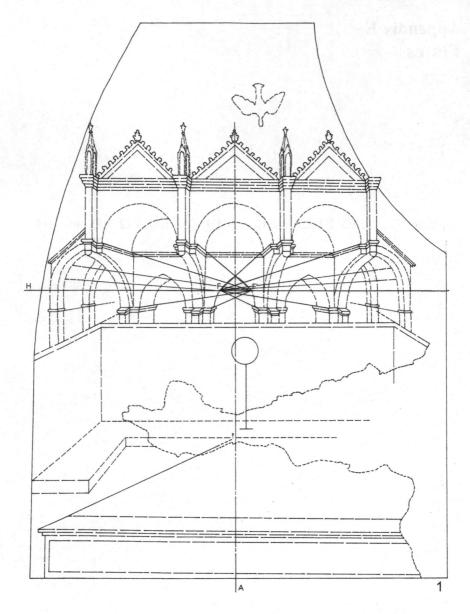

Plate E1 (No. 1) Isaac Master, *The Pentecost*, ca. 1295, Perspective (author's drawing)

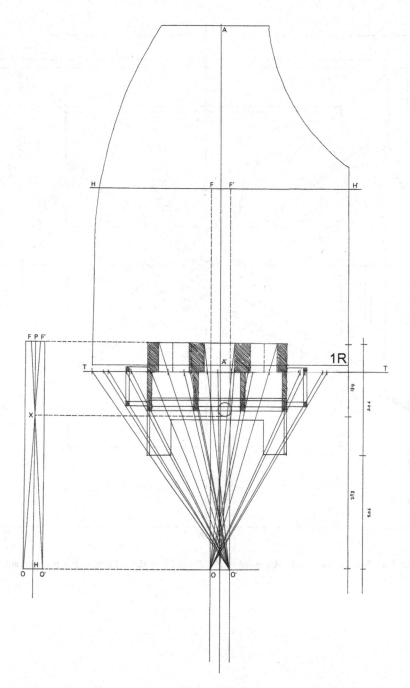

Plate E2   (No. 1R) Isaac Master, *The Pentecost*, ca. 1295, Reconstructed plan (author's drawing)

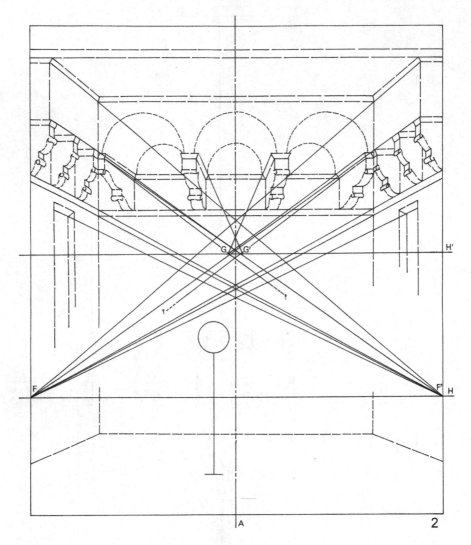

Plate E3 (No. 2) Giotto, *The Approval of the Franciscan Rule*, 1296–9, Perspective (author's drawing)

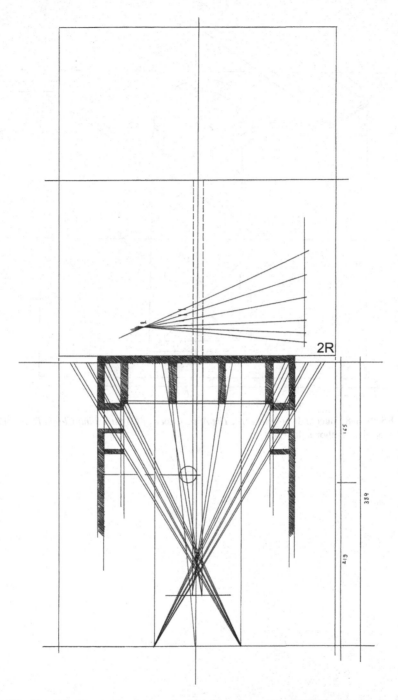

Plate E4 (No. 2R) Giotto, *The Approval of the Franciscan Rule*, 1296–9, Reconstructed plan (author's drawing)

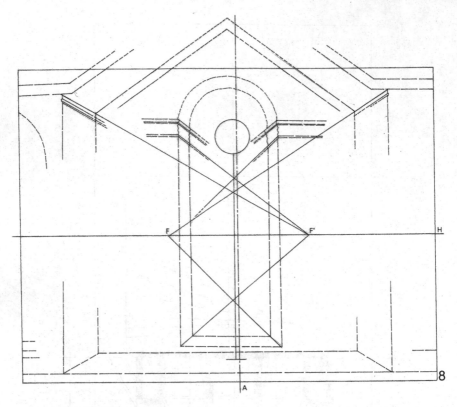

Plate E5 (No. 8) Duccio di Buoninsegna, *The Appearance of Christ behind Closed Doors*, 1308–11, Perspective (author's drawing)

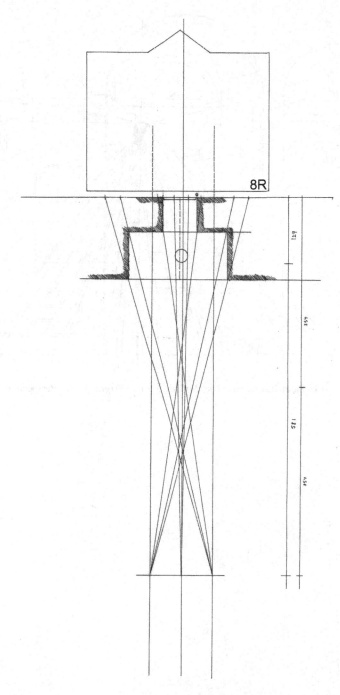

Plate E6 (No. 8R) Duccio di Buoninsegna, *The Appearance of Christ behind Closed Doors*, 1308–11, Reconstructed plan (author's drawing)

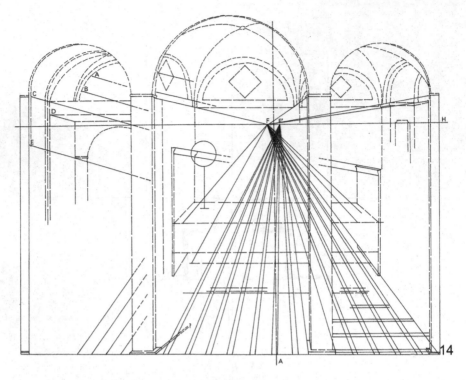

Plate E7 (No. 14) Pietro Lorenzetti, *The Nativity of the Virgin*, 1342, Perspective (author's drawing)

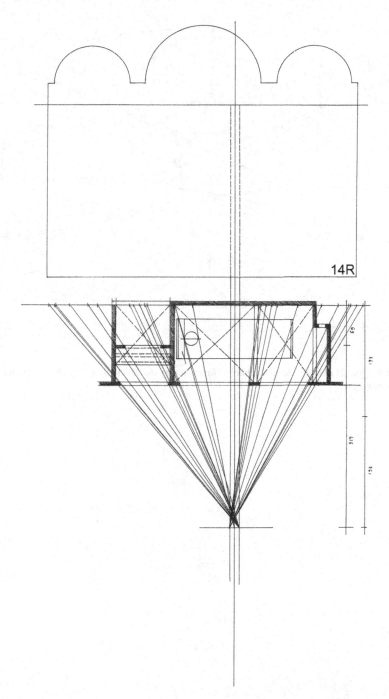

Plate E8 (No. 14R) Pietro Lorenzetti, *The Nativity of the Virgin*, 1342, Reconstructed plan (author's drawing)

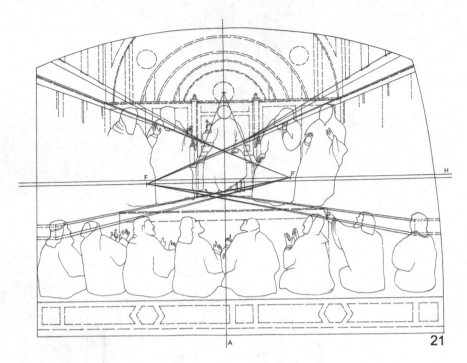

Plate E9 (No. 21) Giusto de' Menabuoi, *Christ among the Doctors*, 1376–78, Perspective (author's drawing)

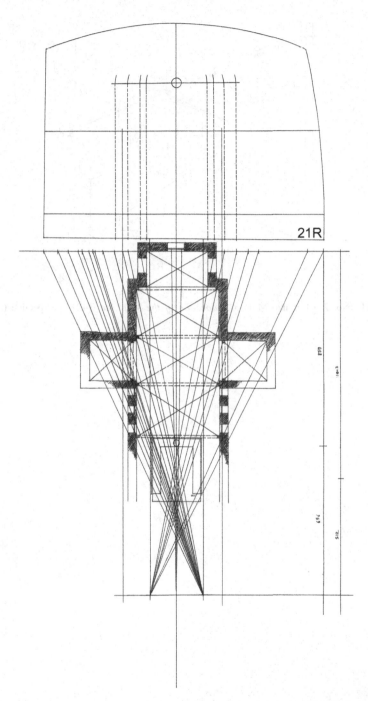

Plate E10 (No. 21R) Giusto de' Menabuoi, *Christ among the Doctors*, 1376–78, Reconstructed plan (author's drawing)

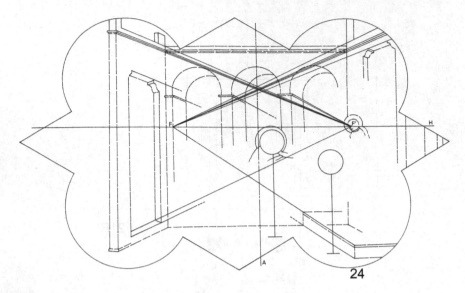

Plate E11 (No. 24) Lorenzo Monaco, *Adoration of the Magi*, ca. 1421, Perspective (author's drawing)

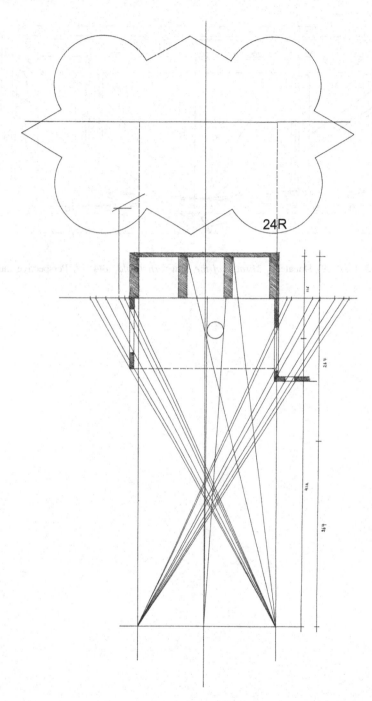

24R

Plate E12 (No. 24R) Lorenzo Monaco, *Adoration of the Magi*, ca. 1421, Reconstructed plan (author's drawing)

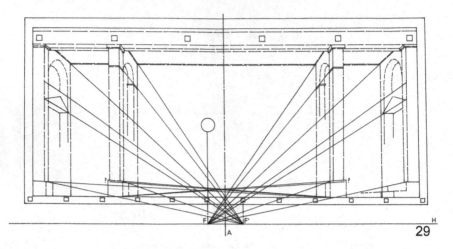

Plate E13 (No. 29) Donatello, *Miracle of the Newborn Child*, 1447–8, Perspective (author's drawing)

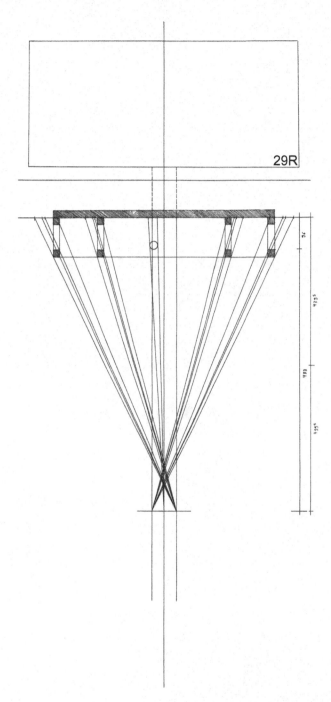

Plate E14 (No. 29R) Donatello, *Miracle of the Newborn Child*, 1447–8, Reconstructed plan
(author's drawing)

Bibliography

Aaen-Stockdale, Craig R. 2008. Ibn al-Haytham and psychophysics. *Perception* 37: 636–638.

Accolti, Pietro. 1625. *Lo inganno degl'occhi, prospettiva pratica di P.A., gentilhuomo fiorentino e della Toscana Accademia del Disegno.* Firenze: appresso P. Cecconcelli.

Aguilonius, Franciscus. 1613. *Opticorum libri sex.* Antwerp: ex. off. Plantiniana.

Aiken, Jane A. 1995. The perspective construction of Masaccio's Trinity fresco and medieval astronomical graphics. *Artibus et Historiae* 31: 171–187.

Alberti, Leon Battista. 1950. *Della Pittura, edizione critica a cura di Luigi Mallè.* Firenze: Sansoni editore: G.C.

Alberti, Leon Battista. 1972. *On Painting and On Sculpture. The Latin texts of the 'De pictura' and 'De statua',* edited with translations, introduction and notes by Cecil Grayson. London: Phaidon; new edition *On Painting.* London: Penguin Books, 1991.

Alberti, Leon Battista.1992. *De la peinture/De pictura (1435).* Préface, traduction et notes par J.-L. Schefer. Paris: Macula.

Alhacen, see Ibn al-Haytham.

Apollonius of Perga. 1891–1893. *Apollonii Pergaei quae graece extant cum commentariis antiquis,* t. I–II, ed. J.L. Heiberg. Leipzig: B.G. Teubner.

Archimedes. 1970. *De la sphère et du cylindre, La mesure du cercle, Sur les conoïdes et les sphéroïdes,* ed. Charles Mugler. Paris: Les Belles Lettres.

Archimedes. 1971. *Des corps flottants, Stomachion, La méthode, Le livre des lemmes, Le problème des boeufs,* ed. Charles Mugler. Paris: Les Belles Lettres.

Aristotle. 1966. *Posterior Analytics,* ed. H. Tredennick. Cambridge: Harvard University Press.

Aristotle, 1994. *Problèmes, tome III: sections XXVIII–XXXVIII,* texte établi et traduit par P. Louis. Paris: Les Belles Lettres.

Arrighi, Gino. 1967. Un estratto dal 'De visu' di M° Grazia de' Castellani (dal Codice Ottoboniano latino 3307 della Biblioteca Apostolica Vaticana). *Atti della Fondazione Giorgio Ronchi* 22: 44–58.

Arrighi, Gino. 1968. Il 'compasso ovale invention di Michiel Agnelo' dal Cod. L.IV.10 della Biblioteca degl' Intronati di Sienna. *Le Machine* 1: 103–106.

Averlino, Antonio detto Il Filarete .1972. *Trattato di architettura,* a cura di A.M. Grassi e L. Finoli. Milano: Il Polifilo.

Bacon, Francis. 1857–1874. *The Collected Works of Francis Bacon,* ed. James Spedding, Robert L. Ellis and Douglas D. Heath. London: Longmans.

Bacon, Roger. 1940. *Communia mathematica Fratris Rogeri,* ed. R. Steele. Oxford: Clarendon Press.

Bacon, Roger. 1897. *The 'Opus majus' of Roger Bacon,* edited with introduction and analytical table by John H. Bridges, 2 vols. London: Williams & Norgate, reed. Frankfurt am Main: Minerva GmbH, 1964.

Baggio, Luca. 1994. Sperimentazioni prospettiche e ricerche scientifiche a Padova nel secondo Trecento. *Il Santo* 34: 173–232.

© Springer International Publishing Switzerland 2016
D. Raynaud, *Studies on Binocular Vision,*
Archimedes 47, DOI 10.1007/978-3-319-42721-8

Ballardini, Antonella. 1998. Lo spazio pittorico medievale: Studi e prospettive di ricerca. In *La Prospettiva. Fondamenti teorici ed esperienze figurative dall'Antichità al mondo moderno*, ed. Rocco Sinisgalli. Convegno Internazionale di Studi (Roma, 11–14 settembre 1995). 281–292. Fiesole: Cadmo.

Bartoli, Cosimo. 1564. *Del modo di misurare le distantie, le superfitie, i corpi, le piante, le prouincie, le prospettiue, & tutte le altre cose terrene, che possono occorrere a gli huomini.* Venezia: Francesco Sanese.

Basile, Giuseppe. 1996. *Giotto Le storie francescane.* Milano: Electa.

Basile, Giuseppe, ed. 1989. *Pittura a fresco. Tecniche esecutive, cause di degrado, restauro.* Firenze: Le Monnier.

Bassi, Martino. 1572. *Dispareri in materia d'architettura et perspettiva.* Brescia: Francesco & Pie Maria Marchetti Fratelli.

Baur, Ludwig. 1912. Die philosophischen Werke des Robert Grosseteste. *Beiträge zur Geschichte der Philosophie des Mittelalters* 9: 1–778.

Beers, Yardley. 1953. *Introduction to the Theory of Errors*, Cambridge, MA: Addison-Wesley.

Bellosi, Luciano, and Giovanna Ragionieri. 2002. *"Giotto e le storie di San Francesco nella basilica superiore du Assisi"*, Assisi anno 1300, a cura di S. Brufani e E. Menestò. 455–473. Assisi: Edizioni Porziuncola.

Bellucci, Roberto and Frosinini, Cecilia. 2009. 'Cum suis debitis proportionibus': Perspective and Geometry in Sassetta's Borgo San Sepolcro Altarpiece. In *Sassetta: The Borgo San Sepolcro Altarpiece*, M. Israëls, ed. vol. 1, 359–370. Florence: The Harvard University Center for Italian Studies, Villa I Tatti.

Bellucci, Roberto and Frosinini, Cecilia. 2014. Underdrawing in Paintings. In *Science and Art: The Painted Surface*, eds. A. Sgamellotti, B.G. Brunetti and C. Miliani, 269–286. London: The Royal Society of Chemistry.

Beltrame, Renzo. 1973. Gli esperimenti prospettici del Brunelleschi. *Rendiconti dell'Accademia Nazionale dei Lincei, serie VIII* 28(3–4): 417–468.

Benedetti, Giovanni Battista. 1585. *Diversarum speculationum mathematicarum et physicarum liber.* Taurini: apud haeredem Nicolai Bevilaquae.

Benevolo, Leonardo, Stefano Chieffi, and Giulio Mezzetti. 1968. Indagine sul S. Spirito di Brunelleschi. *Istituto di Storia dell'Architettura. Quaderni* 85(90): 1–52.

Bergdolt, Klaus. 1989. *Der dritte Kommentar Lorenzo Ghibertis. Naturwissenschaften un Medizin in der Kunsttheorie der Frührenaissance.* Weinheim: Acta Humaniora.

Besançon, Alain. 1994. *L'Image interdite. Une histoire intellectuelle de l'iconoclasme.* Paris: Fayard.

Bevington, Philip R. 1969. *Data Reduction and Error Analysis for the Physical Sciences.* New York: McGraw-Hill.

Beyen, Hendrick G. 1938 *Die pompejanische Wanddekoration (vom zweiten bis zum vierten Stil)*, 2 vols. Den Haag: Martinus Nijhoff.

Beyen, Hendrick G. 1939. Die antike Zentralperspecktive. *Jahrbuch des deutschen archäologischen Instituts* 54: 47–72.

Blasius of Parma [Biagio Pelacani da Parma]. 2009. *Questiones super perspectiva communi.* édité par Graziella Federici Vescovini et Joël Biard, avec la collaboration de Valeria Sorge, Orsola Rignani et Riccardo Bellè. Paris: Librairie J. Vrin

Boethius. 1995. *De Institutione arithmetica*, ed. J.-Y. Guillaumin. Paris: Les Belles Lettres.

Bøggild-Johanssen, Birgitte, and Marcussen Marianne. 1981. A critical survey of the theoretical and practical origins of the Renaissance linear perspective. *Acta ad Archaelogiam et Artium Historiam Pertinentia* 8: 191–227.

Bonaventure. 1885. *Opera Omnia S. Bonaventurae, 2. Commentaria in librum II Sententiarum.* Quaracchi, Florentiae: ex typographia Collegii S. Bonaventurae.

Boncompagni, Baldassare. 1862. *Scritti di Leonardo Pisani, matematico del secolo decimoterzo*, vol. II. *Leonardi Pisani Practica geometriae ed opuscoli.* Roma: Tipografia delle scienze matematiche e fisiche.

Bonelli, Renato. 1985. Basilica di Assisi: i committenti. *Antichità Viva/Mélanges Luisa Becherucci* 24: 174–179.

Borelli, Giovanni Alfonso. 1672. Observations touchant la force inégale des deux yeux. *Journal des Sçavans* 3: 295–298.

Borsi, Franco. 1992. *Paolo Uccello*. Paris: Hazan; English translation by E. Powell, *Paolo Uccello*. New York, NY: H.N. Abrams, 1994.

Boudon, Raymond. 1990. *L'Art de se persuader des idées douteuses, fragiles ou fausses*. Paris: Fayard; English translation by M. Slater: *The art of self-persuasion: the social explanation of false beliefs*. Cambridge, MA: Polity, 1994.

Bozzolato, Giampiero. 1992. *Il Palazzo della ragione a Padova: Gli affreschi, Istituto poligrafico e Zecca dello Stato*. Roma: Libreria dello Stato.

Brugerolles, Emmanuelle, and David Guillet. 1997. Grégoire Huret, dessinateur et graveur. *Revue de l'art* 117: 9–35.

Bunim, Miriam S. 1940. *Space in Medieval Painting and the Forerunners of Perspective*. New York: Columbia University Press.

Busard, Hubert L.L. 1965. The Practica Geometriae of Dominicus de Clavasio. *Archive for the History of Exact Sciences* 2: 520–575.

Busard, Hubert L.L. 1971. Die Traktate 'De proportionibus' von Jordanus Nemorarius und Campanus. *Centaurus* 15: 193–227.

Busard, Hubert L.L. 1998. *Johannes de Muris. De Arte mensurandi*, Stuttgart, F. Steiner.

al-Būzjānī, Abū al-Wafā', see Krasnova.

Camerota, Filippo. 1998. Misurare 'per perspectiva': geometria pratica e *prospectiva pingendi*. In *La Prospettiva. Fondamenti teorici ed esperienze figurative dall'Antichità al mondo moderno*, ed. Rocco Sinisgalli. Atti del Convegno Internazionale di Studi (Istituto Svizzero di Roma, 11–14 settembre 1995). 293–308. Fiesole: Cadmo.

Camerota, Filippo. 2001. "Brunelleschi's panels," *The 4th International Laboratory for the History of Science*. Florence: May 25th 2001.

Camerota, Filippo. 2001. L'esperienza di Brunelleschi, In: *Nel segno di Masaccio*. 32–33. Firenze: Giunti.

Cardano, Girolamo. 1663. "Problematum medicorum. Sectio Secunda," *Hieronymi Cardani Mediolanensis... Operum tomus secundus*. Lugduni: sumptibus I.A. Huguetan & M.A. Ravaud, 636–642.

Carlevaris, Laura. 1989. La prospettiva nell'ottica antica: il contributo di Tolomeo. *Disegnare* 27: 16–29.

Carter, Bernard A.R. 1987. Perspective. In *The Oxford Companion to Western Art*, ed. Harold Osborne, 840–861. Oxford: Clarendon Press.

Catanco, Pietro. 1567. *L'Architettura*. Venezia: Aldo.

Cecchini, Francesca. 1998. "Artisti, committenti e perspectiva in Italia alla fine del Duecento," R. Sinisgalli, ed., *La prospettiva. Fondamenti teorici ed esperienze figurative dall'Antichità al mondo moderno*, Atti del Convegno Internazionale di Studi (Istituto Svizzero di Roma, 11–14 settembre 1995). 56–74. Fiesole: Cadmo.

Cecchini, Francesca. 2006. "Ambiti di diffusione del sapere ottico nel Duecento," M. Dalai Emiliani, M. Cojannot Le Blanc, P. Dubourg Glatigny, eds, *L'Œuvre et l'artiste à l'épreuve de la perspective*, actes du colloque international (Rome, 19–21 septembre 2002). 19–42. Rome: Publications de l'École française de Rome.

Cennini, Cennino. 1859. *Il Libro dell'arte o Trattato della pittura, a cura di Gaetano e Carlo Milanesi*. Firenze: Le Monnier.

Chasles, Michel. 1837. *Aperçu historique sur l'origine et le développement des méthodes géométriques*. Bruxelles: Hayez.

Chastel, André, and Edi Baccheschi. 1982. *Tout l'œuvre peint de Giotto*. Paris: Flammarion.

Chérubin d'Orleans [Michel Lasséré]. 1677. *La Vision parfaite ou Le concours des deux axes de la vision en un seul point de l'objet*. Paris: Chez Sebastien Mabre-Cramoisy.

Chérubin d'Orleans. 1681. *La Vision parfaite ou La veue distincte par le concours des deux axes en un seul point de l'objet*, tome II. Paris: chez Edme Couterot.

Chuquet, Nicolas. 1979. *La Géométrie. Première géométrie algébrique en langue française (1484)*, introduction, texte et notes par Hervé L'Huillier. Paris: Vrin.

Clagett, Marshall. 1964–1984. *Archimedes in the Middle Ages*. Philadelphia: The American Philosophical Society, vol. 1: 1964; vol. 3: 1978, vol. 5: 1984.

Colombo, Realdo. 1559. *De re anatomica libri XV*. Venezia: ex typ. Bevilacquae.

Corbé, Christian, Jean-Pierre Menu, and Gilles Chaine. 1993. *Traité d'optique physiologique et clinique*. Paris: Doin.

Cook, William R. (ed.). 2005. *The Art of the Franciscan Order in Italy*. Leiden: E.J.Brill.

Cooper, Donal, and Janet Robson. 2003. Pope Nicholas IV and the upper church at Assisi. *Apollo* 157: 31–35.

Curze, Maximilian. 1902. *Der Liber Embadorum des Abraham bar Chijja Savasorda in der Übersetzung des Plato von Tivoli*. Leipzig: B.G.Teubner.

Dalai Emiliani, Marisa. 1990. La question de la perspective. In: *Perspective et histoire au Quattrocento*. 97–117. Paris: Les Éditions de la Passion.

Damianos, 1897. *Damianos Schrift über Optik, mit Auszügen aus Geminos. Griechisch und Deutsch herausgegeben von Richard Schöne*. Berlin: Reichsdruckerei.

Danti, Cristina (ed.). 2002. *La Trinità di Masaccio: il restauro dell'anno duemila*. Firenze: Edifir Edizioni Firenze.

Danti, Egnatio. 2003. *Le Due Regole della prospettiva pratica di M. Iacomo Barozzi da Vignola con i commentarij del R.P.M. Egnatio Danti,* Roma: Francesco Zanetti, 1583; nouvelle édition et trad. fr. P. Dubourg Glatigny. Paris: CNRS.

Darr, Alan P. and Bonsanti, Giorgio, eds. 1986. *Donatello e i suoi. Scultura fiorentina del primo Rinascimento*. Detroit: Founders Society and Detroit Institute of Arts/Firenze, La Casa Usher/Milano: Mondadori editore.

Degenhart, Bernhard, and Annegrit Schmitt. 1968. *Einleitung a Corpus der Italienischen Zeichnungen 1300–1450, I–1*. Sud- und Mittelitalien, Berlin: Gebr: Mann Verlag.

Dubourg Glatigny, Pascal. 1999. La merveilleuse fabrique de l'oeil. *Roma moderna e contemporanea* 7: 369–394.

Dubourg Glatigny, Pascal. 2011. *Il Disegno naturale del mondo. Saggio sulla biografia di Egnatio Danti con l'edizione del Carteggio*. Milan: Aguaplano.

Du Tour, Étienne-François. 1760. "Discussion d'une question d'Optique," *Mémoires de mathématique et de physique présentés à l'Académie royale des Sciences par divers savans* 3: 514–530.

Du Tour, Étienne-François. 1763. "Addition," *Mémoires de mathématique et de physique présentés à l'Académie royale des Sciences par divers savans* 4: 499–511.

Dürer, Albrecht. 1995. *Géométrie*, ed. J. Peiffer. Paris: Éditions du Seuil.

Edgerton, Samuel Y. 1975. *The Renaissance Rediscovery of Linear Perspective*. New York: Basic Books.

Edgerton, Samuel Y. 1991. *The Heritage of Giotto's Geometry. Art and Science on the Eve of the Scientific Revolution*. Ithaca/London: Cornell University.

Edgerton, Samuel Y. 2009. *The Mirror, the Window, and the Telescope*. New York: Cornell University Press.

Elkins, James. 1988. Did Leonardo develop a theory of curvilinear perspective? *Journal of the Warburg and Courtauld Institutes* 51: 190–196.

Elkins, James. 1994. *The Poetics of Perspective*. Ithaca: Cornell University Press.

Eriksson, Ruben. 1959. *Andreas Vesalius' first public anatomy at Bologna, 1540. An eyewitness report by Baldasar Heseler*. Uppsala/Stockholm: Almqvist and Wiksells.

Euclid. 1895. *Euclidis opera omnia*, vol. VII: *Euclidis Optica... Catoptrica cum scholiis antiquis*, ed. J.L. Heiberg. Leipzig: B.G. Teubner.

Euclid. 1983. *The First Latin Translation of Euclid's* Elements *commonly ascribed to Adelard of Bath: Books I–VIII and Books X.36–XV.2* ed. Hubert L.L. Busard. Toronto: The Pontifical Institute of Medieval Studies.

Euclid. 1990–2001. *Les Éléments*, vol. 1, *Livres I–IV: Géométrie plane*, vol. 2, *Livres V–VI: Proportions et similitudes, Livres VII–IX: Arithmétique*, vol. 3, *Livre X: Grandeurs commensurables et incommensurables, Classification des lignes irrationnelles*, vol. IV, *Livres XI–XIII: Géométrie des solides*, traduction et commentaire par Bernard Vitrac. Paris: Presses universitaires de France.

Federici Vescovini, Graziella. 1964. Les questions de 'perspective' de Dominicus de Clivaxo. *Centaurus* 10: 236–246.

Vescovini Federici, Graziella. 1965. Contributo per la storia della fortuna di Alhazen in Italia. Il volgarizzamento del ms. vat. 4595 e il 'Commentario Terzo' del Ghiberti. *Rinascimento* 5: 17–49.

Federici Vescovini, Graziella. 1965. *Studi sulla prospettiva medievale*. Torino: Giappichelli.

Federici Vescovini, Graziella. 1969. "L'inserimento della 'perspectiva' tra le arti del quadrivio," *Arts libéraux et philosophie au Moyen Âge*, Actes du Congrès international de philosophie médiévale. Montréal: Institut d'Études Médiévales/Paris: Librairie philosophie J. Vrin, 969–974.

Federici Vescovini, Graziella. 1998. Alhazen vulgarisé: Le 'De li aspecti' d'un manuscrit du Vatican (moitié du XIVe siècle) et le troisième 'Commentaire sur l'optique' de Lorenzo Ghiberti. *Arabic Sciences and Philosophy* 8: 67–96.

Federici Vescovini, Graziella. 2005. "Image et représentation optique: Blaise de Parme et Léon Baptiste Alberti," *Kora. Revue d'etudes anciennes et médiévales* 3/4: 357–376.

Field, Judith V. 1985. Giovanni Battista Benedetti on the mathematics of linear perspective. *Journal of the Warburg and Courtauld Institutes* 48: 71–99.

Field, Judith V., Roberto Lunardi, and Thomas B. Settle. 1989. The perspective scheme of Masaccio's Trinity fresco. *Nuncius* 4: 31–118.

Field, Judith V. 1997. Alberti, the abacus and Piero della Francesca's proof of perspective. *Renaissance Studies* 11(2): 61–88.

Field, Judith V. 1997. *The Invention of Infinity: Mathematics and Art in the Renaissance*. Oxford: Oxford University Press.

Field, Judith V. 2001. "What mathematical analysis can tell us about a fifteenth-century picture?" *Art, Science and Techniques of Drafting in the Renaissance, 4th ILabHS*, working paper, Florence, 24 May–1 June 2001.

Field, Judith V. 2005. *Piero della Francesca, A Mathematician's Art*. New Haven: Yale University Press.

Flocon, André, and André Barre. 1968. *La Perspective curviligne. De l'espace visuel à l'image construite*. Paris: Flammarion.

Folkerts, Menso. 1970. *«Boetius» Geometrie II: Ein mathematisches Lehrbuch des Mittelalters*. Wiesbaden: Steiner Verlag.

Folkerts, Menso. 1996. Piero della Francesca and Euclid. In *Piero della Francesca tra arte e scienza*, eds. M. Dalai Emiliani e P. Curzi, 293–312. Venezia: Marsilio.

Francesca, Piero della. 1970. *Trattato d'abaco*, ed. G. Arrighi, Pisa, Domus Galileiana.

Francesca, Piero della. 1984. *De Prospectiva Pingendi, edizione critica a cura di Giusta Nicco-Fasola*. Firenze: Casa editrice le Lettere.

Francesca, Piero della. 1995. *Libellus de quinque corporibus regularibus*, ed. F.P. di Teodoro. Firenze: Edizione nazionale degli scritti di Piero della Francesca.

Frommel, Christoph L. 1973. *Der Römische Palastbau der Hochrenaissance, 3 Bde*. Tübingen: Verlag E. Wasmuth.

Galassi, Maria Clelia. 1998. *Il disegno svelato. Progetto e immagine nella pittura italiana del primo Rinascimento*. Nuoro: Ilisso.

Galen, Claudius. 1854. *Oeuvres anatomiques, physiologiques et médicales de Galien*, ed. Charles Daremberg. Paris: J.B. Baillière.

Gallay, Antoine. 2013. *The Quest for Perfect Vision. Chérubin d'Orléans' optical instruments and the development of theories of binocular perception in late 17th-century France*, MPhil Dissertation, supervision: Prof. Simon Schaffer, Cambridge.

Genet, Jean-Philippe. 2009. Revisiter Assise: la lisibilité de l'image médiévale. In *Itinéraires du savoir de l'Italie à la Scandinavie (Xe–XVIe siècle). Études offertes à Élisabeth Mornet*, ed. Corinne Péneau, 391– 419. Paris: Publications de la Sorbonne.

Gessner, Samuel. 2006. *Les Mathématiques dans les écrits d'architecture italiens, 1545–1570*, thesis, Université Paris VII.

Ghiberti, Lorenzo. 1947. *I Commentari*, a cura di O. Morisani. Napoli: R. Ricciardi editore.

Gibson, James J. 1979. *The Ecological Approach to Visual Perception*. Boston: Houghton Mifflin.

Gioseffi, Decio. 1957. *Perspectiva artificialis. Per la storia della prospettiva, spigolature e appunti*. Trieste: Università degli Studi di Trieste, Facoltà di Lettere e Filosofia.

Goffen, Rona (ed.). 1998. *Masaccio's Trinity*. Cambridge: Cambridge University Press.

Gorman, Michael J. 2003. Mathematics and modesty in the Society of Jesus: the problems of Christoph Grienberger. In *The New Science and Jesuit Science. Sevententh Century Perspective*, ed. Mordechai Feingold, 1–120. Dordrecht: Kluwer.

Grabmann, Martin. 1906. *Die philosophische und theologische Erkenntnislehre des Kardinal Matthaeus von Acquasparta*. Wien: Verlag von Mayer & Co.

Guillaume, Jean. 1991. Désaccord parfait: ordres et mesures dans la chapelle des Pazzi. *Annali di Architettura* 2: 9–23.

Hahnloser, Hans R. 1935. *Villard de Honnecourt, Kritische Gesamtausgabe des Bauhüttenbuches*. Wien: akademische Druck- und Verlagsanstalt.

Hauck, Guido. 1879. *Die subjektive Perspektive und die horizontalen Curvaturen des dorischen Styls*. Stuttgart: Wittwer.

Hecht, Konrad. 1976. Maßverhältnisse und Maße der Cappella Pazzi. *Architectura* 6: 148–174.

Helmholtz, Hermann von. 1867. *Handbuch der physiologischen Optik*. Leipzig: L. Voss; French translation *Optique physiologique*. Paris: Masson et fils.

Helmholtz, Hermann von, and Ernst W.von Brücke. 1878. *Principes scientifiques des beaux-arts. Essais et fragments de théorie. L'optique et la peinture*. Paris: Baillière et Cie.

Herdman, William G. 1853. *A Treatise on the Curvilinear Perspective of Nature*. London: John Weale.

Hering, Ewald (1861–1864) *Beiträge zur Physiologie*, I. *Vom Ortsinne der Netzhaut*, II. *Vom den identischen Netzhautstellen*, III. *Vom Horopter*. Leipzig: W. Engelmann.

Hero of Alexandria. 1900. *Heronis Alexandrini opera quae supersunt omnia*, vol. II. *Mechanica et Catoptrica*, ediderunt L. Nix and W. Schmidt. Stuttgart: B.G. Teubner.

Hoffmann, Volker. 1996. Masaccios Trinitätsfresko: Die perspektivkonstruktion und ihr Entwurfsverfahren. *Mitteilungen des Kunsthistorischen Institutes in Florenz* 40: 42–77.

Hoffmann, Volker. 2001. "The Trinity of Masaccio: perspective construction—isometric transformation—coordinate system," *Art, Science and Techniques of Drafting in the Renaissance, 4th ILabHS*, working paper, Florence, 24 May–1 June 2001.

Hoffmann, Volker. 2001. "Brunelleschi's invention of linear perspective: The fixation and simulation of the optical view," *Art, Science and Techniques of Drafting in the Renaissance, 4th ILabHS*, working paper, Florence, 24 May–1 June 2001.

Howard, Ian P., and Brian J. Rogers. 1995. *Binocular Vision and Stereopsis*. Oxford: Oxford University Press.

Howard, Ian P. 1996. Alhazen's neglected discoveries of visual phenomena. *Perception* 25: 1203– 1217.

Hugonnard Roche, Henri. 1984. "La classification des sciences de Gundissalinus et l'influence d'Avicenne. In *Études sur Avicenne* eds. Jean Jolivet et Roshdi Rashed, 41–63. Paris: Les Belles Lettres.

Huret, Grégoire. 1670. *Optique de portraiture et de peinture*. Paris: chez l'Autheur.

Huygens, Christiaan. 1704. *Opuscula posthuma quae continent Dioptricam...* Lugduni Batavorum: Boutesteyn.

Ibn al-Haytham. 1971. *Al-Shukūk 'alā Baṭlamyūs (Dubitationes in Ptolemaeum)*, ed. A.I. Sabra and N. Shehaby. Cairo: National Library Press.

Ibn al-Haytham. 1572. *Opticae Thesaurus Alhazeni Arabi libri septem...*, ed. F. Risner. Basileae: Per Episcopios. Reprint with an introduction by D.C. Lindberg. New York: Johnson Reprint Corporation, 1972.

Ibn Isḥāq, Ḥunayn. 1928. *The Book on the Ten Treatises on the Eye Ascribed to Hunain ibn Is-hâq (809–877 A.D.), The Earliest Existing Systematic Text-Book of Ophtalmology*, edited by Max Meyerhof. Cairo: Government Press.

Janson, Horst W. 1967. Ground plan and elevation of Masaccio's Trinity fresco. In *Essays in the History of Art presented to Rudolf Wittkower*, eds. D. Fraser and al., 83–88. London: Phaidon.

Jolivet, Jean. 1997. Classification des sciences. In *Histoire des sciences arabes*, eds. Roshdi Rashed et Régis Morelon, vol. 3, 255–270. Paris: Éditions du Seuil.

Jordanus de Nemore. 1991. *De Elementis arithmetice artis. A medieval treatise on number theory*, ed. Hubert L.L. Busard. Stuttgart: Franz Steiner.

Kemp, Martin. 1978. Science, non-science and non-sense: The interpretation of Brunelleschi's perspective. *Art History* 1: 134–161.

Kemp, Martin. 1985. Geometrical perspective from Brunelleschi to Desargues: a pictorial means or an intellectual end? *Proceedings of the British Academy* 70: 89–132.

Kemp, Martin. 1990. *The Science of Art. Optical Themes in Western Art from Brunelleschi to Seurat*. New Haven: Yale University Press.

Kepler, Johannes (1604) *Ad Vitellionem paralipomena, quibus astronomiae pars optica traditur*, Francofurti, apud Claudium Marnium & Hæredes Ioannis Aubrii.

Kern, Guido J. 1913. Das Dreifaltigkeitsfresko von S. Maria Novella. Eine perspektivisch-architektur geschichtliche Studie. *Jahrbuch der königlich preussischen Kunstsammlungen* 24: 36–58.

Kern, Guido J. 1913. Die Anfänge der zentralperspektivischen Konstruktion in der italienischen Malerei des 14. Jahrunderts. *Mitteilungen des Kunsthistorischen Instituts in Florenz* 2: 39–65.

Kheirandish, Elaheh. 1999. *The Arabic Version of Euclid's Optics: Kitāb Uqlīdis fi ikhtilāf al-manāẓir*, Edited and Translated with Historical Introduction and Commentary. New York: Springer, 2 vols.

Krasnova, Svetlana A. 1966. Abu-l-Vafa al-Buzdjani, Kniga o tom, chto neobhodimo remeslenniku iz geometricheskih postroenij, *Fiziko-Matematicheskie Nauki v Stranah Vostoka* 1: 56–130 (text), 131–140 (commentary).

Krautheimer, Richard and Krautheimer-Hess, Trude. 1956. *Lorenzo Ghiberti*. Princeton, N.J.: Princeton University Press, 2 vols.

Kurz, Otto. 1960. Dürer, Leonardo and the invention of the ellipsograph. *Raccolta vinciana* 18: 15–25.

La Hire, Philippe de. 1730. "Traité sur les differens accidens de la veüe" (1694), *Mémoires de l'Académie royale des Sciences* 9: 530–634.

Langenstein, Henricus de. 1503 "Questiones super perspectivam," *Preclarissimum mathematicarum opus*, ed. Tomás Durán.Valencia: Juan Jofré, fol. 47r–65v.

Laurent, Roger. 1987. *La place de J.-H. Lambert (1728–1777) dans l'histoire de la perspective*. Paris: Cedic-Nathan.

Le Cat, Claude Nicolas. 1744. *Traité des sens*. Amsterdam: chez J. Wetstein.

Le Clerc, Sébastien. 1679. *Discours touchant le point de veue, dans lequel il est prouvé que les choses qu'on voit distinctement, ne sont veuës que d'un oeil*. Paris: chez Thomas Jolly.

Le Clerc, Sébastien. 1712. *Système de la vision fondé sur de nouveaux principes*. Paris: chez Florentin Delaulne.

Le Grand, Yves. 1948–1956. *Optique physiologique*, tome. I: *La dioptrique de l'oeil et sa correction*; tome II: *Lumière et couleurs*, tome III: *L'espace visuel*. Paris: Éditions de la Revue d'Optique.

Lejeune, Albert. 1958. Les recherches de Ptolémée sur la vision binoculaire. *Janus* 47: 79–86.

Lejeune, Albert. 1989. *L'Optique de Claude Ptolémée dans la version latine d'après l'arabe de l'émir Eugène de Sicile*. Leiden: E.J. Brill.

Lindberg, David C. 1967. Alhazen's theory of vision and its reception in the West. *Isis* 58: 326–337.

Lindberg, David C. 1970. *John Pecham and the Science of Optics. Perspectiva Communis*. Edited with an Introduction, English Translation, and Critical Notes. Madison: The University of Wisconsin Press.

Lindberg, David C. 1971. Lines of influence in the thirteenth century optics: Bacon, Witelo, and Pecham. *Speculum* 46: 66–83.

Lindberg, David.C. 1975. *A Catalogue of Medieval and Renaissance Optical Manuscripts*. Toronto: The Pontifical Institute of Mediaeval Studies.

Lindberg, David C. 1976. *Theories of Vision from al-Kindi to Kepler*. Chicago and London: The University of Chicago Press.

Lindberg, David C. 1996. *Roger Bacon and the Origins of Perspectiva in the Middle Ages*. A Critical Edition and English Translation of Bacon's *Perspectiva*, with Introduction and Notes. Oxford: Oxford University Press.

Lindberg, David C. 1998. *Roger Bacon's Philosophy of Nature*. A Critical Edition, with English Translation, Introduction and Notes, of *De multiplicatione specierum* and *De speculis comburentibus*. South Bend: St. Augustine Press.

Listing, Johann Benedikt. 1845. *Beitrag zur physiologischen Optik*. Göttingen: Vandenhoeck und Rupercht.

Little, Alan M.G. 1936. Scaenographia. *The Art Bulletin* 18: 407–418.

Little, Alan M.G. 1937. Perspective and scene painting. *The Art Bulletin* 20: 487–495.

Livesey, Steven J. 1990. Science and Theology in the Fourteenth Century: the Subalternate Sciences in Oxford Commentaries on the Sentences. *Synthese* 83: 273–292.

Long, James. 1979. *Bartholomaeus Anglicus, De proprietatibus rerum, Books 3–4: On the Properties of Soul and Body*. Toronto: The Pontifical Institute of Mediaeval Studies.

Ludi, Jean-Claude. 1989. *La Perspective "pas à pas". Manuel de construction graphique de l'espace et tracé des ombres*. Paris: Dunod.

Luneburg, Rudolf.K. 1947. *Mathematical analysis of binocular vision*. Princeton: Princeton University Press.

Lunghi, Elvio. 1996. *La Basilica di San Francesco di Assisi*. Scala: Antella.

Madre, Alois. 1965. *Nikolaus von Dinkelsbühl, Leben und Schriften, Beiträge zur Geschichte der Philosophie und Theologie des Mittelalters, 40/4*. Münster: Äschendorffsche Verlagbuchhandlung.

Maginnis, Hayden B.J., and Andrew Ladis. 1998. Assisi today: the upper church. *Source* 18: 1–6.

Mancha, José Luís. 1989. Egidius of Baisiu's theory of pinhole images. *Archive for History of Exact Sciences* 40: 1–35.

Manetti, Antonio di Tuccio. 1970. *The Life of Brunelleschi, by Antonio di Tuccio Manetti/Vita di Filippo di Ser Brunelleschi*, Introduction, notes and critical text edited by Howard Saalman, English translation by Catherine Engass. University Park: Pennsylvania State University Press.

Mariotte, Edme. 1668. "La scénographie ou perspective," *Procès verbaux de l'Académie royale des Sciences, Registre de Mathématiques*, tome 4 (14 avril–24 décembre 1668), séance du 20 juin 1668, fol. 62r–73r.

Martindale, Andrew. 1988. *Simone Martini*, Complete ed. Oxford: Phaidon.

Marzinoto, Marica. 2006. "Filippo Gagliardi e la didattica della prospettiva nell'accademia di San Luca a Roma, tra XVII e XVIII secolo," M. Dalai Emiliani, M. Cojannot Le Blanc, P. Dubourg Glatigny, eds, *L'Œuvre et l'artiste à l'épreuve de la perspective*, actes du colloque international (Rome, 19–21 septembre 2002), 153–177. Rome: Publications de l'École française de Rome.

Matteo d'Aquasparta. 1961. *Quaestiones disputatae de anima XIII*, ed. A.J. Gondras. Paris: Librairie philosophique J. Vrin.

Mesa Gisbert, Andrés de. 1989. El 'fantasma' del punto de fuga en los estudios sobre la sistematización geométrica de la pintura del siglo XIV. *D'Art* 15: 29–50.

Molk, Jules. 1906–1911. *Encyclopédie des sciences mathématiques pures et appliquées*, tome 1, vol. 4: *Calcul des probabilités, théorie des erreurs*. Sceaux: Jacques Gabay, 1993.

Montaiglon, Anatole de. 1875–1878. *Procès-verbaux de l'Académie royale de Peinture et de Sculpture, 1648–1793*, tome I: *1648–1672*; tome II: *1673–1688*. Paris: J. Baur.

Monte, Guidobaldo del. 1600. *Perspectivae libri sex*. Pesaro: apud Hieronymum Concordiam.

Müller, Johannes Peter. 1826. *Zur vergleichenden Physiologie des Gesichtssinns des Menschen und der Thiere*. Leipzig: Cnobloch.

Narducci, Enrico. 1871. Nota intorno a una traduzione italiana fatta nel secolo decimoquarto del trattato d'ottica d'Alhazen, matematico del secolo undecimo e ad altri lavori di questo scienziato. *Bollettino di bibliografia e di storia delle scienze matematiche e fisiche* 4: 1–40.

al-Nayrīzī, 1899. *Anaritii in decem libros priores Elementorum Euclidis commentarii ex interpretatione Gherardi Cremonensis, edidit M. Curze*. Leipzig: B. G. Teubner.

Naẓīf, Muṣṭafā. 1942. *Al-Ḥasan ibn al-Haytham: buḥūthuhu wa-kushūfuhu al-baṣariyya (Al-Ḥasan Ibn al-Haytham: His Research and Discoveries in Optics)*, 2 vols. Cairo: Nūrī Press. Reprint with an Introduction by Roshdi Rashed. Beirut: Markaz dirāsāt al-waḥdah al-'arabiyyah, 2008.

Nicéron, Jean-Pierre. 1731. *Mémoires pour servir à l'histoire des hommes illustres dans la République des Lettres*. Paris: chez Briasson.

Ogle, Kenneth Neil. 1950. *Researches in Binocular Vision*. Philadelphia/London: Saunders, 2nd edition 1964.

Pacioli, Luca. 1494. *Summa de aritmetica, geometria, proportione et proportionalita*. Venezia: Paganino de Paganini.

Pacioli, Luca. 1509. *Divina proportione*. Venezia: Paganino de Paganini.

Pagnoni-Sturlese, Maria Rita, Rudolf Rehn, and Loris Sturlese. 1985. *Dietrich von Freiberg. Opera Omnia, IV. Schriften zur Naturwissenschaft*. Hamburg: Felix Meiner.

Palladio, Andrea. 1570. *I Quattro Libri de architettura*. Venezia: Domenico dei Franceschi.

Panofsky, Erwin. 1991. (1924/5) Die Perspektive als "symbolische Form", *Vorträge der Bibliothek Warburg* 4: 258–331; French translation *La Perspective comme "forme symbolique" et autres essais*. Éditions de Minuit, 1975; Paris: English translation by Christopher S. Wood, *Perspective as Symbolic Form*. New York: Zone Books.

Panofsky, Erwin. 1976. *(1960) Renaissance and Renascences in Western art*. Stockholm: Almqvist and Wiksell. French translation *La Renaissance et ses avants-courriers dans l'art d'Occident*. Paris: Flammarion.

Pansier, Pierre. 1909–1933. *Collectio ophtalmologica veterum auctorum*, fasc. 7: Ḥunayn Ibn Isḥāq, *Liber de oculi*; Galien, *Littere Galieni ad corisium de morbis oculorum et eorum curis*. Paris: J.B. Baillière.

Panum, Peter Ludwig. 1858. *Physiologische Untersuchungen über das Sehen mit zwei Augen*. Schwer: Kiel.

Pappus. 1876–1878. *Pappi Alexandrini Collectionis quae supersunt*, ed. F. Hultsch. Berlin: Weidman.

Paravicini Bagliani, Agostino. 1975. Witelo et la science optique à la cour pontificale de Viterbe (1277). *Mélanges de l'École française de Rome, Moyen Âge/Temps modernes* 87: 425–453.

Parronchi, Alessandro. 1957. Le fonti di Paolo Uccello: I prospettivi passati. *Paragone* 89: 3–32.

Parronchi, Alessandro. 1957. Le fonti di Paolo Uccello: I filosofi. *Paragone* 95: 3–33.

Parronchi, Alessandro. 1958. Le due tavole prospettiche del Brunelleschi. *Paragone* 107: 3–32.

Parronchi, Alessandro. 1964. *Studi sulla dolce prospettiva*. Milano: Aldo Martello.

Parronchi, Alessandro. 1974. *"Prospettiva e pittura in Leon Battista Alberti"*, Convegno internazionale indetto nel V Centenario di Leon Battista Alberti (Roma-Mantova-Firenze, 25–29 aprile 1972), 213–232. Roma: Accademia nazionale dei Lincei.

Parsey, Arthur. 1836. *Perspective Rectified*. London: Longmans.

Pecham, John. 1972. *Tractatus de perspectiva*, edited with an introduction and notes by David C. Lindberg. St Bonaventure, N.Y: The Franciscan Institute.

Pedretti, Carlo. 1957. *Studi vinciani, documenti, analisi e inediti leonardeschi.* Genève: Librairie E. Droz.

Pedretti, Carlo. 1963. Leonardo on curvilinear perspective. *Bibliothèque d'Humanisme et Renaissance* 25: 69–87.

Pedretti, Carlo. 1983. *Léonard de Vinci architecte.* Milan/Paris: Electa.

Pedretti, Carlo. 2003. Leonardo discepolo della sperientia, In: *Nel segno di Masaccio.* 184–185. Firenze: Guinti.

Pérez, José-Philippe. 2000. *Optique, Fondements et applications.* Paris: Dunod.

Pigassou-Albouy, Renée (1991/2) *Les strabismes*, vol. 1: *Les divergences oculaires*, vol. 2: *Les convergences oculaires.* Paris: Masson.

Pirenne, Maurice. 1952. The scientific basis of Leonardo da Vinci's theory of perspective. *British Journal for the Philosophy of Science* 3: 165–185.

Pirenne, Maurice. 1970. *Optics, Painting and Photography*, Cambridge: Cambridge University Press.

Platter, Felix. 1583. *De corporis humani structura et usu.* Basileae: ex officina Frobeniana.

Polzer, Joseph. 1971. The anatomy of Masaccio's Holy Trinity. *Jahrbuch der Berliner Museen* 13: 18–59.

Porta, Giambattista della. 1593. *De refractione optices parte libri novem.* Neapoli: apud Io. Iacobum Carlinum & Antonium Pacem.

Porterfield, William. 1759. *A Treatise on the Eye. The Manner and Phœnomena of Vision*, 2 vols. Edinburgh: G. Hamilton and J. Balfour.

Pozzo, Andrea. 1698. *Perspectiva pictorum et architectorum*, vol. 2. Romae: typis Joannis Jacobi Komarek.

Ptolemy, Claudius. 1562. *Liber de analemnate, Ejusdem Federici Commandini liber de horologiorum descriptione*, Rome; Italian translation by R. Sinisgalli and S. Vastola, *L'analemma di Tolomeo.* Firenze: Cadmo, 1992.

Rashed, Roshdi. 1984. *Entre arithmétique et algèbre.* Paris: Les Belles Lettres.

Rashed, Roshdi. 1990. A Pioneer in Anaclastics. Ibn Sahl on Burning Mirrors and Lenses. *Isis* 81: 464–491.

Rashed, Roshdi. 1992. *Optique et mathématiques. Recherches sur l'histoire de la pensée scientifique en arabe.* Aldershot: Variorum Reprints.

Rashed, Roshdi. 1993. *Géométrie et Dioptrique au Xe siècle: Ibn Sahl, al-Qūhī et Ibn al-Haytham.* Paris: Les Belles Lettres.

Rashed, Roshdi. 1993–1996. *Les Mathématiques infinitésimales du IXe au XIe siècle, I: Fondateurs et commentateurs*, al-Furqān: London 1996; *II: Ibn al-Haytham.* London: al-Furqān, 1993.

Rashed, Roshdi. 1996. *Œuvres philosophiques et scientifiques d'al-Kindī: L'optique et la catoptrique.* Leiden: E.J. Brill.

Rashed, Roshdi and Régis Morelon, eds. 1997. *Histoire des sciences arabes*, vol. 2: *Mathématique et physique*, Paris: Éditions du Seuil.

Rashed, Roshdi. 2003. Al-Qūhī et al-Sijzī: sur le compas parfait et le tracé continu des sections coniques. *Arabic Sciences and Philosophy* 13: 9–43.

Rashed, Roshdi. 2005. *Geometry and Dioptrics in Classical Islam.* London: al-Furqān.

Raynaud, Dominique. 1998. Perspective curviligne et vision binoculaire. *Sciences et Techniques en Perspective* 2(1): 3–23.

Raynaud, Dominique. 1998. *L'Hypothèse d'Oxford. Essai sur les origines de la perspective.* Paris: Presses universitaires de France.

Raynaud, Dominique. 2001. Effets de réseau dans la science pré-institutionnelle: le cas de l'optique médiévale. *European Journal of Sociology* 42(3): 483–505.

Raynaud, Dominique. 2001. "Perspectiva naturalis," "Ibn al-Haytham, Tavoletta binoculare," "Le fonti ottiche di Lorenzo Ghiberti". In *Nel Segno di Masaccio,* ed. F. Camerota, 11–14, 79–81. Firenze: Giunti.

Raynaud, Dominique. 2001. La faveur de l'optique à Oxford. *Llull* 24: 727–754.

Raynaud, Dominique. 2003. Linear perspective in Masaccio's Trinity fresco: demonstration or self-persuasion? *Nuncius* 17: 331–344.

Raynaud, Dominique. 2003. Ibn al-Haytham sur la vision binoculaire: un précurseur de l'optique physiologique. *Arabic Sciences and Philosophy* 13: 79–99.

Raynaud, Dominique. 2003. Understanding Errors in Perspective. In *The European Tradition in Qualitative Research*, eds. R. Boudon, M. Cherkaoui and P. Demeulenaere, vol. 1, 147–165. London: Sage.

Raynaud, Dominique. 2004. Une propriété mathématique de la perspective synthétique réfutant son existence médiévale, 1295–1450. *Zeitschrift für Kunstgeschichte* 67(4): 449–460.

Raynaud, Dominique. 2004. Une application méconnue des principes de la vision binoculaire: Ibn al-Haytham et les peintres italiens du Trecento. *Oriens/Occidens* 5: 93–131.

Raynaud, Dominique. 2005. "L'émergence de l'espace perspectif: effets de croyance et de connaissance," A. Berthoz et R. Recht, eds., *Les Espaces de l'homme*, Symposium international du Collège de France (Paris, 14–15 octobre 2003). 333–354. Paris: Odile Jacob.

Raynaud, Dominique. 2006. Le traité sur la quadrature des lunules attribué à Leon Battista Alberti. *Albertiana* 9: 31–68.

Raynaud, Dominique. 2006. "La théorie des erreurs et son application à la reconstruction des tracés perspectifs," M. Dalai Emiliani, M. Cojannot Le Blanc, P. Dubourg Glatigny, eds, *L'Œuvre et l'artiste à l'épreuve de la perspective*, actes du colloque international (Rome, 19–21 septembre 2002), 411–430. Rome: Publications de l'École française de Rome.

Raynaud, Dominique. 2006. "Milieux et réseaux sociaux de diffusion de la perspective," M. Dalai Emiliani, M. Cojannot Le Blanc et P. Dubourg Glatigny, eds., *L'Artiste et l'Œuvre à l'épreuve de la perspective* (Rome, 19–21 septembre 2002), 5–17. Rome: Publications de l'École française de Rome.

Raynaud, Dominique. 2007. Le tracé continu des sections coniques à la Renaissance. Applications optico-perspectives, héritage de la tradition mathématique arabe. *Arabic Sciences and Philosophy* 17: 299–345.

Raynaud, Dominique. 2008. Geometric and Arithmetical Methods in Early Medieval Perspective. *Physis* 45: 29–55.

Raynaud, Dominique. 2009. Las primeras perspectivas de los siglos XIII y XIV según el enfoque del modus operandi. In *Ars, Techné, Technica. A fundamentação teórica e cultural da perspectiva*, ed. M. Mello, 41–62. Belo Horizonte: Argumentum.

Raynaud, Dominique. 2009. La perspective aérienne de Léonard de Vinci et ses origines dans l'optique d'Ibn al-Haytham (De aspectibus, III, 7). *Arabic Sciences and Philosophy* 19: 225–246.

Raynaud, Dominique. 2009. Why Did Geometrical Optics Not Lead to Perspective in Medieval Islam? In *Raymond Boudon. A Life in Sociology*, eds. M. Cherkaoui and P. Hamilton, vol. 1, 243–266. Oxford: Bardwell Press.

Raynaud, Dominique. 2010. Les débats sur les fondements de la perspective linéaire de Piero della Francesca à Egnatio Danti: un cas de mathématisation à rebours. *Early Science and Medicine* 15: 475–504.

Raynaud, Dominique. 2012. Abū al-Wafā' Latinus? A Study of Method. *Historia Mathematica* 39: 34–83.

Raynaud, Dominique. 2013. "As redes universitárias de difusão das ciências matemáticas como fator de desenvolvimento da perspectiva", M. Mello, ed. *A Arquitetura do engano*. 71–86. Belo Horizonte: Fino Traço Editora

Raynaud, Dominique. 2013. "Optics and Perspective prior to Alberti," *The Springtime of the Renaissance. Sculpture and the Arts in Florence, 1400–60*, ed. Beatrice Paolozzi Strozzi and Marc Bormand. 165–171. Florence: Manragora

Raynaud, Dominique. 2013. "Une application de la méthode des traceurs à l'étude des sources textuelles de la perspective," *Vision and Image-Making: Constructing the Visible*, Colloque international, Studium CNRS et Centre d'Études Superieures de la Renaissance, Tours, 13–15 septembre 2013.

Raynaud, Dominique. 2013. Leonardo, Optics, and Ophthalmology. In *Leonardo's Optics. Theory and Practice*, eds. F. Fiorani and A. Nova, 255–276. Venezia: Marsilio Editori

Raynaud, Dominique. 2014. *Optics and the Rise of Perspective. A Study in Network Knowledge Diffusion*. Oxford: Bardwell Press.

Raynaud, Dominique. 2014. A Tentative Astronomical Dating of Ibn al-Haytham's Solar Eclipse Record. *Nuncius* 29: 324–358.

Raynaud, Dominique. 2014. Building the Stemma Codicum from Geometric Diagrams. A Treatise on Optics by Ibn al-Haytham as a Test Case. *Archive for History of Exact Sciences* 68: 207–239.

Raynaud, Dominique. 2015. Un fragment du *De speculis comburentibus* de Regiomontanus copié par Toscanelli et inséré dans les carnets de Leonardo (Codex Atlanticus, 611rb/915ra). *Annals of Science* 72: 306–336.

Raynaud, Dominique, ed. 2015. *Géométrie pratique. Géomètres, ingénieurs et architectes, XVIe–XVIIIe siècle*. Besançon: Presses universitaires de Franche-Comté.

Raynaud, Dominique (in press) "L'ottica di al-Kindī e la sua eredità latina. Una valutatione critica," S. Ebert-Schifferer, P. Roccasecca e A. Thielemann, eds., *Lumen, Imago, Pictura*, Atti del convegno internazionale di studi (Rome, Bibliotheca Herziana, Max-Planck Institut, 12–13 aprile 2010).

Raynaud, Dominique (in press) "A Hitherto Unknown Treatise on Shadows Referred to by Leonardo da Vinci," S. Dupré and J. Peiffer, eds., *Perspective as Practice. An International Conference on the Circulation of Optical Knowledge in and Outside the Workshop*, Max Planck Institut für Wissenschaftsgeschichte (Berlin, 12–13 octobre 2012).

Recht, Roland. 1999. *Le croire et le voir*. Paris: Gallimard.

Roccasecca, Pietro. 1993. Il modo optimo di Leon Battista Alberti. *Studi di Storia dell'Arte* 4: 245–262.

Roccasecca, Pietro. 1998. "Tra Paolo Uccello e la cerchia sangallesca: la costruzione prospettica nei disegni di mazzocchio conservati al Louvre e agli Uffizi," R. Sinisgalli, ed., *La prospettiva. Fondamenti teorici ed esperienze figurative dall'antichità al mondo moderno*, Atti del Convegno internazionale di studi, Istituto Svizzero di Roma (Roma 11–14 settembre 1995). 133–144. Firenze: Cadmo.

Roccasecca, Pietro. 1999. Punti di vista non punto di fuga. *Invarianti* 33(99): 41–49.

Roccasecca, Pietro. 2001. "La finestra albertiana," *Nel segno di Masaccio. L'invenzione della prospettiva*, a cura di F. Camerota. 64–69. Firenze: Giunti Editore.

Roccasecca, Pietro. 2011. La prospettiva lineare nel Quattrocento: dalle proporzioni continuata e ordinata alla proporzione degradatta. In *Proportions. Science – Musique – Peinture & Architecture*, eds. S. Rommevaux, P. Vendrix and V. Zara. 277–297. Turnhout: Brepols.

Roccasecca, Pietro. 2013. Dalla prospettiva dei pittori alla prospettiva dei matematici. In *Enciclopedia italiana di scienze, lettere ed arti. Il Contributo italiano alla storia del pensiero*, 137–144. Roma: Istituto della Enciclopedia italiana.

Romanini, Angiola Maria. 1997. Arnolfo pittore: pittura e spazio virtuale nel cantiere gotico. *Arte medievale* 11: 3–33.

Rösch, Jean. 1943. *Physiologie et géométrie de la vision binoculaire et des mesures stéréoscopiques*. Paris: Hermann.

Rose, Paul L. 1970. Renaissance Italian methods of drawing the ellipse and related curves. *Physis* 12: 371–404.

Russell, Gül. 1994. The anatomy of the eye in 'Alī Ibn al-'Abbās al-Majūsī: A textbook case. In *Constantine the African and 'Alī Ibn al-'Abbās al-Majūsī*, eds. Charles Burnett, and Danielle Jacquart, 247–265. Leiden: E.J. Brill.

Russell, Gül. 1997. La naissance de l'optique physiologique. In *Histoire des sciences arabes*. ed. Roshdi Rashed 2: 319–354. Editions du Seuil, Paris.

Sabra, Abdelhamid I. 1966. Ibn al-Haytham's criticism of Ptolemy's optics. *Journal of the History of Philosophy* 4: 145–149.

Sabra, Abdelhamid I. 1971. The Astronomical Origin of Ibn al-Haytham's Concept of Experiment. In *Actes du XIIe Congrès International d'Histoire des Sciences*, tome III.A. 133–36. Paris: A. Blanchard.

Sabra Abdelhamid I. 1978. Sensation and inference in Alhazen's theory of visual perception. In *Studies in Perception*, eds. Peter K. Machamer, and Robert G. Turnbull, 160–185. Columbus: Ohio State University Press.

Sabra, Abdelhamid I. 1983. *The Optics of Ibn al-Haytham, Books I–III: on Direct Vision*, Arabic text, Edited and with Introduction, Kuwait, National Council for Culture, Arts and Letters, 1983; English Translation and Commentary. London: The Warburg Institute, 1989, 2 vols.

Sabra, Abdelhamid I. 2002. *The Optics of Ibn al-Haytham, Books IV–V: On Reflection and Images Seen by Reflection*. Kuwait: The National Council for Culture, Arts and Letters.

Saint Aubin, Jean-Paul. 1992. Photogrammétrie et étude des ordres: le Louvre de Lescot. In *L'Emploi des ordres à la Renaissance*, ed. J. Guillaume, Actes du colloque de Tours (9–14 juin 1986), 219–226. Paris: Picard.

Sanpaolesi, Piero. 1962. *Brunelleschi*. Milano: Edizioni per il Club del Libro.

Saraux, Henry, and Bertrand Biais. 1983. *Physiologie oculaire*. Paris: Masson.

Scheiner, Christoph. 1619. *Oculus, hoc est fundamentum opticum...* Oeniponti (Innsbruck): apud Danielem Agricolam; London: J. Flesher, 1652.

Scholz-Hansel, Michael (1992/3) Las obras de Pellegrino Tibaldi en el Escorial, *Imafronte* 8/9: 389–401.

Schramm, Matthias. 1963. *Ibn al-Haythams Weg zur Physik*. Wiesbaden: F. Steiner.

Sergescu, Pierre. 1952. *Léonard de Vinci et les mathématiques, Léonard de Vinci et l'expérience scientifique*, 73–88. Paris: CNRS.

Sgarbi, Vittorio. 1979. *Carpaccio*. Bologna: Casa Editrice Capitol.

Simi, Annalisa, and Toti-Rigatelli, Laura. 1993. Some 14th and 15th century texts on practical geometry. In *Vestigia Mathematica. Studies in Medieval and Early Modern Mathematics in Honour of H.L.L. Busard* M. Folkerts and J.P. Hogendijk, 453–470. Amsterdam/Atlanta: Rodopi.

Simi, Annalisa. 2000. "Problemi caratteristici della geometria pratica nei secoli XIV–XVI," *Scienze mathematiche e insegnamento in epoca medioevale*, a cura di P. Freguglia, L. Pellegrini e R. Paciocco, 153–200. Napoli: ESI.

Simi Varanelli, Emma. 1989. Dal Maestro d'Isacco a Giotto. Contributo alla storia della 'perspectiva communis' medievale. *Arte medievale 2. Ser.* 3: 115–143.

Simi Varanelli, Emma. 1995. *Artisti e dottori nel Medioevo. Il campanile di Firenze e la rivalutazione delle arti belle*. Roma: Istituto poligrafico e Zecca dello Stato.

Simi Varanelli, Emma. 1996. *"Arte della memotecnica e primato dell'imagine negli ordines studentes"*, *Bisancio e l'Occidente: arte, archeologia, storia. Studi in onore di Fernanda De' Maffei*, 505–525. Roma: Viella.

Simon, Gérard. 1988. *Le Regard, l'être et l'apparence dans l'optique de l'Antiquité*. Paris: Éditions du Seuil.

Simon, Gérard. 1997. La psychologie de la vision chez Ptolémée et Ibn al-Haytham. In *Perspectives arabes et médiévales sur la tradition scientifique et philosophique grecque*, eds. Ahmad Hasnawi, Abdelali Elamrani-Jamal et Maroun Aouad, 189–207. Louvain/Paris: Peteers/Institut du Monde Arabe.

Simon, Gérard. 2001. Optique et perspective: Ptolémée, Alhazen, Alberti. *Revue d'Histoire des Sciences* 54: 325–350.

Simon, Gérard. 2003. *Archéologie de la vision. L'optique, le corps, la peinture*. Paris: Éditions du Seuil.

Sindona, Ennio. 1970. Una conferma uccellesca. *L'Arte* 9: 67–107.

Sinisgalli, Rocco. 1978. *Per la storia della prospettiva, 1405–1605*. Roma: L'Erma.

Smith A. Mark. 1996. *Ptolemy's Theory of Visual Perception*, An English Translation of the Optics with Introduction and Commentary. Philadelphia: The American Philosophical Society.

Smith, A. Mark. 2001. *Alhacen's Theory of Visual Perception*, A Critical Edition, with English Translation and Commentary, of the First Three Books of Alhacen's *De Aspectibus...* Philadelphia: The American Philosophical Society.

Smith, A. Mark, and Bernard R. Goldstein. 1993. The medieval Hebrew and Italian Versions of Ibn Muʿādh's 'On twilight and the rising of clouds'. *Nuncius* 8: 633–639.

Spielmann, Annette. 1991. *Les strabismes. De l'analyse clinique à la synthèse chirurgicale*. Paris: Masson.

Squadrani, Irenaeus. 1932. Tractatus de luce fr. Bartholomaei de Bononia. *Antonianum* 7: 201– 238, 465–494.

Stidwill, David, and Fletcher, Robert. 2010. *Normal Binocular Vision. Theory, Investigation and Practical Aspects*. London: Wiley.

Ten Doesschate, Gezienus. 1940. *De derde Commentaar van Lorenzo Ghiberti, in verband met de middeleeuwsche optiek*. Utrecht: Drukkerij Hoonte.

Ten Doesschate, Gezienus. 1964. *Perspective. Fundamentals, Controversials, History*. Nieuwkoop: B. de Graaf.

Theisen, Wilfred R. 1979. Liber de visu: The Greco-Latin Tradition of Euclid's Optics. *Mediaeval Studies* 41: 44–105.

Theon of Alexandria. 1895. *Euclidis opera omnia*, vol. VII: *Opticorum recensio Theonis*, ed. J.L. Heiberg. Leipzig: B.G. Teubner.

Tobin, Richard. 1990. Ancient Perspective and Euclid's Optics. *Journal of the Warburg and Courtauld Institutes* 53: 14–41.

Thoenes, Christoph and Roccasecca, Pietro. 2002. Per una storia del testo de Le due regole della prospettiva pratica. In *Jacopo Barozzi da Vignola*, eds. Richard J. Tuttle, B. Adorni, C.L. Frommel, and C. Thoenes, 367–368. Milano: Skira.

Thomas, Ch. 1971. La physiologie de la vision binoculaire. *Archives d'Ophtalmologie* 31: 189– 206.

Tversky, Amos, and Daniel Kahneman. 1973. Availability: a heuristic for judging frequency and probability. *Cognitive Psychology* 5: 207–232.

Ulrich, Gerhard, and Anton Vieth. 1818. Ueber die Richtung der Augen. *Gilbert's Annalen der Physik* 58: 233–253.

Vagnetti, Luigi. 1979. De naturali et artificiali perspectiva. *Studi e Documenti di Architettura* 9 (10): 3–520.

Vagnetti, Luigi. 1980. "La posizione di Filippo Brunelleschi nell'invenzione della prospettiva lineare", *Filippo Brunelleschi. La sua opera e il suo tempo*. Atti del convegno internazionale di studi (Firenze, 16–22 ottobre 1977). 1: 279–306. Florence: Centro Di.

Vasari, Giorgio. 1647. *Vite de' più eccelenti Pittori, Scultori et Architettori*. Bologna: presso gli heredi di E. Dozza; reed. a cura di R. Bertanini e P. Barocchi. Firenze: Sansoni editore, 1967; reed. a cura di Luciano Bellosi e Aldo Rossi. Torino: Einaudi, 1986.

Verstegen, Ian. 2010. A classification of perceptual corrections of perspective distortions in Renaissance painting. *Perception* 39: 677–694.

Verstegen, Ian. 2011. Viewer, Viewpoint, and Space in The Legend of St. Francis: A Viennese-Structural Reading, preprint.

Vesalius, Andreas. 1543. *De humani corporis fabrica*. Basileae: Oporinus.

Victor, Stephen K. 1979. *Practical Geometry in the High Middle Ages. "Artis cuiuslibet consummatio" and the "Pratike de Geometrie,"* edited with translation and commentary. Philadelphia: The American Philosophical Society.

Vinci, Leonardo da. 1882. *Das Buch von Malerei, nach dem Codex Vaticanus (Urbinas 1270)*, ed. H. Ludwig. Wien: W. Braumuller.

Vinci, Leonardo da. 1924. *Libro della pittura*, a cura di A Borzello. Lanciano: R. Carabba.

da Vinci, Leonardo. 1970. *The Notebooks of Leonardo da Vinci*, Compiled and Edited from the Original Manuscripts by Jean-Paul Richter, vol. I. New York: Dover.

Vitruvius. 1487. *L. Vitruuii Polionis ad Cesarem Augustum de architectura libri decem*, Rome: Johannes Sulpicius. English translation by Frank Granger, *On Architecture*, Cambridge: Harvard University Press, 1956; French translation by Philippe Fleury, *De l'architecture*. Paris: Les Belles Lettres, 1990.

Wade, Nicholas J. 1996. Descriptions of visual phenomena from Aristotle to Wheatstone. *Perception* 25: 1137–1175.

Wade, Nicholas J. 1998. Early studies of eye dominances. *Laterality* 3: 97–108.

Wade, Nicholas J., and Hiroshi Ono. 2012. Early studies of binocular and stereoscopic vision. *Japanese Psychological Research* 54: 54–70.

Weisheipl, James A. 1965. Classification of the sciences in medieval thought. *Mediaeval Studies* 27: 54–90.

Weisheipl, James A. 1978. The nature, scope, and classification of the sciences. In *Science in the Middle Ages*, ed. Davic C. Lindberg, 461–482. Chicago: University of Chicago Press.

Wheatstone, Charles. 1838. Contributions to the physiology of vision. – Part the First. On some remarkable, and hitherto unobserved, phænomena of binocular vision. *Philosophical Transactions of the Royal Society of London* 128: 371–394.

White, John. 1949. Developments in Renaissance Perspective: I. *Journal of the Warburg and Courtauld Institutes* 12: 58–79.

White, John. 1951. Developments in Renaissance Perspective: II. *Journal of the Warburg and Courtauld Institutes* 14: 42–69.

White, John. 1956. *Perspective in Ancient Drawing and Painting*. London: Society for the Promotion of Hellenic Studies.

White, John. 1967. *The Birth and Rebirth of Pictorial Space*. London: Faber and Faber; French translation *Naissance et renaissance de l'espace pictural*. Paris: Adam Biro, 1992.

Wickersheimer, Ernest. 1926. *Anatomies de Mondino dei Luzzi et de Guido da Vigevano*. Paris: E. Droz.

Witelo. 1572. *Opticae Thesaurus... Item Vitellonis Thuringopoloni libri decem*, ed. F. Risner, Basileae: Per Episcopios. Reprint with an introduction by David C. Lindberg. New York: Johnson Reprint Corporation. 1972.

Witelo. 1977. *Witelonis Perspectivæ liber primus. Book I of Witelo's Perspectiva*, an English translation with introduction and commentary... by Sabetai Unguru. Wroclaw/Warszawa/Kraków: Ossolineum.

Wolf, Robert E. 1972. "La querelle des sept arts libéraux dans la Renaissance, la Contre-Renaissance et le Baroque," *Renaissance, Maniérisme, Baroque*, Actes du XIe stage international de Tours, 259–288. Paris: Librairie philosophique J. Vrin.

Zanardi, Bruno, Chiara Frugoni, and Federico Zeri. 1996. *Il Cantiere di Giotto. Le storie di San Francesco ad Assisi*. Milano: Skira.

Zanardi, Bruno. 2002. *Giotto e Pietro Cavallini: La questione di Assisi e il cantiere medievale della pittura a fresco*. Milano: Skira.

Zupko, Ronald E. 1981. *Italian Weights and Measures from the Middle Ages to the Nineteenth Century*. Philadelphia: The American Philosophical Society.

Index Nominum

Index Rerum

A

Abacus, arch, 21, 53, 55, 56, 61, 62, 65, 67, 196, 212

Abacus, arith, 7, 20, 25, 40, 43, 191, 217, 220, 221

Abacus, ophthalm, 89, 91

Aberration, 00
chromatic, 230
geometric, 75, 168, 230
spherical, 75, 87, 168, 186, 231, 235
see also: Astigmatism, Coma, Distortion, Field curvature., 00

Academies, 11, 15, 50, 115, 116, 117–122, 122–129, 218–219, 223

Academic curriculum, 11, 21, 35, 50

Accuracy, 00
in metrology (both precision and trueness), 230
perspective, also correctness, exactness, 16, 19, 21–22, 30, 37–39, 232, 234, 41, 44–46, 48, 51, 53, 55, 58, 62–64, 67, 128, 139, 200, 218, 226, 228–229

Adherence to a belief, a knowledge, a rule, etc, 22, 38, 51, 121, 136, 172, 230

Anachronism, 20, 35, 58, 102, 106, 191, 194, 196

Anatomy, anatomists, 5, 32–33, 53, 72, 78, 86, 104, 108, 116–118, 215

Angle(s), 1, 2, 10, 38, 40, 58, 60–61, 63, 82–84, 86–88, 91, 153–154, 170, 231–232, 238–240, 243
inscribed angle theorem, 83–84
axiom of, 163
visual, 158, 178

Antique, 7

Alternatives, 35, 62, 161–162, 188

Application, 11, 20–22, 24, 29–30, 43, 50]51, 54, 95, 114, 135, 138, 155, 158, 186–187, 189, 191, 198, 215

Approximation, 233, 237
paraxial, 75, 231

Architectural, 00
design, 197
element, 155, 194, 212, 225
framework, 43, 135–136, 138, 143, 150–152, 154, 164, 187, 225
model, 133
pattern, 62

Architecture and architects, 1, 3, 17–19, 22, 31, 35, 37, 48, 51, 54, 62–63, 82, 84, 115, 133–134, 143–144, 158, 161, 165, 175, 197, 203, 218, 220–221

Argument from authority, 22, 84, 119

Arithmetic, 2, 3, 43, 221

Arithmetic method, 158, 159, 191–213, 216, 220

Arricio (plaster underlayer), 193

Ars mensurandi, 11, 34

Art(s), 00
liberal, 4, 6
mechanical, 3, 4, 6, 134

Artisans, crafsmen, 6, 16, 194–196, 202, 204, 217, 220–221

Artists and artistic milieux, 20, 24, 33, 38–39, 43, 49, 51–52, 114, 120–122, 128, 134, 138, 141–142, 162, 167, 171, 178, 187, 191, 194, 202, 217–219

De aspectibus (optics), 3, 17, 72–73, 96, 98–101, 103, 147–150, 153, 213, 216, 218

Astigmatism, 75, 168, 231, 235

Astragal, arch, 24

Astrolabe, 7, 11, 29, 54, 223

© Springer International Publishing Switzerland 2016
D. Raynaud, *Studies on Binocular Vision*,
Archimedes 47, DOI 10.1007/978-3-319-42721-8

Printed in the United States
By Bookmasters